CHEMICAL PROCESSES IN
WASTE WATER TREATMENT

CHEMICAL PROCESSES IN WASTE WATER TREATMENT

W. J. EILBECK
Safety and Radiation
Protection Adviser
University College of
North Wales

and

G. MATTOCK
Waste Water
Treatment and
Resource Recovery
Consultant

ELLIS HORWOOD LIMITED
Publishers · Chichester

Halsted Press: a division of
JOHN WILEY & SONS
New York · Chichester · Brisbane · Toronto

First published in 1987 by
ELLIS HORWOOD LIMITED
Market Cross House, Cooper Street,
Chichester, West Sussex, PO19 1EB, England
The publisher's colophon is reproduced from James Gillison's drawing of the ancient Market Cross, Chichester.

Distributors:

Australia and New Zealand:
JACARANDA WILEY LIMITED
GPO Box 859, Brisbane, Queensland 4001, Australia

Canada:
JOHN WILEY & SONS CANADA LIMITED
22 Worcester Road, Rexdale, Ontario, Canada

Europe and Africa:
JOHN WILEY & SONS LIMITED
Baffins Lane, Chichester, West Sussex, England

North and South America and the rest of the world:
Halsted Press: a division of
JOHN WILEY & SONS
605 Third Avenue, New York, NY 10158, USA

© 1987 W.J. Eilbeck and G. Mattock/Ellis Horwood Limited

British Library Cataloguing in Publication Data
Eilbeck, W.J.
Chemical processes in waste water treatment. —
(Ellis Horwood books in water and wastewater technology)
1. Sewage — Purfication
2. Factory and trade waste
I. Title   II. Mattock, G.
628.5′4   TD745

Library of Congress Card No. 86–27798

ISBN 0–85312–791–3 (Ellis Horwood Limited)
ISBN 0–470–20800–7 (Halsted Press)

Phototypeset in Times by Ellis Horwood Limited
Printed in Great Britain by Butler & Tanner, Frome, Somerset

Contents

6   Contents

Preface

There are few industrial waste water treatment operations that do not incorporate chemical processes of one kind or another as part of the overall scheme: their indisputable significance therefore makes it all the more surprising that no reference work exists that comprehensively covers them. There are of course academic texts and reviews in the field of pure chemistry, but these are not directed to the specific needs of waste water treatment. Likewise, the numerous engineering manuals on waste water treatment make scant reference either to the principles of chemical reactions that can be employed or to their optimisation and control. This is the gap we have sought to fill with this book, which was conceived and structurally planned by Gerry Mattock: it is hoped that it will serve as a reference source for design engineers, a groundwork for sanitary and environmental engineering students, a guide to treatment options for those concerned with regulatory control, and as a perspective for research and development workers. We have tried to combine the theoretical with the practical, to provide an understanding of the basis of widely used chemical reactions, how and why they are selected, sensible advice on how they can be applied, and just as importantly, how they can be controlled.

These objectives have dictated the structural approach that has been adopted. There is a primary emphasis on the reactions, leading to their applications, rather than the more common organisation based on waste water treatment problems involving a review of the available process options. (This latter form is suitable for texts dealing with, for example, effluents arising from individual industries, but does not lend itself to thoroughness in dealing with fundamentals. There are in any case several texts of this kind, which fulfil a different function: Appendix 3 gives a bibliography of some of the publications available. The index also provides a guide to textual discussions on specific treatment problems.) Chemical processes are much more widely used for industrial waste water treatment than they are in sewage treatment, so this has directed the main thrust of the book; but sewage chemical treatments are covered as appropriate.

There are four loosely classifiable sections. Chapter 1 reviews generally how chemical processes are applied in terms of engineering requirements and automatic control opportunities, the basic principles of chemical treatment plant design, and approaches to water and materials' conservation. Chapter 2 surveys pH adjustment processes, including measurement and control of pH in waste water treatment situations. Oxidation–reduction reactions inevitably constitute a major section, and these are covered in three chapters, split for convenience into oxidation (Chapter 3), reduction (Chapter 4) and electrochemical (Chapter 5) processes. Finally, Chapters 6 and 7 deal with processes where phase separations occur as a specific objective of treatment, by precipitation, colloid destabilisation, adsorption on to precipitated flocs, and demulsification. Throughout there are theoretical introductions, followed as relevant by reviews of applications, with attention to practical aspects and the reaction monitoring and control techniques currently realisable.

In order to keep the book to a reasonable size we have had to exclude some important processes that arguably could have been included, notably ion exchange, solvent extraction, and adsorption, e.g. by activated carbon. (It is also arguable that ion exchange and solvent extraction are physico-chemical concentration techniques rather than chemical treatment processes; and that activated carbon treatment veers more towards the physical than the chemical—and that it is covered in several existing publications.) However, as with purely physical operations such as settlement, flotation, and the various forms of filtration that are certainly beyond the defined scope, we have made passing references to these topics where omission would clearly have led to imbalance or incompleteness.

It will be apparent from a close reading of the text that there are many areas where more investigation is needed. Also, some previously unpublished work is included, much of it of an exploratory nature. From both of these we hope that not only will the book be of reference value but that it will also act as a spur to further research.

It would be naïvely optimistic to expect that the text is error-free; but we do hope that any mistakes will be obvious enough to prevent the confusion that arises from subtle muddling. We would appreciate being informed of them, as we would of important oversights.

John Eilbeck
Department of Chemistry
University College of North Wales
Bangor
Gwynedd LL57 2UW
Wales

Gerry Mattock
Brook Cottage
Ayleford
Soudley
Nr Cinderford
Glos. GL14 2UF
England

July 1986

Acknowledgments

No volume of this kind can be prepared without the assistance of many people, and we are pleased to acknowledge the help we have received from all directions.

We begin where most authors finish, gratefully recording the care and unstinting willingness of our patient typists: notably we thank Louie Tallis and Sally Eilbeck for their generous work. No less significant has been Peter Holt's careful contribution in preparing the diagrams.

A number of colleagues were kind enough to read and offer constructive comments on some parts of the book. Our thanks here go to Eric Entwistle of Risefield Ltd (Chapter 5),Maurice Hillis of the Electricity Council Research Centre (Chapter 5), and Mike Wilson of Aston University Chemical Engineering Department (Chapter 1). The painstaking care taken by Alan Wheatland of the Water Research Centre, who read the entire text in draft form, is particularly appreciated. Alan spotted several errors, inconsistencies and grammatical infelicities with his eagle eye, and offered many valued criticisms and constructive suggestions from his omnigenous experience.

Various diagrams and tables have been derived from previously published material, and we thank the publishers concerned for their permission to use the source data:

Academic Press (Fig. 7.5) (from P. Sherman (ed.), *Emulsion Science* (1968))
American Chemical Society (Fig. 2.6)
Ann Arbor Science (Table 7.1)
Ellis Horwood (Fig. 3.3 and Table 3.1)
Her Majesty's Stationery Office (Table 2.3)
Institute of Metal Finishing (Figs. 5.11, 5.12)
International Ozone Association (Table 3.2)
International Union of Pure and Applied Chemistry (Table 2.1, © 1985 IUPAC)
John Wiley & Sons/Wiley–Interscience (Appendix 1, Fig. 2.4 (from *pH and pIon Control in Process and Waste Streams*,F. G. Shinsky, Copyright ©1973 John Wiley and Sons Inc. Reprinted by permission of John Wiley & Sons Inc.), and Figs. 6.3 and 6.4, and Table 3.7) (from *Physicochemical Processes for Water*

Quality Control, W. J. Weber, Copyright ©1972 Wiley–Interscience. Reprinted by permission of John Wiley & Sons Inc.)

Pergamon Press (Figs. 6.7, 7.3, and Tables 6.2, 6.3)
Purdue University (Figs. 2.5, 4.1)
Royal Society of Chemistry (Fig. 4.7)
Sijthoff and Noordhoff (Figs. 6.2, 6.6) (from K. J. Ives (ed.) (1978), *The Scientific Bases of flocculation*)
Society of Chemical Industry (Figs. 3.5, 5.8, 5.9, 7.2, and Tables 3.6, 5.1)
Thunderbird Enterprises (Figs. 3.11, 3.12, 4.2, 4.3, 5.13)
Water Pollution Control Federation (Figs. 3.9, 7.4)

We are particularly grateful for the cooperation of Bostock Hill & Rigby, Birmingham, in readily making laboratory facilities available for work that contributed towards Figs. 3.14 and 4.4, and Table 6.4.

Finally, we wish to record the patience and help afforded by Clive Horwood, of our publishers, during the protracted period over which this book was written.

1

The role and application of chemical methods in waste water treatment

CHEMICAL REACTIONS IN POLLUTION TREATMENT

The significance of chemical reactions in pollution control

All pollution treatment operations that are not simply physical separations involve chemical reactions: even biological methods of oxidative degradation are only variations on the means of achieving the chemical reaction. It clearly follows that for a complete understanding of pollution control a good knowledge of these reactions is essential; and certainly no engineering plant design should be undertaken without an awareness of the chemistry that is the foundation of the treatment programme. The contribution that this book seeks to make is therefore to identify and describe chemically induced reactions of significance in pollution control and resource recovery, as the basis for design of treatment plant systems. Aerobic and anaerobic biologically induced processes and engineering practices will not be considered: they are thoroughly reviewed in existing literature, including such texts as those of Schroeder (1977), Barnes and Wilson (1978), White (1978), Metcalf & Eddy, Inc. (1979), Sundstrom and Klei (1979), Benfield and Randall (1980), Curi and Eckenfelder (1980), and Grady and Lim (1980), as well as in the Manuals of Practice published by the Water Pollution Control Federation (1976, 1977), and in any case represent a different line of approach. The scope and opportunities afforded by chemistry are sufficient in themselves to consider in one volume!

Chemical treatment of pollutants, although it has been applied with some success to sewage discharges (see Chapter 6), has been developed far more for industrial waste waters: Table 1.1 presents some of these applications. Industrial waste waters are frequently complex, commonly being variable and high in pollutant loading, and often containing materials toxic or resistant to the organisms on which biological processes depend. These factors, coupled with space restrictions not uncommonly encountered on industrial sites, have encouraged the application of physico-chemi-

Table 1.1 — The application of chemical processes in industrial waste water treatment

Industry	Principal polluting parameters	Typical chemical processes used
Coking (inc. byproducts)	C.O.D., cyanides, sulphur compounds	Oxidation, pH adjustment
Electricity generation	Solids	Flocculation
General engineering (inc. machining, heat treatment, etc.)	Oil, grease, solids, cyanide, surfactants	Emulsion cracking, oxidation, flocculation
Fellmongering, tanning	B.O.D., solids, sulphide, metals, alkalis	Oxidation, flocculation, pH adjustment
Fertiliser	C.O.D.	Oxidation, flocculation
Food processing (inc. abattoirs, meat, fish, fruit and vegetable processing, etc.)	B.O.D., fats, solids	Emulsion cracking, flocculation
Inorganic chemicals	Acids, alkalis, metals	pH adjustment, flocculation
Pesticides, bactericides	C.O.D.	Oxidation, flocculation, pH adjustment
Mining (coal, minerals, etc.)	Solids, metals, acids	Flocculation, pH adjustment
Organic chemicals (inc. plastics)	C.O.D., acids, alkalis	pH adjustment, oxidation, flocculation
Petroleum refinery, petrochemical	C.O.D., oils, cyanide, sulphur compounds, acids, alkalis	Emulsion cracking, oxidation, pH adjustment, flocculation
Pharmaceutical	C.O.D., acids, alkalis	Oxidation, flocculation, pH adjustment
Photographic	Metals, sulphur compounds, acids, alkalis, C.O.D.	Oxidation, reduction, pH adjustment
Pulp and paper	C.O.D., solids, alkalis	Flocculation, oxidation, pH adjustment
Surface treatment (inc. anodising, electroplating, enameling, galvanising, painting, etc., for aircraft, automotive, electronics, domestic goods, ferrous and non-ferrous metals, jewellery, locomotive, etc., industries)	Acids, alkalis, cyanides, metals, surfactants, oils, grease, solids	pH adjustment, oxidation, reduction, emulsion cracking, flocculation
Textile (inc. dyeing)	C.O.D., solids, acids, alkalis, surfactants	Oxidation, flocculation, pH adjustment
Wool processing	B.O.D., grease, solids, alkalis	Emulsion cracking, pH adjustment, flocculation

cal techniques wherever possible. For this reason the main emphasis in this book will be found towards chemical treatments that are applicable to industrial discharges, often as alternatives or supplements to biological ones.

Classification of chemical processes
Granted that a chemical process involves the transformation of a species from one form to another, it is possible to identify, partly for convenience in relation to pollution treatment operations, two main classes of chemical process. The first of these may be termed *transformation reactions*, whereby the species of interest is modified by chemical reaction completely so as to be irreversibly converted to a different compound. An example of this kind of reaction is as follows:

$$CN^- + ClO^- \rightarrow CNO^- + Cl^-$$

A second class of reaction involves not only chemical reaction but is designed also to achieve *phase transfer*. A typical example here would be the precipitation of an insoluble compound by addition of a chemical to a soluble species, as a means of removing the offending material. Another example would be the adsorption of a compound from a solution phase on to a solid phase. The operational importance of this distinction is that in transformation reactions the material so obtained is usually retained in the same phase in which it has been treated (although not necessarily so), while in phase transfer the treatment essentially depends for its efficacy on subsequent physical separation operations. Although the latter operations are not discussed in detail in this text their importance is acknowledged; our concern here, however, is with the chemical procedures that are the precursors of physical ones.

The advantages of chemical treatment and recovery systems
Waste water engineers have become so accustomed to utilising biological treatment systems wherever possible that the opportunities presented by chemically based plants are often overlooked. It is worth identifying some of the advantages.

(1) Inasmuch as chemical treatment systems are more predictable and inherently more subject to control by simple techniques, such plants are usually much more controllable than biologically based ones, and they are frequently amenable to some form of automatic control. They can be operated on an on–off basis without difficulty, and they can be started up rapidly without the need for an acclimatisation time.

(2) Chemical treatments are usually relatively tolerant to temperature changes (although rates for some reactions can be affected, requiring retention times to be based on the slowest conditions).

(3) Chemical treatment plants can be easily designed to accommodate wide variations of input load and substantial variations of flow, a fact of particular importance with industrial waste waters. In contrast to biological systems, where 24 hours per day operation is desirable (although not essential) and loads should preferably be distributed evenly, most chemically based processes can be organised relatively easily to accommodate intermittent and complex discharges. They are of course insensitive to many toxicants such as would affect bacterial action, and therefore although the chemistry must be correctly organised to take account of likely feed compositions, at least the plant is not rendered inoperative by the occasional input of materials that with biological

treatment systems would demand either acclimatisation or a temporary shutdown. Further to this, physico-chemical treatments can achieve excellent results where biological processes either are inapplicable or show poor efficiency, e.g. with toxic discharges, biologically refractory materials, phosphorus removal, and colour removal.

(4) Installation costs and space requirements are usually less for physico-chemical or chemical treatment systems than for biological ones—of particular importance in highly urbanised areas. On the other hand it must be acknowledged that running costs for chemicals are higher than for operating biological treatments: this is, however, normally only significant with large-scale plants.

(5) Chemical plants can generally be rapidly modified or augmented simply by adding a further stage or by changing the nature of the chemical processes. Frequently this does not demand a change in the basic structure of the plant but merely a reorganisation of the chemical operations.

(6) By-products of chemical reaction can often be useful, or at least the process can be designed to provide useful by-products.

There are disadvantages with chemical treatments. For example, compared to alternative biological treatment (where this is possible) chemical treatment results in larger quantities of sludge. Furthermore, where chemical treatment is applied as a precipitative procedure, the concentration of dissolved salts in the discharging effluent is increased (although this is not necessarily the case where metathetical reactions are used).

In practice it is not uncommon to find that integration of biological and chemical treatment processes is the optimum approach. Good examples of this arise where chemical treatments are used to reduce the pollution level of an industrial discharge to make it acceptable to a sewage treatment works: this means that an industrial discharger can apply the advantages of chemical treatments, and the subsequent biological processes can be operated by those who are expert at them, namely sewage plant operators. Furthermore, where biologically intractable materials are present it is essential to use the more robust degradative treatment offered by chemical reaction, or adsorption, or both.

THE USE OF CHEMICAL METHODS IN PLANT SYSTEM DESIGN

The objectives of plant design

These can be summarised as follows:

(1) the minimisation of pollution, with compliance to statutory discharge conditions;

(2) the economical recovery of useful materials where possible;

(3) the provision of operational reliability; and

(4) the achievement of (1), (2) and (3) at minimum cost.

The problem with these criteria is that they are frequently conflicting. Of course, recovery systems should be sought and included wherever possible, but they are not necessarily or always cost-effective. Further, the minimisation of pollution to the theoretical limit, or to the level consistent with the state of the art at any one time, may result in a design that is unnecessarily costly in the context of environmental requirements. Nevertheless the high degree of flexibility offered by chemical reaction-based plant systems offers perhaps more scope for the balanced achievement of all the objectives than does any other.

Evaluation of the pollution-producing processes

As in all cases of plant design it is necessary first to define the requirements and problems. It is particularly important to identify at an early stage when and where chemical methods should be applied, and how these should be integrated with associated physical separations and, as apposite, biological treatments. In order to do this it is essential to undertake a thorough study of the pollution-producing processes and to examine the objectives with the proposed design, and whether these are really appropriate. Frequently dischargers seek to treat their effluents on the basis of minimum cost with little regard to other factors, such as recovery opportunities, and even more importantly, to the possibilities for manufacturing process modifications that would eliminate the need for a good deal of pollution treatment anyway. Since an eye is always cast towards the influence and thinking of the regulatory authority, it is necessary moreover to consider whether the authority's requirements are both reasonable and stable in terms of the lifetime of the proposed plant. It is only by a judicious balancing of these various factors that the true objectives of plant design can be realised to the best interests of the discharger, the regulatory authority, and society.

It is commonly assumed that a necessary precursor for plant design must be a series of sampling and analytical operations to define the character and composition of the effluents. This may have its relevance in the identification of problems associated with an existing treatment plant, but by no means should it necessarily be used as the sole basis for new plant design, the starting point for which should be an initial study of the manufacturing operations and the associated materials' balances. It is important to study first whether the pollution-producing processes can advantageously be modified to minimise the materials lost in the first place. The second step should be the organisation of essential discharges into streams that are defined by the treatment to be employed in relation to the compositions, with an eye both to the efficiency of treatment and to economies that can simultaneously be achieved. The idea of basing treatment designs on chemical or other analyses of existing different streams is often irrelevant to the procedure for efficient plant design: it should be a cardinal rule that one does not design on history, or even on the present circumstances of a discharge, but only after an examination and if necessary a modification either of the effluent-producing process or of the stream segregation system used to convey discharges, or both. With a recognition of this different emphasis much of the apparent significance of aggregated analytical data disappears; although of course where separations of the right kind can be made and real plant samples relevant to a proposed design can be secured, then these constitute good checks on calculations made from materials' balances. It is not the place here to discuss the general and statistical aspects of representative sampling and of effluent testing, however, but simply to identify the correct first stage.

The lesson from this approach to design is that the manufacturing processes cannot be regarded as separate entities from the pollution prevention programme.

This has a further importance when it is recognised that much pollution from industry arises simply from poor housekeeping and through inadequate operational control of production processes.

The influence of required discharge limits
The standards to which a treatment and recovery plant must perform exert a profound influence on both the design of the plant and on its costs. It is therefore crucial that before any considerable design thinking has been crystallised a clear idea of the performance required and the likely stability of that performance requirement be assessed. The definition of discharge standards, whether by concentrations or in terms of total loads, must be made in relation to the analytical criteria that will be used to assess performance. And, it can be added, there should be some understanding and acceptance that the standards are technologically feasible.

As a starting point, there must be an understanding of the analytical criteria on which the discharge standards are based, particularly where chemical treatments are being applied to biologically refractory substances, or to a mixture of them with biodegradable species. The standard biochemical oxygen demand (B.O.D.$_5$) test, consisting of incubating a suitably diluted effluent sample at 20°C in the dark for five days, with measurement of the decrease in oxygen concentration over the period (Department of Environment 1981, 1982, APHA/AWWA/WPCF 1985), may be appropriate as a pollution load indicator for sewage, but it can be very misleading with industrial waste waters. Much more appropriate is the chemical oxygen demand (C.O.D.), where a sample of effluent is boiled with sulphuric acid and potassium dichromate in the presence of catalyst for two hours (Department of Environment 1977, APHA/AWWA/WPCF 1985), which seeks (not always entirely satisfactorily) to oxidise all the organic material to provide a measure of the total oxygen demand; this has largely superseded the 'permanganate value' (P.V.), where the oxygen consumed from acidified potassium permanganate is determined (Department of Environment 1983, APHA/AWWA/WPCF 1985), although P.V. is still widely employed as a quality indicator for water supplies. Yet another parameter is total organic carbon (T.O.C.) (Department of Environment 1979a, APHA/AWWA/ WPCF 1985), which is sometimes used as an indicator to assess the performance of waste water treatment plants. The significant matter here is that the performance of a given chemical operation may be better in one of these criteria than another; while another chemical operation may show different performance patterns with the various analytical assessments. This is particularly true of complex industrial organic wastes.

The position is not necessarily clear with what would appear to be simple parameters, such as suspended solids (does this mean total filterable, or settleable, or including colloidal?), metals (total, or precipitable by pH adjustment?), cyanide (total, or liberating HCN on acidification?), colour (values on the Pt–Co scale (see Department of Environment 1979b), or absorbance at given wavelengths?), and oil (free oil, oil as measured by extraction?—and which extraction, subject to which interferences by co-present organic matter?). It is always possible to define by agreement the analytical criteria (although not always so easy to justify their relevance as pollution indicators for a particular receiving watercourse or sewer);

and it is *essential* for this to be done in the treatment of industrial waste waters in order that an appropriate treatment scheme can be devised, and as a means of measuring performance.

CHEMICAL REACTION KINETICS IN WASTE WATER TREATMENT

Reaction rates

All chemical reactions take place at a finite rate that is governed by the nature of the reactants, their concentrations, and the reaction conditions (temperature, pressure, etc.): even ionic reactions, often assumed to be instantaneous, slow down significantly to easily measurable rates at very low temperatures. However, in waste water treatment, simple reactions, such as neutralisation of acids by sodium alkalis, can usually be regarded for practical purposes as being instantaneous, but as the succeeding chapters show, many others of importance are either quite slow in operational terms, or are apparently so from secondary aspects. Even apparently straightforward processes, such as precipitation, may involve ageing phenomena that can take times of the order of hours for effective completion. It is therefore appropriate to mention briefly the salient principles of chemical kinetics insofar as they bear on waste water treatment operations. For a fuller discussion reference should be made to the many excellent texts covering the subject, e.g. Frost and Pearson (1961), Bamford and Tipper (1969), Cooper and Jeffreys (1971), Weber (1972), and Weston and Schwarz (1972).

The mass law concepts introduced over one hundred years ago by Guldberg and Waage still form the fundamental basis of reaction kinetics. For a simple reaction of the type

$$A + B \rightarrow products$$

in which one molecule of A reacts with one molecule of B, it is evident that before A and B can react there must be some form of contact between them, as by a collision. Since the number of collisions in unit time is proportional to the product of the numbers of A and B molecules present, it follows that the rate of reaction between A and B will be proportional to the product of the concentrations of A and B, i.e.

$$rate = k[A][B]$$

where k is a rate constant, being the rate when concentrations [A] and [B] are unity.

Rate equations are frequently written in terms of the rate of disappearance of a reactant or appearance of a product. In the case above, if a single product is formed, its concentration at any given time being c, while the intitial concentrations of A and B are c_A and c_B,

$$\frac{\partial c}{\partial t} = k(c_A - c)(c_B - c)$$

In the special case where $c_A = c_B$, the equation simplifies to

$$\frac{\partial c}{\partial t} = k(c_A - c)^2 = k(c_B - c)^2$$

In general, the rate of a reaction can be expressed as the product of the reactant concentrations, each raised to a power that is determined experimentally, i.e.

$$\text{rate} = kc_A^a c_B^b c_C^c \ldots$$

The sum of these experimentally determined power terms a, b, c, etc. then defines the 'order' of the reaction. The individual values of a, b, c, etc., provide the orders of reaction in the reactants A, B, C, etc., respectively. In some cases the reaction order may be found to correspond to the number of atoms, ions or molecules participating in the reaction, but this is not necessarily so: some reactions can be effectively zero order (particularly in heterogeneous systems), or not of whole integer, which clearly does not correlate with reaction stoicheiometry.

A typical curve showing the progress of a chemical reaction as a function of time is shown in Fig. 1.1 (the exception to this being for zero order reactions, where the

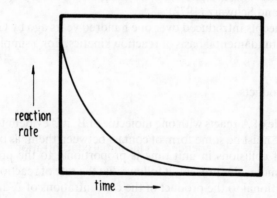

Fig. 1.1 — General form of curve relating degree of chemical reaction with time.

reaction rate is constant). A key feature of this curve is that it is asymptotic with respect to 100% reaction completion, so that in principle it is not possible to achieve complete reaction in a finite time. In practice, the rate constant indicates how closely the reaction can approach effective completion within an operationally practical time. This means that for the execution of a waste water treatment process a knowledge of the rate of reaction is important, particularly where a reactor system receiving a continuous flow of pollutants for treatment is to be used: the reaction rate has an important influence on the reactor retention time (see later in this chapter).

The significance of the order of reaction is shown in Table 1.2, which indicates the

Table 1.2 — Rate laws for some orders of reaction with respect to one reactant

Order	Reaction rate, $\partial c/\partial t$	t	t_{99}	Dimensions of k
0	k	$\dfrac{c_i - c}{k}$	$\dfrac{0.99 c_i}{k}$	concentration $\times$ time^{-1}
1	kc	$\dfrac{1}{k} \log_e \dfrac{c_i}{c}$	$\dfrac{4.6}{k}$	time^{-1}
2	kc^2	$\dfrac{1}{k} \log_e \dfrac{c_i - c}{c_i c}$	$\dfrac{99}{kc_i}$	concentration$^{-1} \times$ time^{-1}
3	kc^3	$\dfrac{1}{2k} \left(\dfrac{1}{c^2} - \dfrac{1}{c_i^2} \right)$	$\dfrac{5000}{kc_i^2}$	concentration$^{-2} \times$ time^{-1}

form and concentration dependences for simple integer orders. It is not uncommon to express reaction rates in terms of the reaction half-time or half-life, $t_{\frac{1}{2}}$, which is defined as the time required for 50% completion of the reaction. For a first order reaction (frequently encountered in waste water treatments), the integrated expression in Table 1.2 shows that

$$t_{\frac{1}{2}} = \frac{1}{k} \log_e 2 = \frac{0.693}{k}$$

indicating that the half-life is independent of the initial concentration. This is not true for the other orders, however.

Of greater practical importance in waste water treatment operations is the time required to complete a reaction to a major or effectively complete degree, e.g. 99%, as for example by lowering the concentration of a pollutant from 100 mg l^{-1} to 1 mg l^{-1}. Table 1.2 shows that the times required for this can be substantially different according to the reaction order, with different influences of the starting concentration. Clearly a knowledge of the kinetic characteristics of a reaction are of importance for the design of waste water treatment plants.

The activation energy
In general, the rates of chemical reactions increase with temperature, as may be expected from a simple collision theory. Experimentally it is found that for many reactions the rate increases by a factor of 2–3 per 10°C rise in temperature, and that generally an equation of the type

$$\log_e k = a - \frac{b}{T}$$

applies, where a and b are constants for the reaction and T is the absolute temperature. This can be written in the alternative form

$$k = A \exp(-E/RT)$$

known as the Arrhenius equation, where A is a constant, R is the gas constant, and E is known as the activation energy of the reaction.

The activation energy is an important parameter in determing the rate of a reaction: the higher the activation energy, the slower the reaction at a given temperature, or the higher the temperature needed to achieve a given rate. A plot of the logarithm of the reactions rate, k, against $1/T$ should, if the Arrhenius equation applies, give a straight line the slope of which is the activation energy, E. As is demonstrated further in Chapter 3, simple energetic considerations based on the driving force available (in terms of free energy) to cause a reaction to take place are not necessarily sufficient to determine whether that reaction will occur at a useful rate: the governing rate factor is the activation energy.

It must be pointed out that the Arrhenius equation does not apply to all chemical reactions: a plot of k against $1/T$ is sometimes curvilinear, and for some complex oxidation processes there may be maxima or minima. Nevertheless, as a comparative guide between reactions the activation energy is a useful indicator to the practical achievement of reactions in a plant system.

THE USE OF INSTRUMENTATION AND AUTOMATIC CONTROL

Instrumentation requirements
Chemical reactions are often amenable to monitoring and control by instrumentation. However, a distinction must be drawn between instruments that are required or designed to monitor certain analytical parameters and those that are intended for control of treatment. Monitoring demands the development of instruments that will give a concentration or equivalent indication of the component level in solution over a fairly broad range. This is a much more demanding specification than applies to control, where relative accuracy is necessary only at the desired control set point. It is equally important to consider what it is one actually wishes to measure. When impurities are being monitored it is the impurity concentration itself or some closely connected derivative that needs to be measured. With treatment control by addition of chemicals, however, it is not the impurity concentration that should necessarily be measured but the presence or absence of the treatment chemical, indicating that completion of reaction has occurred. In an ideal case one would observe both the absence of the impurity and the presence of the treatment agent, but apparent absence of impurity can be misleading because it is then dependent on the inherent discrimination and sensitivity level of the instrument for the particular component that is being measured. Complications can arise when reactions are slow or where side reactions inhibit the full use of the reagent for the purpose intended, thereby

resulting in co-existence of residual impurity and excess reagent. This situation should obviously be avoided if possible, and design thinking must be governed by an understanding of the implications of the kinetics of the chemical reactions as well as by a knowledge of the instrument control options.

Review of the range of instruments used in practice for control of chemical treatment processes reveals that these are relatively few in number. This may partly be attributed to the fact that there are relatively few types of pollution control reactions, namely oxidation–reduction reactions, pH adjustment, emulsion breaking, and precipitation/precipitative adsorption reactions. In the first two of these, the reactions are often adequately controlled by redox or dissolved oxygen and pH measurement systems, respectively, while with the others it is sometimes possible to use a secondary feature of the reaction that permits also the use of these monitors. Flocculation is sometimes monitored and controlled by turbidimetric instruments.

Electrometric sensors, showing a logarithmic response to changes in activity or concentration, have some advantages when it is necessary to observe in a controlled condition a wide range of concentrations, from a large excess of a component to its virtual elimination. For accurate concentration monitoring logarithmic responses are too crude, but for control use their range and relative sensitivity at low concentrations is an advantage. There are a number of analytical alternatives available as a result of the development of specific ion electrodes (see Kakabadse 1978 for a review of applications in waste water treatment), and there is usually a colorimetric analytical reaction that can be followed by means of automated on-line equipment. However, the use of specific ion electrodes in plant application demands sophisticated dosing and standardisation procedures that result in more complex equipment than with simple redox and pH systems, and the same is true for colorimetric equipment used in monitoring and control. The specific nature of this type of equipment, coupled with its sophistication and corresponding expense, has resulted in scant application for control work, although some systems have been used for monitoring waste water discharges and water quality. Conductivity is occasionally used as an indicator and is, for example, used to conserve water usage in rinsing operations involved in metal finishing processes, but while simple to measure, it is usually insufficiently specific for pollution treatment control. It does, however, have some value in water quality monitoring as a rough guide to the levels of total dissolved salts.

For plant control purposes, the primary requirements in an instrument system are relevance to the parameter to be controlled, and simplicity, coupled with a high degree of reliability demanding only simple planned maintenance. Insofar as an automatically controlled plant is critically dependent upon the performance of the control equipment, and even more particularly on the sensor where one is used as the primary element, it can be seen that if this is not trustworthy the whole plant may fail. Lees (1976), in a review of instrumentation reliability in three chemical works, found that for a total of 34 pH meter systems operating over a total of 15.8 instrument years, 93 faults were identified, corresponding to a failure rate of 5.88 faults per year. This in our experience is a relatively good performance, and rather better than may be expected in many pollution treatment plants operated by industrial dischargers. In the simple case of electrical conductivity instruments, where 5 systems were employed over a total of 1.99 instrument years, 33 faults were identified, corresponding to a failure rate of 16.70 faults per year. It is obvious that faults in a control system rendering it inaccurate or inoperative are more serious from the treatment point of view than faults in monitors identifying effluent qualities.

The automatic control loop in chemical treatment practice

It is inappropriate here to review the subject of automatic control loop theory and practice, and there are in any case many texts available (see, e.g., Coxon 1962, Pollard 1971, Shinskey 1973, and Moore 1978 in relation to the present subject). It is, however, useful to highlight elements peculiar to waste water treatment technology, in view of the importance of automatic control in chemical treatments.

Types of control action

The dosing control loop as applied to a chemical reactor is illustrated in its basic form in Fig. 1.2. It consists of the process that is being regulated, the measuring device,

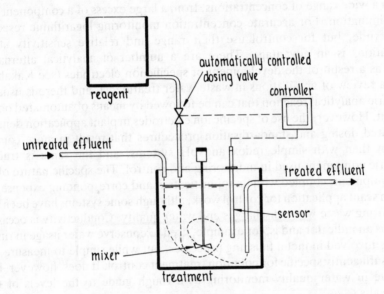

Fig. 1.2 — The automatic dosing control loop applied to a continuous flow stirred chemical reactor (CFSTR).

and the controlling/regulating device. Variations in the effluent quality being measured are sensed by the measuring device, which sends a signal to the controller, which in turn operates a dosing device, e.g. pump or valve, to bring the parameter being measured back to the desired value—so-called 'feedback control'. There are four fundamental types of control action:

On-off (two-step) control

This is very simple, in that when the variable, e.g. pH, moves away from the desired value, the dosing unit introduces reagent at a fixed rate until the desired value is achieved again. A feature of such control action is a sinusoidal response to a step change in the variable, as shown in Fig. 1.3 It is only adequately effective when the changes are small, or when the retention time in the system is sufficient to allow the system to stabilise from a given change. In practice a 'dead band' is established on the controller around the desired value setting to minimise undue hunting.

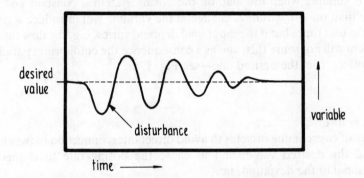

Fig. 1.3 — Characteristic time–response curve of a variable parameter for on–off control action.

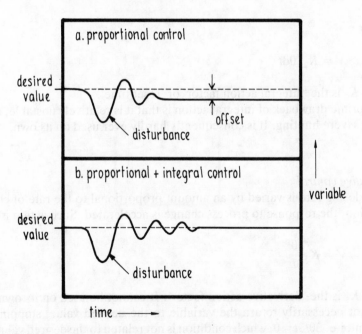

Fig. 1.4 — Characteristic time–response curves of a variable parameter applied to: a. proportional control (showing offset), b. proportional plus integral control.

Proportional control
A more sensitive response is provided when the dosing unit delivers reagent to an extent governed by the magnitude of the deviation from the desired value. If θ is the deviation and V is the dosing rate (say, magnitude of valve opening), then

$$V = K_1\theta$$

where K_1 is the 'proportional action factor', corresponding to the amount of deviation that must occur for dosing to change from zero to maximum rate. This system is suitable when the load on the dosing system is constant and is faster to stabilise than on–off action. A change in the variable will introduce a given load of reagent in unit time, but if the input load demand varies, e.g. the flow rate increases, the system will not sense this, and as a consequence the equilibrium parameter value will be offset from the desired one—see Fig. 1.4a.

Integral control
One way of overcoming offset is to avoid direct interconnection between the dosing rate and the desired value, and to cause the dosing rate to change at a rate proportional to the deviation, i.e.

$$\frac{\partial v}{\partial \theta} = K_2\theta$$

so that

$$V = K_2 \int \theta \mathrm{d}t$$

where K_2 is the 'integral action factor' or 'reset time'.
    A prime drawback of integral action is that it is relatively unstable, and can give rise to severe hunting. It is consequently hardly ever used on its own.

Derivative control
If the dosing rate is varied by an amount proportional to the rate of change of the deviation, the response to process change is accelerated. Such action is derivative:

$$V = K_3\frac{\partial \theta}{\partial t}$$

where K_3 is the 'derivative action factor'. It cannot be used on its own, because it does not necessarily return the variable to the desired value, stopping when θ is constant, i.e. $\partial\theta/\partial t = 0$, which condition is not related to the desired value setting. Its main advantage is its speed of response to changes in the process variable.

Combination control actions
The individual disadvantages of the various control actions can be overcome by combining them, e.g. proportional plus integral (p+i) (see Fig. 1.4b), or proportional plus integral plus derivative (p+i+d). Integral action eliminates the offset

propensity in proportional control, proportional control provides rapid stabilisation, and derivative action gives a fast rate of response to changes. Such combination control systems operate on a variable action dosing unit, e.g. variable setting valve, to provide optimum control conditions. Simpler variations are also used, e.g. so-called 'mark-space' control. This utilises a constant dosing rate, e.g. a dosing valve that is either fully open or fully closed, as in on–off control, but the time for which the dosing occurs is a function of the deviation of the process variable value from the desired value. In practice the dosing operates discontinuously on–off for reduced times as the deviation is reduced. Another alternative that is used is multiple dosing from on–off control, where a series of simple dosing units are used, additional ones operating as the deviation increases. This calls for a controller having several on–off relay settings, rather than a system based solely on the controller identifying the deviation in a continuous manner.

Control response
Within a control loop there are certain inherent time lags, and the magnitudes of these have an important influence on the choice of control action. The overall response time in a loop is governed by these delay times; and to appreciate the relevance of control action choice in waste water treatment it is necessary to summarise these time factors and indicate their orders of magnitude.

Measurement lag
This is the delay in the measuring system responding to a change in the measured parameter, and is an omnibus term for sensor lag, e.g. from a pH or redox or gas membrane electrode, together with instrument response time. In practice the major time delay is usually from the sensor, and it is normally exponential in character. Sensor lag time is not significant with clean pH glass electrodes, the time constant being of the order of seconds (unless the control pH is less than about 2 or higher than about 11), but with redox electrodes it can, for thermodynamically irreversible reactions, be of the order of $\frac{1}{2}$–1 minute, coresponding to approximately 97% total response time of $2\frac{1}{2}$–5 minutes (see, e.g., Chapter 3 for a discussion on this for various redox reactions); while with gas membrane, e.g. oxygen electrodes, the diffusion time constant can again be of the order of up to a minute, depending on the type and thickness of membrane used. If electrodes are dirty, as from the presence of coatings of precipitates or oil, then response lags can be dramatically increased to the order of several minutes. With analysers using additions of chemicals to an effluent sample to provide the means of instrument measurement overall measurement lag can also be quite significant, again of the order of minutes.

Transfer lag
This is the delay between the controller transmitting the signal for correcting action and the arrival of the correcting action to the process (excluding distance–velocity lag, described below). It corresponds to the time for a valve or pump to respond to a signal from the controller and also to the time taken for reagent to pass to the reactor. Transfer lag can arise as a significant factor when, for example, motorised valves are used for reagent dosing (where the response time may be of the order of $\frac{1}{2}$ minute) or when the dosing valve or pump is remote from the reactor vessel. Attention to plant

engineering design should minimise transfer lags from these sources; and the equally undesirable consequences of using long delivery pipes that can drain after dosing action has been stopped should similarly be avoided.

Distance–velocity lag

This is a most important factor in chemical treatment control, and arises from the fact that reagent must be added at a point removed from measurement, simply because it is clearly undesirable for the sensor to observe reagent rather than the result of interaction of the reactants. It includes the time taken for the reagent to reach the sensor within the reactor vessel. In principle it is desirable to keep the time as short as possible without hunting occurring from the influence of mixing and reaction lags (see below), but in waste water treatment control practice it is often by necessity the largest time lag of all. For example, it is a well-known fact that precipitation reactions can cause coatings to deposit on sensor elements immersed in the reactor systems; but it is also generally true that a precipitate that has been formed is less likely to form an adhering coating than a precipitate that is in process of formation. This means that an effective ageing time must be allowed before measurement is made, so a compromise must be accepted; and such ageing times for precipitation reactions are usually of the order of minutes.

Mixing lag

Time is needed to blend the reactants, and this is of course a function of mixing efficiency in relation to the volume of the reactor vessel. Mixing lag is exponential, and a well-designed system should have a mixing time constant of less than a minute. So-called 'flash' mixing is sometimes used as a preliminary to main reactor mixing, but this has little value when it implies a short reactor retention time followed by a long one, where measurement has to be made (for reasons already discussed above) in the larger reactor vessel.

Reaction lag

Although reactants may be blended efficiently together, the overall reaction being monitored may not be fast, being subject to a rate-limiting step. In the case of precipitation reactions there is usually an ageing phenomenon associated with stabilisation of the precipitate. As discussed in Chapter 2, for example, precipitation of metal hydrous oxides by addition of alkali can involve initial occlusion of excess reagent, with subsequent slow release, causing change in the measured variable (pH); and furthermore a change in the chemical composition of the precipitate with time can occur, with release or uptake of ions that may influence the measured variable. The times necessary for reactions to complete effectively are indicated in the reviews of reaction mechanisms in succeeding chapters.

The practice of waste water treatment control

When one considers the time lags peculiar to chemical treatment monitoring and control in relation to available control technologies, some conclusions become apparent that are significant for the design of waste water plants. It is obvious that simple or multiple stage on–off control is applicable, if only because of its simplicity, but what of its proclivity towards hunting? This is only compensated by the use of an extended retention time in the reactor system, and is this acceptable? The answers to these questions lie in recognition of the relative magnitudes of the various control lags in practice. The dominant factors are those from distance–velocity, measurement (sometimes), and reaction lags: the first two exacerbate the hunting, but the last demands the use of extended retention times, which helps to reduce the hunting. The practical consequence is that by necessity a nominal retention time of less than 5 minutes is rarely safe for any form of control, and that 5–30 minutes (depending on the reaction) is usual. This of itself tends to minimise the effects of hunting even in simple on–off control with relatively large variations in input loads (effects which can in any case be mitigated by the use of multiple dosing); at the same time the advantages of using p+i or p+i+d control systems, being primarily ones of speed of response and accommodation of load variations, are diminished.

Another aspect is related to the characteristics of the sensor output where this is of the potentiometric type, e.g. in pH and redox measurements. The response curves of signal output with respect to quantity of reagent dosed (or quantity of contaminant treated) follow the corresponding titration curve, which is generally sigmoid—see, e.g., Fig. 2.1 (p. 45) (fuller discussions are given in Chapters 2 and 3). The sigmoid shape relates to the sensor response being a logarithmic function of the parameter of interest: in the equivalence region there is a high gain in the system, and this diminishes sharply with the appearance of excess reactant of either type. Attempts to control in the equivalence region can therefore lead to instability, and so it is usual practice to set the desired value to a slight excess of reagent, i.e. on the knee of the curve. This is also good practice, because it helps to ensure full treatment.

The logarithmic response character of potentiometric sensors such as pH and redox electrodes can pose some problems in ensuring adequate dosing capability. Theoretically, with proportional control, a logarithmic sensor response demands the use of a logarithmic action dosing unit, but this is not only expensive but complicated as well. Conventional proportional action valves normally have an output dosing range of about 30:1, which is only just over one logarithmic unit change—barely enough for adequate control in many practical situations. The extension is then to provide a 'ganged' system of two valves, giving $(30 \times 30):1$ range. At this point one asks whether this has much advantage over simple on–off multiple dosing. It is certainly possible to ensure a smoother control condition with a given retention time, and where flow rates are high, and the use of large reaction vessels is impractical (say above about $100 \, m^3 \, h^{-1}$) then there are advantages. But otherwise, where smoothing can be provided by slightly longer retention times or, better, multiple reactors (see next section), there is little advantage. Where extremely wide input load variations may occur, multiple on–off controlled reactors with sequential approximations to the desired value are certainly advantageous. Even feedforward control can be used in these circumstances as a preliminary stage (Chang 1964, Shinskey 1973), whereby the incoming water parameter value is measured and used to control dosing downstream. However, in general, feedforward control is so dependent on precise correlation between the measured variable and a corresponding dosing level (virtually impossible in plant practice) that at best it is only a crude method and at worst it can even cause loss of control.

In summary, then, it can be concluded that although p+i and p+i+d control have advantages in principle, in practice the circumstances of waste water treatment, including a need for systems that can be understood by operators who are not instrument engineers, favour on–off control and its variations. Certainly in our experience by far the majority of waste water treatment plants use simple on–off, multiple stage on–off, or mark-space on–off control.

Finally, the reader is recommended to the trenchant and cautionary tales of pH control as relayed by McMillan (1984): there is a vast difference between simple theory and the real world of pH (and redox) control systems.

THE PRACTICE OF CHEMICAL TREATMENT PLANT DESIGN

This section is intended solely to outline the means whereby the chemical processes reviewed in this book may be realised in practice. Many texts are available that discuss the basic engineering principles (see, e.g., Holland and Chapman 1966, Cooper and Jeffreys 1971, Levenspiel 1972, Coulson and Richardson 1979, and Denbigh and Turner 1984): here we are concerned only with the experience of applying chemical reaction data in waste water treatment plant design.

Organisation and use of data

It has been stressed earlier in this chapter that a preliminary to plant design must be the collection of *appropriate* data, and this usually means a study of the processes producing the waste water rather than simply the collection of samples of discharging waste waters. The first steps may be summarised as follows:

(1) The effluent-producing process should be subjected to a series of mass balances for each stage, to determine where the wastes arise, how much there is, and how they are discharged.

(2) The quantities of water used for each individual waste conveyance should be determined, to provide an indication of concentrations of streams at source, i.e. before any admixture has been allowed at a subsequent point.

(3) The individual chemical characteristics should be ascertained for each stream source (before admixture), and studied to determine the best means of treatment. From this there should be an evaluation to determine whether an existing admixture arrangement is desirable or not from a treatment point of view.

(4) At this stage it is always useful to ask whether the waste, or as much of it (including, importantly, water) need be discharged. This is the most constructive approach to conservation, and is nearly always more economic as well as more reliable than recirculation or extraction recovery. It may involve recommendations for modifying the process—calling for enlightened plant managers as well as intelligent and diplomatic approaches by the designer!

(5) Having established the essential discharges and their nature, it is then necessary to settle the degree of stream separation and admixture that may be necessary in terms of the treatments that are being considered. This calls for chemical common sense, with a good understanding of the chemical principles involved, and needs also to be worked out with an eye on the performance that the treatment plant is expected to provide, viz., the discharge quality criteria; there are few benefits from over-designing. Just as an example: it is not chemical common sense to mix cyanide-containing streams, which require oxidative

destruction at high pH, with chromate-containing ones, if low pH chemical reduction treatment is to be applied for these, before the individual treatments have been carried out. Likewise there is no point in allowing relatively unconta-minated water, requiring minimal treatment, to mix with streams that will demand a sequence of other treatment operations; this will merely increase the size and therefore capital costs of the treatment plant (and probably the operating costs as well).

(6) It may of course be necessary to carry out feasibility studies to decide on the best form of treatment, and this may call not only for laboratory process evaluations but also a comparative review of alternative possible unit operations. This review may be confined to a paper study when a new process is being established, and this indeed usually permits evolution of a simpler design than when one is designing for existing discharges. Whatever the route, the objective must be to produce at this stage as simple and as 'clean' a flow diagram as possible, incorporating as many fail-safe features as are economically and technologically sensible; reaction systems, reagents, reagent storage, and dosing systems that are selected with an awareness of health and safety implications; a plant that will not be complicated or costly to operate; and a recognition that God favours His way: so for example it is worth remembering that gravity does not fail, and that pumps are inventions of the Devil. One then passes into the realms of design and contract management, which is another subject (Mattock 1976) . . .

Chemical dosing and reactor systems

Within the flow diagram the various unit operations can be defined and separated conceptually, although of course the impact of each stage upon succeeding ones is not only inevitable but significant in relation to choice between options. For example, one form of precipitation treatment may produce solids that float more easily than settle, whereas another way may cause the reverse (this can happen, for example, in the flocculation of certain types of paint waste). In that case the chemical process cannot be isolated from the subsequent treatment(s) it imposes, and thus it has to be selected with an understanding of the alternative merits of these, and how they can be interpreted. It is certainly not relevant here to cover physical processes such as screening, flotation, settlement and the various forms of filtration, including the very important carbon adsorption—these are reviewed in texts such as those of Purchas (1971), Weber (1972), White (1978), Metcalf & Eddy, Inc. (1979), and Grady and Lim (1980)—but some remarks on the practical engineering of chemical treatment systems are appropriate.

Reactors

The engineering features of these are to some extent controlled by whether the design calls for continuous or batch treatments. In the latter case it is not uncommon to allow for a succession of treatments, including physical operations such as settlement, within the one vessel. With continuous treatments the reactor is usually only called upon to provide the means for one stage.

The basic design requirements of the so-called CFSTR (continuous flow stirred tank reactor) are:

(1) it must incorporate a mixer that will blend all reactants efficiently in minimum time and ensure all the tank contents are so blended;
(2) the flow path through the vessel must be such as to utilise the retention time provided by the vessel volume to a maximum extent, with avoidance of short-circuiting (and this implies also introduction of all reactants at or near the inlet to the vessel, and into the mixing zone);
(3) the retention time must be consistent with the time scale of the reaction(s) to be accomplished (more than one vessel may be used in series to accomplish this); and
(4) where automatic dosing control is being applied provision must be made for the retention time consequence of the control loop.

A diagrammatic representation of a CFSTR is given within Fig. 1.2. For ease of theoretical analysis, equilibrium conditions of steady state mass balance are often assumed as an ideal model. Such a model leads to the relationship

$$\frac{V \partial c_e}{W \partial t} = c_e - c_i \tag{1.1}$$

where W is the flow rate into a reactor of volume V, c_i is the input concentration into the reactor of the species of interest, and c_e is the steady state concentration of that species in the reactor, assumed perfectly mixed. The degree to which conversion from c_i to c_e occurs is a function of the reaction kinetics, which thus govern the design residence time t_r, equal to V/W.

Many reactions, such as those involving simple ionic interaction, are extremely fast, and for these the residence time is mostly governed by other factors, such as control loop lags. Many slower reactions follow first order kinetics, and even where these do not apply in the strict sense, they may do so as pseudo first order, from reactants other than the incoming contaminant being present in effectively constant concentration. This applies in principle where reactants are dosed automatically into the reactor to maintain constant excess concentration conditions. An assumption of effectively first order kinetics is thus reasonable for many systems.

For first order kinetics, as has been shown earlier,

$$\frac{\partial c_e}{\partial t} = kc_e \tag{1.2}$$

where k is the rate constant. This relationship, when substituted into equation (1.1), gives

$$t_r = \frac{1}{k}\left(\frac{c_i}{c_e} - 1\right) \tag{1.3}$$

Where there are n reactors in series, all of equal volume V and nominal retention time t_{rn}, the overall retention time nt_r is given by

$$nt_r = \frac{n}{k}\left[\left(\frac{c_i}{c_e}\right)^{1/n} - 1\right]$$ (1.4)

It is illuminating to compare the retention times needed to achieve a given degree of conversion(c_i/c_e) for simple batch reaction and continuous flow-through reactors, based on first order kinetics. For a first order reaction the time to achieve a given degree of conversion is given by

$$t = \frac{1}{k}\log_e\left(\frac{c_i}{c_e}\right)$$ (1.5)

For the case where 99% conversion is sought, $c_i/c_e = 100$, *and* $t = 4.6/k$. (This corresponds to the behaviour in a plug-flow reactor also.) The same degree of conversion in an ideal single stage CFSTR requires, from equation (1.3), a retention time of $99/k$; while for a two stage reactor the time is $16/k$. — $18/k$

In practice the steady state equilibrium conditions assumed for CFSTR theory do not always apply, particularly where widely varying input loads are being treated with reagents dosed under automatic control. Importantly also, the influence of the various control lags have to be taken into account, and these may introduce time factors that are at least of the same order of magnitude as the reaction ones. This means that even with steady state conditions the retention volume of a reactor has to be calculated on the basis of the longest of the time factors, and not only on the reaction kinetics. Practical experience in some of the more commonly used reaction systems is described in succeeding chapters, to give an indication of the retention times that can be adopted in plant design.

Mixing
It is manifestly important to employ efficient mixing in a CFSTR system. The type of mixer is not particularly important—powered paddle (or some variation on this), air (subject to any oxidation problems that may arise), or pump recirculation, for example—provided the results are achieved with a good power economy. The key requirement is that the inputs be fully blended throughout the entire reactor within the retention time. Of course, the nominal retention time does not represent the time during which *all* inputs are held in the reactor: this only applies in an unstirred plug flow reactor. In practice a proportion of the fluid will pass through a CFSTR in substantially less than the nominal retention time, to a degree influenced by path flow patterns, which are a function of the reactor shape, inlet and outlet arrangements, and of the mixer flow pattern itself. For a good reactor design where path length is maximised, and with orthodox propeller type mixing, a useful guide is to use a mixer having a blend time that is no more than one-tenth of the nominal retention time for the volume being mixed; but in any case where automatically controlled reagent dosing is being applied the blend time must be of no greater magnitude than other control lags.

Mixing theory and a description of the wide variety of mixing systems is beyond the scope of this book: for further information the reader is referred to chemical engineering texts, e.g. Holland and Chapman (1966), Cooper and Jeffreys (1971), and Denbigh and Turner (1984). However, it can be pointed out that the great majority of chemical reactor systems used in waste water treatment employ electrically powered mixers of a propeller or related type, such simple designs providing adequate efficiency for the requirements.

The reactor vessel

The design details of a CFSTR demand attention. As has already been stressed, inlet and outlet arrangements must minimise short-circuiting, and so if the waste water and reagents enter on one side and at the top of the liquid in the vessel, the exit should be diametrically opposite and from the bottom, or vice versa. This can be achieved by the use of a baffle, or with a pipe extending to the bottom of the tank (see Fig. 1.2). Some account must be taken of the circulation pattern imposed by the mixer in this. The usual arrangement for propeller type mixers is as shown in Fig. 1.2, with the paddles at a level some two-thirds to three-quarters of the solution depth.

Although rectangular reactors can be used, and indeed are far more conveniently constructed when a pit system is employed, it is better to use circular vessels to avoid the problem of dead zones in corners. The dimension ratios of the vessel may be influenced to some extent by the type of mixer employed, but it is common practice to arrange for the solution depth to be approximately the same as the vessel diameter or width. A tall small-diameter tank may have the advantage of providing a longer path length, but mixing efficiency may suffer and other inconveniences follow, such as excessive tank heights imposing access difficulties.

The waste water and reagent inlet pipes should enter the vessel in close proximity, and towards the mixing zone. If these inlets are to be on one side of the vessel this implies that the mixer should be off-centre and towards the inlet side, which is in any case desirable to avoid the vortexing that will occur with central placement (unless baffles are constructed on the vessel walls). Too great a displacement from centre may give rise to significant loss of mixing efficiency, and so a compromise is necessary: placing a paddle mixer about one-quarter to one-third of the diameter from the inlet is usually satisfactory.

As has been indicated earlier, the position of the sensor probe when used is of importance in the context of automatic control lags. The sensor position is also significant in relation to the extent that the sensor may become coated where precipitates are a reaction product. (This is discussed further for pH control in Chapter 2 under 'The Sensor System', the considerations however being general.) Once again a compromise is needed between placement close to the mixing zone/ solutions' inlets to provide a rapid response with minimum distance–velocity lag and the alternative of location at the outlet to reduce hunting and the extent of precipitate coatings. A good position is towards the outlet on the opposite side of the mixer to the inlets, with the sensor of the probe at about the level of the mixer paddles; with deeper tanks (from larger volumes) this, however, may imply an inconveniently long probe, which is to be avoided if possible for reasons of maintenance convenience.

It will be evident from the foregoing that the practice of reactor design is one of compromise. Provided principles are recognised, rules can be bent in the interests of practical engineering.

Dosing systems

Any dosing system to be applied in a chemical reactor must have an output capability that is consistent with the chemical demands that will be made; thus the required mass input of reagent(s) as dictated by the mass input of reactant to be treated must be available within the time scale dictated by the retention time of the reactor. This means that the maximum demand rate in a varying input load rate must be available, and for proportional control and the associated control actions normally used with it the variable output dosing valve or pump must be sized accordingly. For valves the potential output is giverned by the flow factor, k_v, defined as the flow rate of a fluid of specific gravity unity through a valve across which unit pressure difference is applied. The relationship between k_v, the flow rate W, the specific gravity γ, and the pressure drop Δp is given by

$$k_v = W\sqrt{\left(\frac{\gamma}{\Delta p}\right)}$$

Where there are n valves in series the resultant flow factor is related to the individual ones by

$$\sum \frac{1}{k_v^2} = \frac{1}{k_{v1}^2} + \frac{1}{k_{v2}^2} + \cdots \frac{1}{k_{vn}^2}$$

For valves in parallel the resultant k_v is simply the sum of the individual flow factors.

In practice it is desirable to provide valve or pump outputs in excess of theoretical, and to adjust as necessary by means of flow restrictors. This is particularly important in the application of on–off control dosing systems, where mass output rates have a significant effect on the general controllability. Inadequate dosing capability can lead to loss of control in high load demand conditions, but too high a dosing rate can cause hunting. Obviously much depends on the input load variations: if occasional peak demands well in excess of a normal level are allowed (or anticipated), this will require the setting of a relatively high dosing rate, which in normal conditions will give rise to control instability. A rough guide is that the dosing rate output should provide 'on' time to an 'off' time in the ratio range 1:3 to 1:10. If the load variations are greater than this ratio range can accommodate it is better to provide load smoothing by means of a larger retention time in the reactor, or to use multiple on–off dosing, with spaced control set point values to bring in each dosing stage in a pseudo-proportional manner, rather than to increase the dosing output rate. Indeed much can be achieved by merely adjusting the reactor retention time (which is more satisfactory than providing a balancing tank upstream), or by employing multiple reactors (for which CFSTR theory shows that it is then possible to reduce the overall retention time).

The dosing time ratio itself is not the only factor to be taken into account. Too long an absolute dosing time can lead to insensitivity to load changes, with the risk of over-run, while too short a time can result in hunting. A rule of thumb for many continuous reactors is that the 'non-dosing' time should be no more than about one-fifth of the nominal reactor retention time. Application of dosing devices must be made with recognition of transfer lag, which should be minimised by ensuring that the time taken to transport reagent from the dosing device to the reactor must be minimised. One technique is by providing a circulating loop of reagent, with short tee connections for the control valves located at the reactor entry point. Short delivery runs after valves are necessary to avoid drainage effects following valve closure; pumps can be fitted with loading valves at the discharge point to prevent delivery pipe drainage after pump stoppage.

Practical conclusions

When one takes the various practical factors into account, even with the fastest of reactions it is usually imprudent to use nominal retention times of less than 5 minutes for any individual reactor stage, including systems of multiple reactors.

Where a single control system is applied to a series of reactors, this should preferably be at the first stage: subsequent stages are then essentially smoothing devices to balance out the hunt arising during control of the first stage. If mixing is not employed in the smoothing stages, efficiency is reduced, often to the extent that the smoothing time needs to be increased by 100% with respect to stirred reactors.

CONSERVATION AS A DESIGN APPROACH

Water and materials as cost factors

It has been stressed that conservation is better than treatment, and indeed is better than recycling or re-use, accepting that the concept of zero discharge is fundamentally meaningless and inconsistent with industrial processes as they are currently realised. Apart from the environmental factors, however, it is always important in the preliminary analysis for plant design to recognise that discharges should be regarded as a form of inefficiency in the production process: thus the materials that are discharged and the water that is used to convey them represent costs on the production process that should be reduced if not eliminated. The equation of cost must therefore include the costs of the materials being treated as well as the costs of treatment. Additionally, there is the close correlation between plant capital costs and water flow rates, which often follows a pattern expressed by

$$\frac{\text{Cost for flow } W_1}{\text{Cost for flow } W_2} = \left(\frac{W_1}{W_2}\right)^{0.60 \pm 0.05}$$

(while recognising that as flows decrease one eventually reaches an irreducible minimum associated with constant cost components, such as instrumentation and dosing equipment). On this basis alone it is clearly preferable to conserve to achieve a lower flow rate than to treat at a higher rate and then recycle at the same flow rate, provided the manufacturing process can tolerate this. So far as materials are concerned, the costs here normally relate to reagent demands which are in the main independent of flows. The ambition here must therefore be to reduce discharges at source: it is in any case simpler and usually more reliable to treat a small volume of concentrated discharge than a large volume of dilute one, because it brings about the possibilities for batch treatment. Also, concentrated solutions are frequently more amenable to recovery, or even to offsite treatment elsewhere—a parochial form of zero discharge!

Methods for conservation

A particularly important opportunity here, but outside the scope of this book, is through use of concentrating devices, whereby solutions are treated by separators to provide a concentrate that may be directly re-used or may approximate sufficiently to the original material to enable purification to be made for subsequent re-use or recycle. Particular examples of concentrator systems are evaporative devices, reverse osmosis, and ultrafiltration. Ion exchange does not truly concentrate, in that the composition of the eluate differs from that of the original solution, but it does provide a more concentrated version of the original that in some circumstances can be accepted for re-use. It can also be used to good effect to purify dilute aqueous discharges to provide high quality water that can be recirculated back to the process stations, and indeed it is in this respect that ion exchange finds most application in conservation.

Given that water conservation must always have a high priority, attention must be paid in plant design to the technique of achieving it. Where in industrial practice water is used to clean or otherwise rinse the material being processed, it is much more effective to use a counterflow of water with respect to the work flow in a series of discrete stages than to use one water station (see Fig. 1.5). If steady state

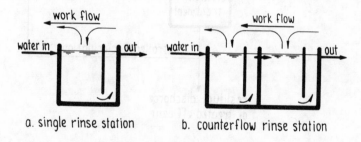

a. single rinse station   b. counterflow rinse station

Fig. 1.5 — Types of continuous water rinsing.

equilibrium conditions apply, the concentration c_{1e} of contaminant in water leaving a single rinse tank to which water is fed at a flow rate of W is given by

$$c_{1e} = \frac{c_{1i}D + c_o W}{W + D}$$

where c_{1i} is the contaminant concentration in solution introduced from a work article at a rate D, and c_o is the contaminant concentration in the water.

Where two rinse stages are used in series, as in Fig. 1.5b, the equation becomes

$$c_{2e} = \frac{c_{1i}D^2 + c_o W(W + D)}{W^2 + D(W + D)}$$

Steady state rinsing theory of this kind has been reviewed by Kushner (1949, 1951) and others. For non-steady state conditions, other factors enter the equation, such as volume of the wash tank (see the discussion by Pinner 1967): and where incomplete mixing occurs the dragover from the rinse tank may be substantially greater than steady state theory predicts, as has been shown by the experiments of Tallmadge and Buffham (1961). In this case the contaminant concentrations in the rinse water are of course less than predicted.

Further savings can thereafter be achieved by recirculation of chemically treated rinse water, e.g. by returning treated water back to the first rinse stage of a counterflow system at an increased rate, thus allowing a reduction of fresh water input (see Fig. 1.6). With this arrangement the water flow rate required to achieve

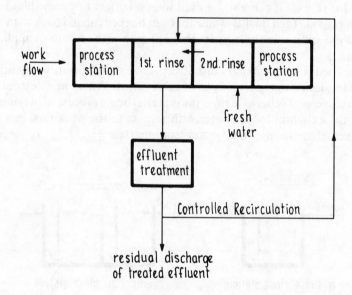

Fig. 1.6 — Water conservation in counterflow continuous rinsing systems by Controlled Recirculation.

the same concentration in the second rinse stage as with normal counterflow operation, i.e. having the same rinsing efficiency, is less, according to the relationship (Mattock 1971)

$$\frac{W_r^2 + W_r D}{W^2 + WD} = \frac{1}{1+r}$$

where W_r is the new input rate, W is the normal counterflow rate, r is a recirculation ratio factor, and rW_r is the recirculation flow rate (assuming the input water has zero contamination). Other related recirculation techniques have been applied for particular industrial processes, e.g. metal finishing, for which the so-called 'integrated' treatment system developed by Lancy (1957) in the United States can be applied. Closed loop treatment is applied to water recirculating through a first rinse stage, permitting the overall use of less water in a second (separate) rinsing stage (for full details see the review by Pinner 1967).

Quite apart from restraint and recirculation measures, water consumption can also be minimised by more efficient drainage of processed work before subsequent operations are applied, thus reducing the carry-over of process solution from one stage to the next. There are several means by which such intrinsic materials' conservation measures can be realised, most of which are used in economy programmes for the surface treatment industry: see Mattock (1979) for a summary.

As is so often said, both pollution and water waste are indicators of inefficiency; but at the same time it must be remembered that even efficiency has its price, and that economy measures can increase capital outlay. A cost–benefit analysis should therefore always be carried out on all design schemes, and particularly on conservation and recovery systems, to ensure optimisation of the total production and waste treatment programme.

REFERENCES

APHA/AWWA/WPCF (1985) *Standard Methods for the Analysis of Water and Wastewater,* 16th edn, American Public Health Association, American Water Works Association, Water Pollution Control Federation.

Bamford, C. H., and Tipper, C. F. H. (eds.) (1969) *Comprehensive Chemical Kinetics,* Vols. 1 and 2, Elsevier.

Barnes, D., and Wilson, F. (1978) *Chemistry and Unit Operations in Sewage Treatment,* Applied Science.

Benfield, L. D., and Randall, C. W. (1980) *Biological Process Design for Wastewater Treatment,* Prentice-Hall.

Chang, J. W. (1964) *Proc. Conf. Automatic Control in Chemical Process and Allied Industries,* Society of Chemical Industry.

Cooper, A. R., and Jeffreys, G. V. (1971) *Chemical Kinetics and Reactor Design,* Oliver & Boyd.

Coulson, J. M., and Richardson, J. F. (1979) *Chemical Engineering,* Vol. III, 2nd edn, eds. Richardson, J. F., and Peacock, D. G., Pergamon.

Coxon, W. F. (1962) *Automated Process Control,* Heywood.

Curi, K., and Eckenfelder, W. W., Jr. (1980) *Theory and Practice of Biological Wastewater Treatment,* Sijthoff and Noordhoff.

Denbigh, K. G., and Turner, J. C. R. (1984) *Chemical Reactor Design,* 3rd edn, Cambridge University Press.

Department of Environment (1977) *Methods for the Examination of Waters and Associated Materials. Chemical Oxygen Demand (Dichromate Value) of Polluted and Waste Waters,* Her Majesty's Stationery Office.

Department of Environment (1979a) *Methods for the Examination of Waters and Associated Materials. The Instrumental Determination of Total Organic Carbon, Total Oxygen Demand and Related Determinands,* Her Majesty's Stationery Office.

Department of Environment (1979b) *Methods for the Examination of Waters and*

Associated Materials. Colour and Turbidity of Waters, Her Majesty's Stationery Office.

Department of Environment (1981, 1982) *Methods for the Examination of Waters and Associated Materials. Biochemical Oxygen Demand*, Her Majesty's Stationery Office.

Department of Environment (1983) *Methods for the Examination of Waters and Associated Materials. The Permanganate Index and Permanganate Value Tests for Waters and Effluents*, Her Majesty's Stationery Office.

Frost, A. A., and Pearson, R. G. (1961) *Kinetics and Mechanisms*, Wiley.

Grady, C. P. L., Jr., and Lim, H. C. (1980) *Biological Wastewater Treatment*, Marcel Dekker.

Holland, F. A., and Chapman, F. S. (1966) *Liquid Mixing and Processing in Stirred Tanks*, Reinhold.

Kakabadse, G. (ed.) (1978) *Chemistry of Effluent Treatment*, Applied Science.

Kushner, J. B. (1949) *Plating* **36** 789; 915.

Kushner, J. B. (1951) *Metal Finish.* **49** (11) 59; (12) 58; 67.

Lancy, L. E. (1957) *Electroplat. Metal Finish.* **10** 251.

Lees, F. P. (1976) *Chem. Ind. (London)* 195.

Levenspiel, O. (1972) *Chemical Reaction Engineering*, 2nd edn, Wiley.

McMillan, G. (1984) *Intech* **31** 69.

Mattock, G. (1971) *Chem. Ind. (London)* 46.

Mattock, G. (1976) *Chem. Ind. (London)* 918.

Mattock, G. (1979) *Finish. Ind.* **3** 24.

Metcalf & Eddy, Inc. (1979) *Wastewater Engineering. Treatment, Disposal and Re-Use*, 2nd edn, McGraw-Hill.

Moore, R. L. (1978) *Neutralisation of Waste Water by pH Control*, Instrument Society of America.

Pinner, R. (1967) *Electroplat. Metal Finish.* **20** 208; 248; 280.

Pollard, A. (1971) *Process Control*, Heinemann.

Purchas, D. B. (1971) *Industrial Filtration of Liquids*, Godwin.

Schroeder, E. B. (1977) *Water and Wastewater Treatment*, McGraw-Hill.

Shinskey, F. G. (1973) *pH and pIon Control in Process and Waste Streams*, Wiley.

Sundstrom, D. W., and Klei, H. E. (1979) *Wastewater Treatment*, Prentice-Hall.

Tallmadge, J. A., Jr., and Buffham, B. A. (1961) *J. Wat. Pollut. Control Fed.* **33** 817.

Water Pollution Control Federation (1976) *Operation of Wastewater Treatment Plants. Manual of Practice No. 11.*

Water Pollution Control Federation (1977) *Wastewater Treatment Plant Design. Manual of Practice No. 8.*

Weber, W. J., Jr. (1972) *Physicochemical Processes for Water Quality Control*, Wiley–Interscience.

Weston, R. E., Jr., and Schwarz, H. A. (1972) *Chemical Kinetics*, Prentice-Hall.

White, J. B. (1978) *Wastewater Engineering*, 2nd edn, Edward Arnold.

2

pH Adjustment

THE NATURE AND SIGNIFICANCE OF pH

The pH concept

The properties of all but the most concentrated aqueous solutions reflect the unusual characteristics of the solvent water. Water is a weak electrolyte, having a high dielectric constant (approximately 80), and it dissociates according to the nett equation

$$H_2O \rightleftharpoons H^+ + OH^-$$

the hydrogen and to a lesser extent the hydroxyl ions themselves being surrounded by water molecules, i.e. they are hydrated. The degree of water dissociation, K_w, from the above equation, defined by

$$K_w = \frac{a_{H^+} \cdot a_{OH^-}}{a_{H_2O}}$$

is quite small, K_w being of the order of 10^{-14} at room temperature. In pure water, a_{H_2O}, the activity of the solvent water, is effectively constant, and is defined as unity; so for these circumstances each species has an equal activity of $a_{H^+} = a_{OH^-} = \sqrt{K_w} = 10^{-7}$. This corresponds to a neutral condition. However, it should be noted that because K_w is temperature-dependent, so too is the pH, reducing from 7.5 at 0°C, to 7.0 at 25°C, and 6.5 at 60°C.

When an electrolyte is dissolved in water, this equivalence or neutrality may be disturbed, as for example by the introduction of hydrogen ion (proton)-donating species, which gives an 'acid' solution, or of a proton-accepting species, giving an 'alkaline' solution. The former results in an excess of hydrogen over hydroxyl ions, and the latter the opposite.

The term 'pH' is used to indicate the degree of acidity of aqueous solutions, and derives from an old definition that sought to relate an electrochemically determined quantity, pH_c, to the hydrogen ion concentration, c_{H^+}, by

$$pH_c = -\log c_{H^+}$$

In fact this 'definition' cannot generally be realised in practice with any acceptable degree of accuracy for fundamental thermodynamic reasons, and pH as a usable concept is nowadays based on operational standards.

This may be demonstrated by consideration of the two primary types of electrochemical cell that may be employed for pH measurement. The more common uses two half-cells, of which one responds in electrical potential to changes in hydrogen ion activity, a_{H^+} (equal to hydrogen ion concentration, c_{H^+}, times an activity coefficient, f), while the other exhibits a constant potential (at constant temperatures); the two are connected by means of a liquid junction. The primary form of hydrogen ion-responsive half-cell is a platinum–hydrogen gas electrode immersed in the electrolyte (the more commonly used glass electrode approximates in behaviour to this), while a mercury–mercurous chloride half-cell (the so-called 'calomel' electrode) may be used as the constant potential reference, being in contact with a constant activity of chloride ions at any given temperature. A representation

Pt; H$_2$	|	HX	||	satd. KCl	|	Hg$_2$Cl$_2$; Hg
hydrogen electrode		electrolyte		reference electrolyte		calomel electrode

$$\underbrace{\qquad\qquad\qquad\qquad}_{\text{hydrogen half-cell}} \qquad\qquad \underbrace{\qquad\qquad\qquad}_{\text{calomel half-cell}}$$

of the complete cell is thus the twin vertical lines representing the liquid–liquid junction or interface. The open circuit e.m.f. of this cell at an absolute temperature T is given by the difference between the two half-cell potential values, from the relevant Nernst equations:

$$E = E_{cal} - E_H$$

$$= \left(E_{cal}^{\ominus} - \frac{2.3026RT}{F} \log a_{Cl^-} \right) - \left(E_H^{\ominus} + \frac{2.3026RT}{F} \log a_{H^+} \right) + E_j$$

where R is the gas constant; F is the Faraday; $E_{cal}^{\ominus}$ is the standard potential of the calomel electrode at temperatue T, i.e. the value of E_{cal} when $\log a = 0$; $E_H^{\ominus}$ is the standard potential of the hydrogen electrode; and E_j is the so-called 'liquid junction potential' that arises at the boundary between the two dissimilar electrolytes. By electrochemical convention, $E_H^{\ominus} = 0$, so that

$$E = E_{cal}^{\ominus} - \frac{2.3026RT}{F} (\log a_{H^+} + \log a_{Cl^-}) + E_j$$

$$= E_{cal}^{\ominus} - \frac{2.3026RT}{F} (\log c_{H^+} f_{H^+} + \log c_{Cl^-} f_{Cl^-}) + E_j$$

It can be seen that $\log c_{H^+}$, and thus pH_c as defined above, cannot be calculated from this relationship, because the ionic activity coefficient f_{H^+} is not known, nor is f_{Cl^-}, nor is E_j. It is not even possible to calculate pH precisely in practical terms from the ideal relationship

$$pH = -\log a_{H^+}$$

because it is not possible to evaluate a_{Cl^-} and E_j precisely.

The second type of cell does not use a liquid junction. It is typified by a cell such as

$$Pt; H_2|HCl|AgCl; Ag$$

where the e.m.f. is given by

$$E = E^{\ominus}_{Ag; AgCl} - \frac{2.3026RT}{F}(\log a_{H^+} + \log a_{Cl^-})$$

Here again a problem arises in that there are two variables, namely a_{H^+} and a_{Cl^-}, neither being independently determinable.

If, therefore, pH is to be a practical parameter, it has to be based on certain extra-thermodynamic assumptions, concerning values of activity coefficients, and, if cells with liquid junctions are to be employed, concerning liquid junction potentials. These assumptions can then be used to define a set of solution reference standards whose pH values provide a scale spanning the range 0–14. Two systems have been widely used, namely the National Bureau of Standards (USA) scale and the British Standards Institution pH scale. These are based on different premises, and so differ slightly, although not materially except for the most precise work. Thus the NBS scale is based on measurements made on electrochemical cells without liquid junction of the type mentioned above, and extra-thermodynamic assumptions are made to calculate values for a_{Cl^-}, subsequently to provide operational activity pH definitions for the standard solutions. The BSI scale makes no attempt at activity pH definition: a single standard reference solution is *defined* as having a given pH, and all other solutions are then assigned pH values from the results of experimental observations on cells with liquid junction of the type referred to above. The standard pH reference values for the two scales are thus slightly different. A key difference between the two scales is that no liquid junction potential term appears in the NBS scale, but that E_j factors are included in the BSI reference values. According to the British Standard, the difference in pH between a test solution X having pH = pH_x and a standard solution S having a defined pH = pH_s is measured in open circuit conditions at constant temperature using a cell of the type

$$Pt; H_2|X \text{ or } S\|\text{reference electrode}$$

with observation of the e.m.f. of the cell with solution X ($= E_x$) and of the e.m.f. of the cell with solution S ($= E_s$). Then

$$(pH_x - pH_s) = (E_x - E_s)\bigg/\frac{2.3026RT}{F}$$

It is possible to apply NBS standard values here, assuming that the liquid junction potential error introduced in the measurement with solution S is cancelled by the presence of an identical potential when solution X is used. This is broadly reasonable to within ± 0.02 pH unit, provided $(\text{pH}_x - \text{pH}_s)$ is not greater than ± 3, that the pH is not at the scale extremes of 0–2 and 12–14, and provided no other unusual solution conditions prevail.

Recently a compromise agreement accommodating the two standards was reached (Covington *et al.* 1983, 1985) within the framework of the International Union of Pure and Applied Chemistry (IUPAC), whereby a reference value pH standard (0.05 mol kg^{-1} potassium hydrogen phthalate) was defined, against which a series of operational pH standards was also defined. The relevant pH values for these solutions are given in Table 2.1; they correspond also to the British Standard, BS 1647 (1984).

pH as measured is an indication of the hydrogen ion 'level', but it is not simply related to hydrogen ion concentration, and is only approximately related to a hydrogen ion activity. This must be borne in mind when attempts are made to calculate concentration levels of species in solution from pH measurements, the significance of which will be apparent in later sections of this chapter. pH is important in pollution treatment, but perhaps more as a number than as a hydrogen ion indicator. Also, pH in this context does not usually demand measurement precision to better then ± 0.05 unit, so although limitations on interpretation must be recognised, they should not be regarded as inhibitory to the use of 'pH'.

For comprehensive discussions on the theoretical basis and practical application of the pH concept, the reader is referred to Bates and Guggenheim (1960), Mattock and Band (1967), Bates (1973, 1981, 1982), Covington (1981), and Covington *et al.* (1983, 1985).

Acid–base titration curves and buffer solutions
The operational hydrogen ion activities that can be calculated from pH measurements from the relationship

$$\text{pH} = (-\log a_{\text{H}^+})_{\text{op}}$$

are not necessarily linearly related to the total acidity or basicity of an electrolyte solution; such a relationship could only in principle apply where there is complete dissociation of acids into anions and hydrogen ions, and complete uptake by bases of hydrogen ions to leave cations and hydroxyl ions. Thus, for example, if a plot is made of the concentration of an acid solution against the antilog of the corresponding pH values the resulting curve will not necessarily be linear. It does not even occur with a 'strong' acid, such as hydrochloric, where total dissociation is often assumed and linearity of the plot may therefore be naïvely anticipated. The non-linearity here arises from the fact that the derived a_{H^+} calculation includes a variable activity coefficient, f_{H^+}, from the relationship

$$a_{\text{H}^+} = c_{\text{H}^+} f_{\text{H}^+}$$

Table 2.1 — pH values for some operational standard reference buffer solutions (after Covington et al. 1985)

Solution	Temperature, °C														
	0	5	10	15	20	25	30	37	40	50	60	70	80	90	95
(1) 0.05 mol kg⁻¹ potassium tetroxalate	—	—	1.638	1.642	1.644	1.646	1.648	1.649	1.650	1.653	1.660	1.671	1.689	1.72	1.73
(2) 0.05 mol kg⁻¹ potassium hydrogen phthalate	4.000	3.998	3.997	3.998	4.000	4.005	4.011	4.022	4.027	4.050	4.080	4.116	4.159	4.208	4.235
(3) 0.025 mol kg⁻¹ disodium hydrogen phosphate + 0.025 mol kg⁻¹ potassium dihydrogen phosphate	6.963	6.935	6.912	6.891	6.873	6.857	6.843	6.828	6.823	6.814	6.817	6.830	6.86	6.90	6.92
(4) 0.05 mol kg⁻¹ disodium tetraborate	9.475	9.409	9.347	9.288	9.233	9.182	9.134	9.074	9.051	8.983	8.932	8.898	8.88	8.84	8.89
(5) 0.025 mol kg⁻¹ sodium hydrogen carbonate + 0.025 mol kg⁻¹ sodium carbonate	10.273	10.212	10.154	10.098	10.045	9.995	9.948	9.889	9.866	9.800	9.753	9.728	9.725	9.75	9.77
(6) Saturated calcium hydroxide	13.360	13.159	12.965	12.780	12.602	12.431	12.267	12.049	11.959	11.678	11.423	11.192	10.984	10.80	10.71

Preparation

(1) Dissolve 12.710 g $KH_2C_4O_8 \cdot 2H_2O$ in 998.2 g distilled water, or 12.62 g in 1 litre solution at 20°C.

(2) Dissolve 10.211 g $C_6H_4(CO_2H)(CO_2K)$ in 1000 g distilled water, or 10.138 g in 1 litre solution at 20°C.

(3) Dissolve 3.549 g Na_2HPO_4 and 3.402 g KH_2PO_4 in 1000 g distilled water, or 3.54 and 3.39 g respectively in 1 litre solution at 20°C.

(4) Dissolve 19.068 g $Na_2B_4O_7 \cdot 10H_2O$ in 991.0 g distilled water, or 19.01 g in 1 litre solution at 20°C.

(5) Dissolve 2.100 g $NaHCO_3$ and 2.650 g Na_2CO_3 in 1000 g carbon dioxide-free distilled water, or 2.10 g and 2.64 g respectively in 1 litre solution at 20°C.

(6) Use at least 1.6 g $Ca(OH)_2$ per litre solution at 20°C. (Excess solid should be present at usage temperature.)

Use high quality analytical reagent chemicals. For the most precise work, purification should be undertaken according to BS 1647 (1984). Other reference solutions are also given in this Standard.

while the acid concentration corresponds to c_{H^+}, when total dissociation is assumed, such that

$$c_{HX} = c_{H^+} = c_{X^-}$$

One interpretation of this non-linearity may be that ion association occurs in such solutions, resulting in deviations of f_{H^+} from unity, implying that complete dissociation is a concept of dubious value. This underlines the fundamental unreliability of the calculations of acid or base concentrations sometimes crudely made from pH measurements.

A different situation can be mechanistically recognised, however, with for example 'weak' acids, where only incomplete dissociation of the acid HX may be considered to occur:

$$HX \rightleftharpoons H^+ + X^-$$

definable by a dissociation constant

$$K_{HX} = \frac{a_{H^+}.a_{X^-}}{a_{HX}}$$

Here it can be understood that greater departures from linearity will occur in a plot of acid concentration against antilog pH, the deviation increasing with increasing values of K_{HX}.

The same essential arguments apply to pH changes that occur in titrations involving acids with bases, except that the situation here is rendered more complex by the co-presence of two species whose dissociative/associative characteristics may be intrinsically different and be further modified by interactions that may occur in solution between the various species. The form of an acid-base pH titration curve depends primarily on the intrinsic strengths of the acid and base involved. Fig. 2.1 shows the familiar pH curves for the titrations of strong acid by strong base, weak acid by strong base, weak acid by weak base, and dibasic acid (representative of the general class of polybasic acids) by strong base. (Formally similar curves result from titration of a strong base or a weak base by a strong acid, or of a weak acid by a weak base, or of a polyacidic base by a strong acid.) The example of polybasic acid being titrated by a strong base, it may be noted for later reference, is exemplified in the titration of a hydrolysable polyvalent metal ion by a strong base, where the hydrated metal ion is behaving as a polybasic acid. (For theoretical analyses of titration curves see, for example, Ricci (1952) and Stumm and Morgan (1981), although it must be pointed out that these authors use concentration rather than activity terms.) The important features of these curves in the present context is that in certain regions they show an insensitivity in pH to the degree of titration. Such a solution whose pH varies only slightly on addition of small quantities of acid or base is said to be a 'buffered' solution. Two types of solution show pH buffer qualities:

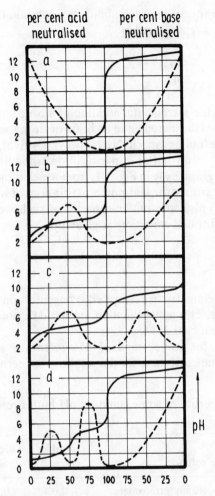

Fig. 2.1 — pH titration curves and buffer capacity curves (broken lines) for titration of acids by alkalis. a. Strong acid plus strong base. b. Weak acid plus strong base. c. Weak acid plus weak base. d. Dibasic acid plus strong base.

(1) Mixtures of a weak acid and its salt with a strong base, or of a weak base and its salt with a strong acid, or of a mixture of a weak acid and a weak base.
(2) Mixtures of strong acids or of strong bases, not more dilute than about 0.01 M, with small quantities of strong bases or strong acids respectively, and solutions of strong acids or bases alone that are not more dilute than about 0.01 M.

The reasons for the buffer properties differ in cases (1) from cases (2). In the latter, buffer characteristics arise simply from the logarithmic relationship of pH to hydrogen ion activity (the latter being crudely indicative of the acid or base concentration), this relationship being relatively insensitive at low and high pH values (0–2, 12–14). In cases (1) it is due to the ability of weak acid anion or a weak base cation to accept or donate hydrogen ions respectively.

The quantitative expression of the ability of a solution to behave as a buffer to acid or base addition is represented by

$$\beta = \frac{\partial c_b}{\partial pH} = -\frac{\partial c_a}{\partial pH}$$

where ∂c_b represents the incremental addition of strong base in gram-equivalents/litre, and ∂c_a represents the incremental addition of strong acid in gram-equivalents/litre. β is known as the 'buffer capacity' (it is always positive); its reciprocal is known as the 'buffer index'. The buffer capacities corresponding to various titration conditions are shown graphically in Fig. 2.1, from which it can be seen that fairly flat maxima are exhibited around the mid-points of titrations of weak acids or bases. To a first approximation (only), the buffer capacity of such weak acid or weak base systems in these half-titrated conditions is given by

$$\beta \approx \frac{2.3c}{4}$$

where c is the stoicheiometric (pre-titrated) concentration of the acid or base in gram-equivalents/litre. This approximation, it should be noted, can only be applied to solutions up to about $c = 0.25\,M$.

Buffer capacities for some standard pH reference solutions are given in Table 2.2. The importance of these for the solutions concerned is that pH reference

Table 2.2 — Properties of some standard pH buffer reference solutions at 25°C

Buffer solution	Buffer capacity	Dilution value	Salt effect
0.05 mol kg^{-1} potassium tetraoxalate	0.070	+0.186	—
Satd. (at 25°C) potassium hydrogen tartrate	0.027	+0.049	—
0.05 mol kg^{-1} potassium hydrogen phthalate	0.016	+0.052	−0.01(KCl)
0.025 mol kg^{-1} potassium dihydrogen phosphate 0.025 mol kg^{-1} disodium hydrogen phosphate	0.029	+0.080	−0.01(NaCl)
0.01 mol kg^{-1} borax	0.020	+0.010	−0.01(NaCl)
Satd. (at 25°C) calcium hydroxide	0.090	−0.280	—

standards (Table 2.1) desirably should have relatively high buffer values for practical reasons of stability with respect to accidental contamination by acid or base. In a wider sense, the buffer characteristics of solutions in general are of considerable importance in both waste/water treatment operations, as will be shown later, and in stabilising environmental conditions.

Another term of significance here is the 'dilution value', $\Delta pH_{\frac{1}{2}}$, defined by Bates (1954) as being the increase in pH exhibited by a solution when it is diluted with an equal volume of water:

$$\Delta pH_{\frac{1}{2}} = (pH)_{\frac{1}{2}c} - (pH)_c$$

Table 2.2 shows some dilution values for various buffer solutions. It can be appreciated that pH reference standards should desirably have low dilution values.

Introduction of salts into buffer solutions can have an effect on their pH values. This can loosely be regarded as the opposite of a dilution effect, and can arise in practice, for example, through leakage of salt bridge solution into the buffer. Table 2.2 gives an indication of the orders of magnitudes for three primary buffers, written in terms of pH/salt molarity, up to 0.05 M salt strengths.

pH in the context of ecology and pollution control

A small change in pH corresponds to a large change in hydrogen ion activity, and it is perhaps surprising that the tolerance often shown by living systems in aquatic environments to changes in pH between approximately 6.5 and 8.5, corresponding to a change of two orders of magnitude of hydrogen ion activity, is so great. Nevertheless it is generally true that it is the pH rather than the titratable acidity or basicity that is of importance in aquatic systems, and consequently the process of pH adjustment constitutes one of the key operations in pollution control. The actual absorptive capacity of a receiving watercourse for acids or bases is dependent on its buffer capacity, and for discharges representing a relatively high proportion of the watercourse flow, on its dilution value also. The dilution value of the discharging solution is also of importance. In some cases the buffer capacities of fresh water streams or lakes may be quite low; and sea water also has a relatively low buffer capacity—a fact which makes relatively confined volumes of sea water particularly susceptible to pollution by acids and alkalis.

The achievement of an acceptable pH for discharge is not, however, the sole reason for the importance of pH in waste water treatment. Many treatment processes, e.g. those involving oxidation–reduction reactions, or those involving precipitation of polluting compounds for subsequent removal, are critically dependent on careful adjustment to a specific pH not necessarily in the neutral region. This chapter examines those processes where pH adjustment constitutes the primary treatment procedure; in succeeding chapters the influence of pH is demonstrated in the relevant sections.

pH MEASUREMENT

Principles of measurement

The hydrogen ion-responsive system for pH measurement is usually one of three main types:

(1) the platinum–hydrogen gas electrode, of fundamental importance, because its standard potential, $E_H^\ominus$, in the Nernst equation

$$E = E_H^\ominus + \frac{2.3026RT}{F} \log a_{H^+}$$

is conventionally defined as being zero, making it a true reference electrode, and of practical importance because of the high precision with which measurements can be made with it;

(2) the glass electrode, which shows a similar response characteristic to the platinum–hydrogen gas electrode for much of the pH range; and

(3) various metal–metal oxide systems which exhibit pH response over certain sections of the practical pH range.

Of these the glass electrode is by far the most commonly employed, being convenient, simple, and adequately precise for the majority of requirements. The hydrogen gas electrode is essentially a research and reference tool, while of the metal–metal oxide systems only the antimony electrode finds occasional application, e.g. in fluoride-containing solutions that attack glass electrodes.

The properties of glass electrodes have been fully described in the literature, and both their advantages and their limitations have been exhaustively discussed; likewise reference electrodes and their properties have been thoroughly examined: see, for example, Ives and Janz (1961), Mattock (1961), Eisenman (1967), and Bates (1973). Because of the high electrical resistance of glass electrodes (usually in the range 50–500 megohms), glass-reference electrode cell e.m.f. values must be measured using high impedance systems capable of discriminating to the necessary degree of replicability (0.1–0.5 mV, equivalent to approximately 0.02–0.1 pH) at very low cell currents corresponding closely to ideal open circuit conditions. Such instruments constitute the pH meters commercially available for day-to-day plant and laboratory pH measurements.

The constituent components of glass and reference electrodes and of the measuring system are represented in Fig. 2.2. The total cell e.m.f., E, can be written as the sum of all the potential components:

inner ref. electrode	inner ref. solution	glass membrane	solution	salt bridge solution	external ref. solution	external ref. electrode
$-E_i$		$-E_{ib} + E_a$	$+k T \mathrm{pH}$	$+E_{j1} + E_{j2}$		$+E_e$

The e.m.f. response E_t for the test solution is compared with the response E_b for a standard buffer solution. Provided the temperatures of the two solutions are the same, the assumption can usually be made that E_i, E_{ib}, E_a, E_{j1}, E_{j2} and E_e remain constant in the two measurements, and so cancel, so that

$$E_t - E_b = kT(\mathrm{pH}_t - \mathrm{pH}_b)$$

where ($k = 2.3026R/F$). When the temperatures are not the same the situation is

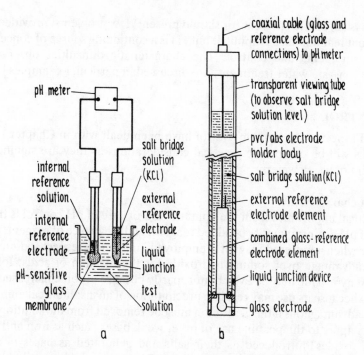

Fig. 2.2 — The pH measurement cell. a. Diagrammatic representation, showing component elements. b. Typical industrial (plant) cell assembly.

more complicated, but suitable instrumental correction can usually be applied either manually or automatically (e.g. with the aid of a resistance thermometer incorporated in the cell) to allow for variations in E_i, E_{ib} and E_e. The glass asymmetry potential, E_a, and liquid junction potential, E_{j1}, are only likely to vary when the pH difference between test and standard solution is marked (>7 pH), or when the temperature difference is marked (>20°). Commercial instruments provide variable offset voltages to balance out $(- E_i - E_{ib} + E_a + E_{j1} + E_e)$, thus to display a pH reading corresponding to the standard solution value when the standardisation is effected. The scale calibration is based on the slope factor kT, modified by zero shifts that occur with temperature according to the types of inner and external reference electrodes employed.

Practical factors
Three main problem areas, or sources of error, can be identified in pH measurement with glass electrodes;

(1) deterioration of, or coatings on, the glass surface, and blockage of the liquid junction system, which can have a marked influence on the apparent pH;
(2) the high electrical impedance of the pH cell, which demands a high and maintained standard of electrical insulation in the measuring system; and
(3) stability and reliability of the pH meter, which is obviously of prime importance.

Nowadays (2) and (3) present (or should present) few problems, provided adequate equipment maintenance is applied, but (1) is a continuing source of concern in plant pH measurements and control. The characteristic difficulties observed in (1), relevant to waste water treatment, are discussed in particular examples later.

pH CONTROL

General aspects of automatic control have been dealt with in Chapter 1. Here the emphasis will be on specific practical considerations as they are significant in the treatment of waste waters.

Reagent characteristics

Of practical importance from the point of view of ease of pH control is the titration curve of pH against quantity of added reagent, and the choice of reagent. pH control is easier when a buffered reagent is employed, and in this respect sodium carbonate has an advantage over sodium hydroxide for the titration of acids, for example. Calcium hydroxide may be even better, partly because of some increased buffering, but also because its use may result in precipitation of an insoluble calcium salt, which may be advantageous if the acid anion must be removed from the solution. The same remarks apply to the possible use of other weak bases, such as iron and aluminium hydrous oxides, introduced as their salts and generated as bases *in situ* by the addition of an alkali to the acid–salt mixture. Magnesium hydroxide has both buffer-forming characteristics and less propensity towards forming insoluble salts than calcium hydroxide; it is, moreover, more expensive, although weight for weight more effective. Ammonia has occasionally been used as a neutralising base, but its use can bring unwanted side effects, such as the introduction of ammoniacal nitrogen into the effluent, or the complexing of heavy metals, so it is not generally to be recommended.

It is possible to calculate the curve characteristics from a knowledge of the dissociation constant, K, and concentrations of both the acid being titrated and of the base being used as reagent, by making the assumption that activity coefficients are constant, and that $pH = pH_c = -\log c_{H^+}$. These assumptions are not really justifiable, as has been shown above, and it must therefore be understood that calculations made therefrom can be regarded as indicative only, and that practical curves will differ from the calculated ones. Of particular importance are the curves obtaining with strong acids where any one of sodium hydroxide, sodium carbonate or calcium hydroxide is used as neutralising reagent, exemplified in Fig. 2.3 for a monobasic acid, e.g. hydrochloric acid, or sulphuric acid (which, although dibasic, acts effectively monobasically through both hydrogen ions dissociating together). Sodium carbonate behaves as a diacidic base in this context:

$$CO_3^{2-} + H^+ \rightleftharpoons HCO_3^-$$
$$HCO_3^- + H^+ \rightleftharpoons H_2CO_3 \rightleftharpoons H_2O + CO_2$$

This has a flattening effect on the titration curve, a feature of some importance in pH control.

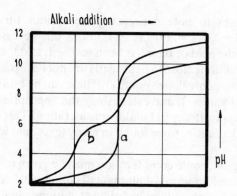

Fig. 2.3 — pH titration curves for the addition of alkali to hydrochloric acid.
a. NaOH. b. Na_2CO_3.

Where a strong base is added to a 'stronger' weak acid such as acetic ($pK_a = -\log K_a = 4.8$, where K_a = dissociation constant for

$$CH_3COOH \rightleftharpoons CH_3CO_2^- + H^+)$$

or to intermediate strength acids such as carbonic (pK_a for

$$H_2CO_3 \rightleftharpoons HCO_3^- + H^+$$

being 6.5), the titration curves are flattened even more; while with addition to very weak acids, such as boric (pK_a for

$$H_2CO_3 \rightleftharpoons H_2CO_3^- + H^+$$

being 9.2), there is no inflection in the curve.

Another case is where a polybasic acid is being neutralised by a strong base. To take a dibasic acid, if the first dissociation

$$H_2X \rightleftharpoons HX^- + H^+$$

is strong, and the second weak, as in the case of sulphuric acid, only one inflection is observed: otherwise two inflections will be seen. With phosphoric acid, three dissociation steps can be identified:

$$H_3PO_4 \rightleftharpoons H_2PO_4^- + H^+ \qquad pK_1 = 4.5$$
$$H_2PO_4^- \rightleftharpoons HPO_4^{2-} + H^+ \qquad pK_2 = 9.5$$
$$HPO_4^{2-} \rightleftharpoons PO_4^{3-} + H^+ \qquad pK_3 = 14.5$$

In practice, the third inflection will be difficult to observe, and in the case where calcium hydroxide is being used as reagent, is further complicated by the precipitation of insoluble calcium phosphates, e.g. $CaHPO_4 \cdot 2H_2O$ and $Ca_3(PO_4)_2$.

An important point to note concerning the various titration curves is that neutralisation to pH $= \sqrt{K_w} = 7.0$ (at 25°C) does not necessarily mean that the equivalence point of the acid or base has been reached, and therefore that the ionic species present in conditions close to neutral pH are not necessarily correspondent to those to be found at the equivalence point(s). These can be established by noting the equivalence point pH value(s), and establishing the appropriate dissociation equations that therefore apply at any pH under consideration. This exercise is a piece of standard chemical calculation, to be found in many texts, and will not be considered here further.

In choosing a reagent, some consideration must be given to the buffer characteristics of the reagent itself, in terms of ultimate pH obtainable versus the desired pH. Thus, for example, although sodium hydroxide solutions can provide a high pH (M NaOH, i.e. 4%, has a pH of approximately 13.7 at 20°C), M Na_2CO_3, i.e. 5.3% at 20°C has a pH of only 10.9, while saturated lime slurry has a pH of 12.6 at 20°C. At some point, excessive amounts of reagent are needed to raise the pH by only a small amount, rendering the operation uneconomic. However, the buffer properties of some materials can sometimes be used to advantage to assist in stabilising the pH: for instance, aluminium sulphate, used as an emulsion-cracking acid, will at the same time provide a measure of pH control by its buffer properties.

The neutralisation of alkalis arises less frequently in pollution treatment, but the control problems are similar. It is customary to employ either sulphuric or hydrochloric acids as reagents (where other waste acids are not available), and of course it is necessary to study the titration curve characteristics of actual samples to determine the control modes to be adopted. Carbon dioxide has been used as an acidifier, since it offers the advantage of providing a buffered pH control curve, and is sometimes relatively inexpensive when available in waste flue gases.

Acid is used in pH control for secondary purposes, e.g. in oil emulsion cracking (see Chapter 7) and in chromate reduction (see Chapter 4).

Reaction times

Attention must be paid in reactor design to reaction and control lags, as already outlined in Chapter 1. The glass electrode sensor and associated measuring equipment normally has an adequately fast response, and the neutralisation process is normally regarded as being instantaneous. However, the *effective* reaction rate can be controlled by other factors. For example, calcium hydroxide is normally dosed as a slurry, and a limiting factor on reaction rate is then the rate of solution of the hydrated lime particles (which is itself a function of their degree of sub-division), the solution pH, and the temperature (calcium hydroxide exhibits a negative solubility–temperature relationship), as well, of course, as mixing efficiency. Reactions between sodium carbonate and acid involve the formation of carbonic acid, then release of carbon dioxide from the H_2CO_3 molecule, and then gaseous evolution of the CO_2, the last of which is usually rate-determining:

$$CO_3^{2-} + H^+ \rightarrow HCO_3^- \xrightarrow{H^+} H_2CO_3 \rightarrow H_2O + CO_2 \rightarrow CO_2 \uparrow$$

The time reaction characteristics for interaction of hydrochloric acid with lime slurry are illustrated graphically in Fig. 2.4, taken from Shinskey's monograph (1973). Shinskey has suggested that for single stage continuous neutralisation with lime slurry a nominal residence time of 15 minutes should be used when the influent pH is 3 or above. He also recommends the use of a two stage reactor system for more acid conditions or where a reduction in total residence time is required. If, as Shinskey suggests, the time delay is due to the presence of calcium carbonate in lime, then the reaction of the calcium carbonate must be considered. This will not react as a base above pH values 9–10, although it will at pH values below about 8. The time scales are consistent with other experiments that were carried out by Hoak *et al.* (1947) on the relative efficacies of different types of lime, which additionally illustrate a faster response for high calcium lime reagents than for those containing magnesium, e.g. dolomitic products, when sulphuric acid or waste pickle liquor are being neutralised. The different rates have been explained in terms of the lower pH of the magnesium mixtures, which provide a weaker driving force for reaction completion.

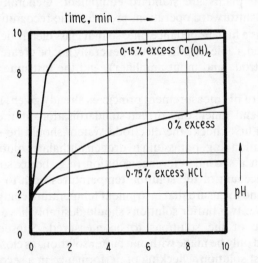

Fig. 2.4 — Reaction time characteristics for neutralisation of hydrochloric acid by lime slurry (reproduced from Shinskey 1973).

In our experience the use of two stage lime slurry dosing in an on–off mode in a single reactor is quite effective using nominal residence times of 5–15 minutes, even when influent pH values are less than 3. Of rather more importance than influent pH is the rate and degree of its variation, and whether or not precipitation occurs in the reaction, e.g. with treatment of sulphuric acid to form calcium sulphate. The latter tends to form coatings of insoluble calcium sulphate on the lime slurry particles, inhibiting further reaction; not only does it slow down the effective rate of reaction, but it may also result in pH drifts subsequent to the reactor stage. There are several ways of mitigating this problem apart from increasing the residence time; for example, more efficient mixing, the use of a highly dispersed lime slurry (note that dry solids should not be used directly), and a fine particle size of slaked lime.

Sodium carbonate can present a greater problem from the time required in practice to disperse and remove the carbon dioxide formed, although 10–20 minutes residence time should suffice, with two stage dosing to aid matters if necessary. It may seem superficially that sodium hydroxide, with its faster response time in relation to the other two reagents, would present an attractive alternative, but the benefits are illusory. The control hunting problem that can arise with high pH reagents when used for control in neutral pH zones where buffer capacity is weak can be greater than reaction lag difficulties; this may be seen from an inspection of the pH titration curves shown in Fig. 2.3. It is often better to try to avoid using sodium hydroxide where little buffer capacity is available within the solution to be treated. The difficulty is not necessarily overcome by the employment of more sophisticated control systems, because although these may theoretically reduce minimum residence times, the other limiting factors discussed are of greater significance.

The sensor system
The pH electrode assemblies and instrumentation used for pH monitoring and control of treatment plants are standard equipment. Generally speaking, their installation is a straightforward operation, provided due recognition is made of the fact that the system is a high impedance one: so that glass electrode connection cable must be well insulated, cable lengths should generally not be greater than about 50 m (between glass electrode and input amplifier), and the instrument should be well shielded electrically.

From the review of pH measurement principles already given, it can be seen that reliable results depend much on correct standardisation of the glass–reference electrode cell. As a first check, the electrode system should be standardised in a reference buffer solution (e.g. potassium hydrogen phthalate solution of pH≈4) and then performance in a second reference solution (e.g. borax solution of pH≈9) noted, taking due account of the solution temperatures both to define the buffer solution pH values and to ensure that instrument temperature compensation is being applied properly. The two buffer solutions should desirably have pH values straddling the pH range of the solutions to be measured. Subsequently, periodic standardisation need only be made with one buffer solution, as close as possible in its pH to that of the test solution, checking of performance in a second standard only being necessary when measurements are suspect or the electrodes have been subjected to cleaning or some other surface action. The frequency of standardisation is difficult to define in general terms. For simple applications, e.g. water measurement, weekly checks may be sufficient for ±0.1 pH replicability. Where coating of the glass electrode or clogging of the liquid junction, or even of the reference electrode, may occur (which can result in cell e.m.f. drift or shortening of scale length response, i.e. observed pH difference betweeen two standard solutions being less than the reference pH values require), standardisation may be necessary as much as once or twice daily, preceded by cleaning procedures. Much depends both on the application and on the siting of the electrode system.

Possibly the greatest practical difficulties encountered in pH control of processes derive from the formation of precipitates on the responsive glass surface or in the liquid junction devices, or from the coating of the surfaces by oil or grease. The effects of these are, in the case of the glass electrode, a sluggishness or even a loss of pH response, while with blockages of the reference electrode liquid junction, erratic readings can result. There is no simple answer to this problem, although many approaches have been tried, ranging from periodic flushing with solubilising reagent at the electrodes, to mechanical wiping devices, to ultrasonic agitators. The last of these seem to work best where crystalline deposits tend to form, but are less effective on slimy coatings, where wiping is usually more efficient. On the other hand, mechanical wipers may merely spread the deposit around the surface rather than clean it, and then the solubilising method must be tried. This can be arranged to be automatic, and *in situ*, or may alternatively be achieved by the regular manual withdrawal of the electrode system for chemical cleaning.

An important distinction must be drawn between conditions where precipitates are forming and where coagulation has already occurred. Serious deposition, i.e. that which cannot be removed by solution agitation or by fast flows, rarely occurs in solutions containing formed precipitates; the problems usually arise when the electrodes are in the reagent–solution reaction zone. The lesson from this is that siting of electrodes is of great importance: the optimum position is as far from the reaction zone as possible without introducing undue distance–velocity control lag. A desirable feature is the use of fast flowing conditions past the electrode surfaces, which minimises coating: location of electrodes in relatively sluggish flow zones or in dead pockets should be avoided. Despite all this, the necessity for regular maintenance cleaning and electrode standardisation remains, and indeed it is the key to success. Cleaning by chemicals will depend on the nature of the coating. Hydrochloric acid may sometimes be sufficient, or a complexing agent, such as EDTA, can be tried. Very aggressive agents, such as chromic acid and strong alkalis, should desirably be avoided.

Oil and grease coatings on a glass electrode are inimical to the response, and problems can be expected with solution containing oil dispersions, even in the form of emulsions. If the oil or grease tends to separate, every effort should be made to avoid locating the sensor element in the oily layer or causing the element to be moved through the layer. Detergent solutions are usually the most effective agents for cleaning electrodes that have been coated with oil or grease, being relatively harmless to the glass surface. Electrodes can be cleaned with *water miscible* solvents, e.g. acetone, but non-polar organics should be avoided except in extreme circumstances.

The other factor, that of blockage of liquid junction devices such as ceramic plugs, is best offset by the use of a positive head of salt bridge solution from the reference electrode against the liquid junction with respect to the solutions being measured. A number of commercially available devices, e.g. those using gel salt bridges sealed to obviate the need for salt bridge replenishment, are satisfactory when used with clear solutions and in systems producing soluble reaction products, but are sometimes less satisfactory in conditions where precipitation is occurring or the temperature is varying. Where blockages develop in a liquid junction, or back-diffusion into the salt bridge solution degrades the reference electrode, there really is no effective alternative to the use of a relatively higher pressure of salt bridge solution, e.g. approximately 0.5 m water gauge differential head with respect to the solution being measured.

There can be other problems connected with the sensor system. For example, prolonged immersion of glass electrodes in very high or very low pH solutions can lead to a response sluggishness, making pH control at extreme values (less than 2 or higher than 11) more difficult. This is of more importance in control of pH for associated treatment procedures as, for example, described in Chapters 3 and 4, but should not be overlooked in the general context.

Self-regulation of pH

Mention may be made here of a technique that is self-regulating in pH, eliminating the need for control equipment. If an insoluble reagent forms a soluble reaction product of suitable characteristics for direct discharge, this permits the employment of a column of the reagent through which the effluent is passed. Flow may be either downwards or upwards, the latter being advantageous in promoting solution–solid contact when a fluidised condition for the solid is achieved, either fully or partially. The fluidised condition, although preferable, is more difficult to maintain, demanding stabilisation of the solution flow and an adequate mass of solid reagent. Because the reagent will be consumed continuously when reactant passes through the column, total fluidisation is rarely practical, and the compromise of partial achievement is the better objective. On the other hand, downward flow may be satisfactory provided compacting does not occur and providing there is no build-up of reaction solids on the reagent surface.

An important criterion for establishing this form of self-regulating control is that the reaction products should have a pH that is acceptable for direct discharge. This means the careful selection of neutralising reagent. A suitable solid reagent may be magnesium oxide, MgO, which forms soluble salts with many acids, the equilibrium pH values of which are usually about 10.

Wheatland and Borne (1926) investigated the use of a bed of calcined magnesite for the neutralisation of sulphuric acid, operated in upward flow, and found that the effluent pH depends on the rate of flow through the bed, and, in their experiments, on the bed depth. Lime, $Ca(OH)_2$, or limestone, $CaCO_3$, are less satisfactory with acids, such as sulphuric, that form insoluble calcium salts: these tend to coat the reactive particles and inhibit further reaction. Gehm (1944) and Reidl (1947) sought to overcome this disadvantage by using a fast upward flow through a bed of limestone grit so as to expand the particles, but coating still occurs if the acid strength is high (more than 5000 mg l^{-1} for sulphuric acid). Eden and Truesdale (1950) found that the limiting concentration at which inactivation occurs is temperature-dependent. These workers also established that under fast flow conditions dissolved iron did not form coatings of hydrous oxides on the particles.

As a treatment technique the method is useful in circumstances where concentrations of acid and dissolved heavy metals are relatively low, so that coating does not occur; but the effluent pH tends to be variable unless conditions are rigidly controlled.

THE ISOELECTRIC pH IN TREATMENT AND RECOVERY

The isoelectric pH and the point of zero charge

A large number of substances exist that are capable of both proton-donating (acidic) and proton-accepting (basic) functions, and water itself is an obvious example of such an amphiprotic compound. Other examples of importance in pollution problems are certain of the metal hydrous oxides, where the same group (in this case of the type $[M\!-\!OH]^{(m-1)+}$, $[M(OH)_{m+1}]^-$, $[M_n(OH)_m]^{n(m-1)+}$, or other complex ionic species) is responsible for both properties, e.g. zinc, aluminium and tin hydrous oxides. In others the acidic and basic groups are different and separately present, and are exemplified by the amino acids that form key components within proteins, and with certain types of detergent, e.g. aminobenzene sulphonic acid, and monosodium N-lauryl-β-β'-iminodipropionate:

$$C_{12}H_{25}\!-\!N\!\!\begin{array}{c} \diagup CH_2CH_2CO_2H \\ \diagdown CH_2CH_2CO_2Na \end{array}$$

The general representation of such species may be by NH_2RSO_2H or NH_2RCO_2H, where R is either an aliphatic or an aromatic group. An important property of these amphoteric species is their migration behaviour consequent upon the application of an electric field. In acid solutions they move towards the cathode, while in alkaline solutions they migrate to the anode. This behaviour is explained by proton uptake in the acid environment resulting in the formation of a positively charged entity, and proton release (or hydroxyl ion uptake, which in aqueous medium is thermodynamically indistinguishable from proton release) in the alkaline condition giving rise to a negatively charged ion. At the so-called isoelectric point there is nett zero charge and thus no movement in an electric field.

It is possible to ascribe the behaviour in the case of some of the aromatic amino acids, such as aminobenzoic acid, and the aminophenols, to the separate acidic and basic characteristics of the groups present, e.g.

$$-NH_2 + H^+ \rightleftharpoons -NH_3^+ \qquad \text{(basic function)}$$
$$-CO_2H \quad \rightleftharpoons -CO_2^- + H^+ \quad \text{(acidic function)}$$

For the aliphatic types and certain of the aromatic ones such as aminosulphonic acids, however, the view is held that they exist as dipolar species, representable as

$$\begin{array}{c} NH_3^+ \cdots \\ | \\ R \\ | \\ CO_2^- \cdots \end{array}$$

and termed 'zwitter' ions.

Their behaviour with strong acids or bases is thus to be identified as

$$+NH_3.R.CO_2^- + H^+ \rightleftharpoons +NH_3.R.CO_2H \qquad \text{(basic function)}$$
$$+NH_3.RCO_2^- \qquad \rightleftharpoons NH_2.R.CO_2^- + H^+ \quad \text{(acidic function)}$$

i.e. the active groups play a role opposite to that assumed for the other types. The dissociation constants for these two equilibria are given by

$$K_1 = \frac{a_{RH^{\pm}} a_{H^+}}{a_{RH_2^+}}$$

$$K_2 = \frac{a_{R^-} a_{H^+}}{a_{RH^{\pm}}}$$

where $RH^{\pm} = {}^+NH_3 . R . CO_2^-$ (the dipolar ion),

$RH_2^+ = {}^+NH_3 . R . CO_2H$, and

$R^- = NH_2 . R . CO_2^-$.

This represents the ampholytes as being dibasic acids. An approximation to permit rough calculation of the isoelectric pH assumes that in this condition $a_{RH_2^+}^+ = a_{R^-}$, so that from the above equations

$$a_{H^+} = \sqrt{(K_1 K_2)}$$

i.e. $pH_{iso} \approx -\frac{1}{2} \log K_1 K_2$

There are other, less well-defined species that may be involved in charge sign change as the pH changes. Good examples, reviewed later in this chapter, are the various metal hydrous oxide species that are formed as the pH of a solution of metal ions is increased: the simple aquo–metal ion loses a proton, and then these species condense to form polymers, developing in size to become colloidal in nature. In lower pH concentrations the colloidal or particulate species are positively charged, being protonated hydroxo or other hydrolysis products, such as $[M_n(OH)_m]^{n(m-1)+}$; as the pH rises, there is progressive proton release (or hydroxyl ion adsorption) to develop negatively charged species such as $[M_n(OH)_{m(n+1)}]^{m-}$ (where m is the valence of the metal and n is the degree of polynucleation). In an intermediate region there is nett zero charge, corresponding to a mean isoelectric pH. The participating ions need not be H^+ and OH^- only: co-present anions may be adsorbed to provide the negative charge. In this case the nature of the anions, and particularly their charge value, will determine the value of the mean isoelectric pH.

A distinction can be drawn between the well-defined amphoteric species, such as amino acids, and including proteins, whose isoelectric pH values are fairly closely definable, and others that exhibit charge sign transitions to some extent as a result of adsorption of co-ions from solution, exemplified by hydrolysing metal ions. In the former case, a knowledge of the isoelectric pH is of some importance and value in separation of the materials , because at the point of zero charge minimum solubility is frequently exhibited. In the latter case, flocculation and coagulation may be maximised at the point of zero charge, but solution conditions may markedly affect the pH at which this occurs.

In Chapter 6 this is considered again in relation to the zero point charge conditions that arise in precipitation processes and in other solution dispersions (e.g. colloids).

Use of the isoelectric pH

At the isoelectric pH the amounts of charged acid and base conjugates co-present with the electrically neutral ampholyte are at a minimum, and for this reason many of the physical properties of the ampholytes are maximised. Particularly important is the solubility, which, as has been mentioned above, is frequently at a minimum at the isoelectric pH. One example where pH adjustment to the isoelectric point has been proposed for waste water treatment is in colour removal from vegetable tanning solutions (Tomlinson *et al*. 1975). The isoelectric pH of such spent liquors is in the region of 2.2, but adjustment to pH just less than 3.0 is satisfactory when cationic polyelectrolyte is added, causing precipitation and colour adsorption. (The cationic polyelectrolyte works more effectively on negatively charged particles, and hence a pH higher than the isoelectric value is used.) Colour reductions of more than 90% are claimed in the studies reported, with concurrent reduction of C.O.D. (50–60%) and suspended solids (85–90%). Studies on fellmongering wastes have shown that adjustment of the pH to 3 of liquors previously aerated in the presence of a manganese(II) catalyst, followed by settlement, enhances the C.O.D. and suspended solids' removals by 55–80% and 80–90% respectively with respect to settlement without such treatment (see Chapter 6, p. 266); this may also be an isoelectric effect. Another example where acidification to pH<2 can reduce C.O.D. levels substantially is in the treatment of waste waters containing phenol–formaldehyde resins used for gluing wood structures (Nieweglowska and Bartkiewicz 1980).

In the case of proteins and of aminoacid species having dipolar ion structures, pH adjustment to the isoelectric point can in principle be applied to achieve precipitative separation for useful preliminary recovery operations. For example, the addition of acid to fish-processing waste causes coagulation and allows recovery of as much as 90% of the proteins, with C.O.D. reductions of the order of 70%, and suspended solids' removals to 80–90%. Coagulation pH values may be as low as 2, or as high as 6 (Civit *et al*. 1982), but in all cases it is necessary to heat as well, to 60–80°C. Similarly, in the treatment of slaughterhouse wastes it is possible to achieve substantial protein removal by adjusting the pH to approximately 4.5. However, in most practical treatment operations it is more usual to add a precipitating agent, which is introduced into the solution that may be held at or near the isoelectric pH, thereby lowering the solubility still further. Such processes are examined in Chapter 6.

NEUTRALISATION AND PRECIPITATION TREATMENT OF ACIDS AND BASES

The process of neutralisation constitutes one of the primary forms of treatment for waste waters, and is practised in some manner in practically all treatment plants. One factor of importance concerns the ease with which the pH adjustment to neutrality can be controlled, discussed earlier. Let it be noted, however, that neutralisation alone of organic acids and bases may not generally constitute an adequate form of treatment when the C.O.D. or B.O.D. or any toxic characteristics of the materials are not also reduced. In these cases it may be important to remove or to modify the offending material (or to reclaim it). Apart from a biological degradation procedure, chemical degradation may be employed, as may precipitation in some cases, this latter being most commonly achieved in the case of acids by the addition of non-toxic heavy metal salts (such as those of iron and aluminium). Although in some respects this can be regarded as a pH titration process, it is more conveniently considered in the context of Chapter 6.

HYDROLYSIS AS A TREATMENT PROCEDURE

Hydrolytic action

The term 'hydrolysis' refers to the reaction of the solvent water with dissolved solutes. In practice this may occur either under acid or alkaline conditions, and so is usually a function of pH. The hydrolytic degradation of toxic or otherwise environmentally undesirable compounds represents a treatment possibility that has found some application, although so far only to a limited extent. It is attractive where the local circumstances provide acid or alkaline reagents in a cheap form, or where other chemical methods such as oxidation are either costly or demanding of large amounts of space, e.g. for biological treatments. The degree of molecular breakdown may not be sufficient from hydrolysis to permit direct discharge, but may constitute useful partial treatment. A few examples will illustrate the theme.

Hydrolysis of condensed phosphates

In recent years the role of phosphorus in contributing to pollution through its nutrifying enrichment of natural waters, causing growths and eventual eutrophication, has been highlighted by a large number of studies. Condensed phosphates such as tripolyphosphate, type $P_3O_{10}^{5-}$, are used in detergents, and concern has been expressed over their appearance in sewage works, where their precipitation has been undertaken by the addition of lime or aluminium or iron salts (reviewed in Chapter 6). Much of the investigative work that has been carried out has, however, been concerned with orthophosphates (PO_4^{3-}, HPO_4^{2-} and $H_2PO_4^-$), whereas actual waste water discharges may contain such entities as tripolyphosphates. The extent of hydrolytic breakdown of these to simpler species is clearly of importance, especially when it is recognised that the relevant hydrolysis reactions are not always fast. Another important factor is that certain condensed phosphates, such as pyrophosphate, $P_2O_7^{4-}$, and polymetaphosphates, $(PO_3)_n^{n-}$, form soluble complexes with heavy metals (indeed they are used industrially for this very property), so that there is a risk of heavy metal solubilisation in sewage systems unless the complexing ions can be transformed to monomeric forms.

The hydrolysis of condensed phosphates in waste waters has been reviewed by Heinke and Norman (1969), but hydrolysis data are scattered. Most work on kinetics has been directed to alkaline solutions (see, e.g., Bell 1947 and Watzel 1942), and nearly all of it using distilled water. Only more recently have hydrolysis studies been made on natural or waste waters (see, e.g., references quoted by Heinke and

Norman) and significantly these suggest different kinetic characteristics from those in distilled water.

Aqueous phosphate chemistry is very complex, and it is possible here only to summarise salient features of interest. The ultimate hydrolysis endpoint is an orthophosphate, but the routes and rates depend on the starting species and the conditions. Pyrophosphates hydrolyse to orthophosphate entities, expressible for different pH regions by the equations

(pH of 1% solution $=$ 4.6) $H_2P_2O_7^- + H_2O \rightarrow 2H_2PO_4^-$

(pH of 1% solution $=$ 7.3) $HP_2O_7^{3-} + H_2O \rightarrow HPO_4^{2-} + H_2PO_4^-$

(pH of 1% solution $=$ 10.2) $P_2O_7^{4-} + H_2O \rightarrow 2HPO_4^{2-}$

Hexametaphosphate (used in sequestering to prevent hardness scaling) transforms by the reactions,

$$3(PO_3)_6^{6-} + 12H_2O \rightarrow 2(PO_3)_3^{3-} + 12H_2PO_4^-$$

$$(PO_3)_3^{3-} + H_2O \rightarrow H_2P_3O_{10}^{3-}$$

$$H_2P_3O_{10}^{3-} + 2H_2O \rightarrow 3H_2PO_4^-$$

Tripolyphosphate hydrolyses by two consecutive mechanisms:

$$P_3O_{10}^{5-} + H_2O \rightarrow P_2O_7^{4-} + PO_4^{3-} + 2H^+$$

$$P_2O_7^{4-} + H_2O \rightarrow 2PO_4^{3-} + 2H^+$$

These reactions show first order rate characteristics, the rates being directly proportional to phosphate concentration, although there are data (Shannon and Lee 1966) to suggest that at very low concentrations zero order may apply.

The effect of pH on the reaction rate constant has been summarised by Griffiths (1959) for distilled water solutions, and confirmed by Heinke and Norman (1969). The correlation for tripolyphosphate in distilled water is shown in Fig. 2.5, demonstrating a 10^3-fold increase in rate as the pH is lowered from 9.5 to 1.0, and 10-fold from pH 5 to pH 2. Significantly, however, Heinke and Norman found that in a sewage waste water the fastest rate of hydrolysis is at pH levels around 7.5, the rates at pH 5.5 and 9.0 being two and four times slower respectively. Data on the rates of hydrolysis of phosphates are sparse, but decomposition of $HP_2O_7^{3-}$ and $H_2P_2O_7^{2-}$ to $H_2PO_4^-$ seems to proceed to a substantial degree at pH 2 within a time of about 30 minutes. This provides a means for breaking down metal–pyrophosphate complexes as a treatment procedure prior to precipitation of the metal by alkali addition.

Temperature has an effect that is dependent on the background nature of the medium (Heinke and Norman 1969). Thus the activation energy for the tripolyphosphate hydrolysis reaction in distilled water is about $100 \, kJ \, mol^{-1}$, but this drops to less than half in natural and sewage waste waters. Likewise the presence in waste

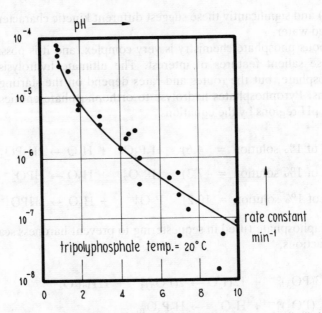

Fig. 2.5 — The effect of pH on the rate of hydrolysis of tripolyphosphate in distilled water
(after Fig. 1 in Heinke and Norman 1969).

waters of various organisms and enzymes can increase the hydrolysis rate by as much as 10^6-fold. Contact with activated sludge may increase this rate still further. A further discussion on phosphate equilibria is given in Chapter 6 (p. 271 *et seq.*).

Hydrolysis of cyanates and cyanides
The destructive oxidation of cyanides by chlorination is widely employed in pollution treatment (see Chapter 3), resulting in the formation of cyanate under usual operating conditions. Concern is sometimes expressed at the presence of cyanate in discharges, and environment legislation has occasionally demanded removal of the species. This is despite the fact that cyanate ions are many orders of magnitude less toxic than cyanide ions, that cyanate shows a negligible complex-forming propensity, and that the likelihood of its reductive reversal to cyanide is extremely remote. Resnick *et al.* (1958) have shown that in sewage systems, for example, the anaerobic decomposition of cyanate does not give rise to cyanides, the products appearing to be ammonium and carboxylate ions; and u.v. irradiation also does not give rise to cyanides. Official concern nevertheless prompts some consideration of means of destroying cyanate.

One method is by further oxidation (see Chapter 3). However, a simpler and often cheaper alternative can be to take advantage of acid hydrolysis. Fig. 2.6 shows the rate of removal of cyanate by hydrolysis at different pH values from data obtained by Resnick *et al.* (1958). It can be seen that at a pH of 2 the reaction rate is fast enough to permit its employment for continuous treatment operations. The effect of increasing pH is quite marked, such that at a pH of 7 the cyanate species is stable for several weeks.

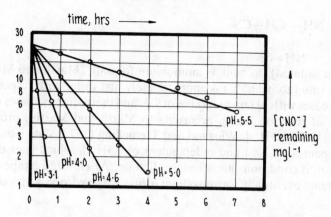

Fig. 2.6 — The effect of pH on the rate of hydrolysis of cyanate (reprinted with permission from Resnick *et al.*, *Ind. Eng. Chem.*, 1958, **50**, 71. Copyright (1958) American Chemical Society).

There is some apparent conflict in the literature over the hydrolysis mechanisms. Jensen (1958) has suggested the following reactions, based on rate studies.

acid medium	$HOCN + H^+$	$\rightarrow \quad NH_4^+ + CO_2$	(fast)
	$HOCN + H_2O$	$\rightarrow \quad NH_3 + CO_2$	(slow)
strongly alkaline medium	$NCO^- + 2H_2O$	$\rightarrow \quad NH_3 + HCO_3^-$	(very slow)

According to Amell (1956), the possible side reaction between NH_4^+ and cyanate ion to form urea is too slow to affect the overall stoicheiometry. He has suggested a mechanism involving the intermediate formation of carbonic acid (at pH = 4) as being the rate-determining reaction, but gives no sound evidence. Resnick *et al.* (1958), from their experimental work, stated that the hydrolysis rate is proportional to $[CNO^-][H^+]^{0.5}$.

The direct hydrolysis of cyanide ion is extremely slow at room temperatures, although in strong hydrochloric (not sulphuric) acid medium there is a catalytic acceleration (Krieble and McNally 1929, Krieble and Peiker 1933) of the hydrolysis reactions:

$$HCN \rightarrow H_2O \rightarrow HCONH_2$$

$$HCONH_2 \rightarrow H_2O \rightarrow HCO_2NH_4$$

The initial reaction to give formamide is much slower than the second step, to ammonium formate. In acid medium the hydrolysis is independent of [HCN]. In alkaline conditions, the decomposition path is different, giving a soluble polymer (probably $NH_2CH(CN)_2$) and a brown precipitate called azulnic acid (from which a tetramer,

$$NH_2-CH-CN,$$
$$|$$
$$NH=CN$$

has been isolated), as well as ammonium formate (Marsh and Martin 1957). The reaction rate (above 65°C) is approximately first order with respect to [HCN]; the rate increases with pH up to a value of 9.3, and thereafter increases only slowly, with the rate of reaction being, according to Marsh and Martin, proportional to the product [HCN][OH⁻]. Wiegand and Tremelling (1972) state it as being first order with respect to [CN⁻] and independent of [OH⁻], which is essentially the same conclusion if conditions are strongly alkaline. At elevated temperatures and pressures, steam eventually forms sodium formate by hydrolysis of sodium cyanide.

Hydrolysis of cellulose derivatives

It is well known that while cellulose is readily biodegradable, even small degrees of substitution into side groups renders it resistant to microbial attack (Kenyon and Gray 1936, Siu 1951). As a consequence, many cellulose-based materials present problems in biological treatment, including textile products (such as cotton cellulose), cellulose nitrate, and cellulosic paints. However, the substituted materials are often susceptible to hydrolytic attack to give products that are biodegradable, so the opportunity develops for combined chemical–biological treatment.

Cellulose nitrate is attacked by acids, and, more readily, by alkalis: the decomposition rate increases with temperature and increasing alkali concentration (Kenyon and Grey 1936), and this hydrolytic approach was examined by Wendt and Kaplan (1976). They digested nitrocellulose at 95°C for 30 minutes, using 3% sodium hydroxide, obtaining a complex but mainly biodegradable product (together with a small residue of resistant material) which was found to be amenable to activated sludge treatment and denitrification (necessary because of the liberation of nitrates).

Cellulose-based paints, used in the automotive industry as finishes, appear as wastes in the recycled water used in the paint spray booths. A paint slime develops that hardens and releases soluble materials into the water that degrade in time to produce noxious smells and an extremely high C.O.D. One treatment procedure is to use alkaline additives to raise the pH of the water to approximately 12, which helps to denature the paint and permits removal of the solids by dissolved air flotation. The subnatant water can then be recycled for a longer period, although regular disposals are still necessary as a consequence of build-up of soluble solvent materials. There are other techniques of flocculation that are rather better, however (see Chapter 6, p. 267).

Decomposition of pesticides

The breakdown of pesticides is often, slow, although aggressive oxidation treatments can be helpful (see Chapter 3), and alkaline hydrolysis alone can produce some decomposition in certain species. Sodium hydroxide degrades malathion, for example (Leigh 1969, Kennedy *et al.* 1969), and also, apparently, lindane and DDT (Kennedy *et al.* 1972). A summary of the applicability of alkaline hydrolysis as a degradation procedure is given in Table 2.3.

Table 2.3 — Summary of the effects of alkaline hydrolysis on some pesticides (after Department of Environment 1980)

Group	Pesticide	Recommended as general method	Degree of degradation	Identity and environmental hazards of degradation products
Insecticides				
Organophosphorus compounds	Naled	Yes	Complete degradation	The alkaline salts of dimethylphosphoric acid, hydrobromic acid and dichlorobromoacetic acid formed are non-toxic
	Diazinon	Yes	Complete degradation	The alkaline salt of diethylthiophosphoric acid and the 2-isopropyl-4-methyl-6-hydroxypyrimide formed are considerably less toxic than diazinon
	Methylparathion (Parathion-methyl)	No	Complete degradation	p-Nitrophenol formed is moderately toxic
	Chlorpyrifos	No	Complete degradation	Toxicity of trichlorohydroxypyridine formed is unknown
	Azinphos-methyl	Yes	Complete degradation	The alkaline salt of dimethyldithiophosphoric acid and the anthranilic acid formed are non-toxic
	Malathion	Yes	Complete degradation	The alkaline salt of dimethyldithiophosphoric acid formed is non-toxic. The diethyl fumarate formed is not a cholinesterase inhibitor
Organochlorine compounds	Methoxychlor	No	Slow reaction	Environmental hazards of the 1,1-dichloro-2,2-bis(p-methoxyphenyl)ethylene formed are not known
	Chlordane	No	Unknown	Partial dechlorination leads to the splitting out of one or two chlorine atoms. Environmental hazards of products unknown
	Toxaphene (Camphechlor)	No	Partial dechlorination	Identity and toxicity of products unknown
Carbamates	Carbaryl	Yes	Complete degradation	1-Naphthol and methylamine formed have low toxicity. Alkaline carbonate formed is non-toxic

Table 2.3 — continued

Group	Pesticide	Recommended as general method	Degree of degradation	Identity and environmental hazards of degradation products
Herbicides				
Phenolics	Pentachlorophenol	No	Neutralisation	Forms stable toxic salt
Phenoxyacids	2,4-D	No	Neutralisation	Forms stable toxic salt which is resistant to further reaction
Substituted ureas	Diuron	No	Complete degradation in boiling caustic alkali	3,4-Dichloroaniline and dimethylamine formed are more toxic than diuron
Triazines	Atrazine	Yes	Complete degradation in strong alkali	Hydroxyatrazine formed is herbicidally inactive and will further decompose in plants to amines and carbon dioxide
Benzoic acids	Chloramben	No	Unreactive	Most benzoic acids will form salts
Dinitroanilines	Trifluralin	No	Unreactive	
Anilides	Alachlor	No	Complete degradation	The secondary amine formed may react with nitrites in the soil to form nitrosoamine, a suspected carcinogen
Others	Picloram	No	Unknown	Decarboxylation and partial dechlorination of picloram. Complete identity and toxicity of products unknown
Fungicides				
Dithiocarbamates	Maneb	No	Unknown	Products unknown
Phthalimides	Captan	Yes	Complete degradation	Tetrahydrophthalimide and alkaline sulphide, chloride and carbonate are formed. Tetrahydrophthalimide is further hydrolysed to phthalic acid

Hydrolysis of nylon wastes
Nylon is hydrolysed in strongly alkaline conditions, and this has been turned to advantage in the treatment of nylon wastes. These are subjected to concentrated sodium hydroxide treatment to yield starting materials that can be recycled for polycondensation.

Treatment of fluoborates
Removal of fluoride associated as the BF_4^- ion has been examined by Korenowski *et al.* (1977), using a two stage process involving hydrolysis in acid conditions in the presence of Ca^{2+} ions:

$$BF_4^- + 3H_2O \xrightarrow[\text{heat}]{H^+, Ca^{2+}} 4HF + H_3BO_3$$

$$2HF + 2OH^- + Ca^{2+} \rightarrow CaF_2 + 2H_2O$$

The calcium ions apparently enhance the hydrolysis rate even in small concentrations, and subsequently precipitate calcium fluoride when conditions are made alkaline. The hydrolysis step is best carried out pH $\leqslant 4$, and the rate is improved substantially by increasing the temperature to $\geqslant 70°C$.

Liberation of free acids and bases
Although not strictly hydrolysis in the sense of molecular rearrangement by reaction with the solvent water, pH adjustment for the liberation of water-immiscible acids and bases is included here for completeness. Water-soluble fatty acids and amines can be liberated from their respective salts by generous additions of acid or alkali respectively, and floated or solvent-extracted from the treated solution. The efficacy of the process is usually improved by the addition of coagulating agents, e.g. ferric salts. Amines can thus be adsorbed on to alkaline flocs of hydrous oxides, while with the acids insoluble salts may be formed, as described in the treatment procedures of Chapter 6.

Gas stripping, an operation by which dissolved volatile materials are removed from aqueous solutions by gaseous displacement, e.g. with air, is often preceded and made chemically feasible by prior addition of acid or alkali. Examples include the important process of ammonia stripping, which is aided by alkaline conditions, and the acid liberation of hydrogen sulphide, mercaptans and hydrogen cyanide. The products may then be recovered or destructively oxidised. As an example, one commercial process that has been used for the treatment of waste concentrated cyanide solutions involves addition of sulphuric acid, followed by catalytic combustion of the evolved hydrogen cyanide to carbon dioxide, nitrogen and water at 300°C (Jola 1970), this being exothermic and thus attractive in energy terms. (High temperature decomposition of cyanides is also described by Hoerth *et al.* (1973) and Schwarzbach (1974).)

pH adjustment to form an undissociated acid or base can sometimes be advantageously employed in the treatment of metal ion complexes. By pH depression or elevation the undissociated acid or base species is formed, the complex being broken, thus providing opportunities for further treatment procedures. An example is the decomposition of concentrated copper ammine solutions arising in printed circuit board manufacture: ammonia is liberated by adding sodium hydroxide and heating to $\sim 80°C$, leaving recoverable cuprous oxide. Other procedures are discussed in practical detail in succeeding sections in this chapter.

PRECIPITATIVE REMOVAL OF METAL IONS BY pH ADJUSTMENT

The precipitation of dissolved heavy metals to form so-called 'hydroxides', by pH adjustment through the addition of alkalis, is the major method for removing toxic metals in pollution treatment.[a] The importance of this treatment process in its application to the removal of heavy metals from discharges merits detailed discussion of important background features, possible mechanisms, and practical aspects connected with control and plant design.

The characteristics of metal ions in solution

Metal ions do not exist as such in aqueous solution, but are always associated with a hydrating sheath of water molecules. This interaction occurs between the positive metal ion and the residual negative charge on the oxygen atom that arises from the dipolar characteristics of the water molecule:

$$M^{z+} + O^{2\delta-} \underset{H^{\delta+}}{\overset{H^{\delta+}}{<}} \quad \rightarrow \quad \left[M^{z+} \ O^{2\delta-} \underset{H^{\delta+}}{\overset{H^{\delta+}}{<}} \right]$$

The extent of hydration is governed by a number of factors, including the charge, radius and coordination characteristics of the metal ion. The coordination involves association of water molecules in a primary hydration sheath, where the binding is at its strongest, and with outer sheaths of less strongly associated water 'clouds'. The inner sheath involves a number of water molecules that is defined by a characteristic coordination number. (For detailed reviews of hydration theories and experimental information, see, e.g., Burgess 1978.)

The association does not stop there. There is a tendency for the metal ion–oxygen atom association to polarise the water molecule further, to a point where hydrogen ions dissociate from the entity—a process also encouraged by hydration of the hydrogen ion. This may be pictured mechanistically by

$$\left[(H_2O)_{n-1}M—O \overset{H}{\underset{H}{<}} \right]^{z+} \rightleftharpoons [(H_2O)_{n-1}M—OH]^{(z-1)+} + H^+aq.$$

This constitutes an induced dissociation of the water molecule by the metal ion. A sequence of monomeric hydrolysis products can be formed by a succession of reversible step-wise reactions such as:

a This process is usually termed hydrolysis, although strictly it is not the degradation of the metal ion but splitting of the water molecule, and should therefore really be termed lysis. However, to conform with convention the term hydrolysis will be employed here.

$$[M(H_2O)_n]^{z+} \rightleftharpoons [M(H_2O)_{n-1}OH]^{(z-1)+} + H^+aq.$$

$$[M(H_2O)_{n-1}OH]^{(z-1)} \rightleftharpoons [M(H_2O)_{n-2}(OH)_2]^{(z-2)+} + H^+aq.$$

$$[M(H_2O)(OH)_{n-1}]^{(z-n+1)+} \rightleftharpoons [M(OH)_n]^{(z-n)+} + H^+aq.$$

It has long been known that these hydrolysis reactions are not the only ones that occur, but that more complex species also exist in solution, involving more than one metal ion per entity, and possibly arising by condensation polymerisation from simpler species. e.g.

$$2[(M(H_2O)_{n-1}OH]^{(z-1)+} \rightarrow [(H_2O)_{n-1}M-O-M(H_2O)_{n-1}]^{2(z-1)+} + H_2O$$

$$2[(M(H_2O)_{n-1}OH]^{(z-1)+} \rightarrow [(H_2O)_{n-1}M \overset{\displaystyle OH}{\underset{\displaystyle OH}{\diagdown\diagup}} M(H_2O)_{n-1}]^{2(z-1)+}$$

Such polymeric products will themselves be involved in equilibria involving hydrogen ions, e.g.

$$[(H_2O)_{n-1}M-O-M(H_2O)_{n-1}]^{2(z-1)+} \rightleftharpoons$$

$$[(H_2O)_{n-1}M-O-M(H_2O)_{n-2}OH]^{2(z-3)+} + H^+aq.$$

and

$$[(H_2O)_{n-1}M-O-M(H_2O)_{n-1}]^{2(z-1)+} + H^+aq. \rightleftharpoons$$

$$[H_2O)_{n-1}M-\overset{\displaystyle H}{\overset{\displaystyle |}{O}}-M(H_2O)_{n-1}]^{(2z-1)+}$$

It can be seen that these reactions involve formation of species with —O— ('oxo') and —OH— ('ol') bridges. A whole range of products can be established with several metal ions per ionic species, which may be cationic or anionic. This picture embraces both metals such as Ni, Cu, and Pb, and those that are amphoteric, such as Al, Zn, and Sn.

The formation of hydrolysed aquo–metal ion species can be followed by an alkali titration curve (Fig. 2.7), which indicates the polyprotic nature of a transition metal acid. It also demonstrates the substantial buffering power of hydrolysing metal ions. Titration curves of this kind can be analysed mathematically to reveal both the nature of the ions formed and their formation constants (although only in well-controlled experimental conditions). Aqueous solutions of metal ions are very complex, and are made more so by effects that may occur from interaction with the anionic species that will be co-present, the strength of whose complexing ability will modify both the degree of hydration and the extent of hydrolysis and polymerisation.

The major work in unravelling the difficult problems of identifying hydrolytic species was carried out by Sillén (1959) and his co-workers (see Sillén's review in Martell 1971); this and subsequent work has been reviewed by Baes and Mesmer (1976), who have also summarised the predominant species believed to exist for

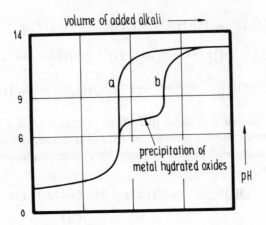

Fig. 2.7 — Characteristic titration curves for addition of alkali to acid solutions.
a. Acid solution alone. b. Acid solution containing heavy metal ions.

various metals. The relative preponderances of each species is a function of the formation constants, of the pH where hydrogen ions are involved in the equilibrium equation, and of the metal ion activity (concentration) where polymeric species are involved. Equilibrium formation/dissociation constant data are to be found in Baes and Mesmer's text.

The precipitation of hydrous metal oxides

The process of precipitation of metal hydrous oxides may be regarded as a sequence of interacting steps that occur as the pH or metal ion activity (or concentration) increases. As hydrolysis occurs, usually through polymerisation, so the sizes of the ionic species increase, resulting eventually in macromolecular or colloidal particles that subsequently themselves coagulate to form relatively insoluble hydrous metal oxide precipitates. Some of the transitional species are thermodynamically unstable, and act as intermediates in the precipitation process. It should be noted that these precipitates do not necessarily correspond to particular stoicheiometric formulations: their structures generally consist of local polynuclear groupings, probably interacting partially directly and partially through the bridging of occluded anions. As the pH of the solution is increased, the positively charged character of the hydrolysis product changes, and may pass through a point of zero charge and on to a negatively charged (anionic) condition. This may derive from an amphoteric property of the metal concerned, or from adsorption on to the solid surface of negatively charged species; in the former case some re-dissolution or peptisation may occur at high pH values. Interaction may also occur between charged species and the water dipole, favouring solubilisation. The point of zero charge corresponds to the isoelectric pH of the mixed species, and is relatively important in the achievement of maximum precipitation. There may be a band of pH over which the charge is effectively zero, and within which good precipitation will occur. The introduction of polyelectrolyte flocculants at this condition can enhance precipitate flocculation; which action is considered in Chapter 6.

The quantitative expression of full precipitation is, in principle, found in the solubility product. Where a pure solid hydroxide is in thermodynamic equilibrium with the constituent free ions in solution, the appropriate equation can be written

$$M(OH)_{z(S)} \rightleftharpoons M^{z+} + zOH^-$$

The corresponding dissociation constant is

$$K_{MOH} = \frac{a_{M^{z+}} \cdot a_{OH^-}^z}{a_{M(OH)_z}}$$

Since the solid reactant has, by definition, an activity of unity, this relationship simplifies to:

$$K_{sp}^0 = a_{M^{z+}} \cdot a_{OH^-}^z$$

where K_{sp}^0 is called the solubility product. From solubility product data it is in principle possible to calculate the residual metal ion activity level from a knowledge of the pH, from the rearrangement

$$K_{sp}^0 = \frac{a_{M^{z+}}}{a_H^z} K_w$$

remembering that pH $= -\log a_{H^+}$. However, matters are by no means as simple as that. Firstly, much of the recorded K_{sp}^0 data are not truly thermodynamic (i.e. based on activity relationships), and the concentration K_{sp}^0 values only apply within the experimental conditions of their determination. There is thus often considerable variation in the literature. Secondly, and even more importantly, very rarely can it be assumed that the solid phase is of a single species, or even of a constant composition. (This is apparent from the numerous species that form during the hydrolysis reactions leading to precipitation.) Thus not only are much of the data very specific to their conditions of determination, but they are probably fundamentally unreliable anyway. The most that can be said for hydroxide solubility product data (and to some extent those of oxides as well) is that they give a semi-quantitative indication as to the level of metal ion concentrations that may apply in given pH conditions. They can also be applied for order of magnitude comparisons against similar data for metal phosphates, carbonates, etc. As a useful guide where direct experimental data are not available on the precipitation characteristics of metals, the graphical compilations of Kragten (1978) can be consulted.

The residual metal ion level in a solution in contact with a precipitate depends on a host of factors as well as pH. Adsorption of metal ion species on to the precipitate takes place, particularly of the polynuclear hydrolysis products (Stumm and Morgan 1962, O'Melia and Stumm 1967, Stumm and O'Melia 1968). This occurs even against electrostatic repulsion, implying that the bonding forces are fairly strong. The adsorption appears to commence abruptly in a narrow critical pH range corresponding roughly to the region where hydrolysis may be said to start. Stumm and O'Melia

(1968) have proposed qualitative explanations for the preferential adsorption of hydrolysed species, based on the idea that the substitution of OH groups for H_2O molecules in the metal ion hydration sheath and aggregation to larger species creates favourable conditions for bonding at the precipitate surface. Leckie and James (1974) have reviewed adsorption theories, and the reader is referred to this summary for further details. In practical pollution treatment terms, adsorption is important for two reasons: firstly, choice of favourable adsorption conditions can minimise residual metal ion concentrations, and secondly, the existing presence of hydrous metal oxides or other insoluble materials (as in settlement tanks and post-settlement filter units) can promote metal removal. Recent studies have indeed shown that adsorption of metals on, for example, ferric and manganese hydroxides, can result in very low residual levels in solution. The effect is, as may be expected, pH-dependent: Benjamin and Leckie (1981) found that increasing pH values were needed for maximum removals of metals in the order Pb, Cu, Zn and Cd. Generally speaking, the degree of adsorption is in the sequence Hg>Pb>Cu>Zn>Cd/Ni>Co, but much depends on the conditions, e.g. the types of anion co-present, the nature of any complexing species, relative metal concentrations, etc. It is significant that even strong complexing agents do not necessarily inhibit adsorption. Benjamin *et al.* (1982) found, for example, that zinc, copper, and cadmium in boiler fly ash leachates were strongly adsorbed on freshly precipitated ferric hydroxide at pH 8–9 from starting concentrations as low as 3×10^{-3} mmol^{-1} even in the presence of EDTA. It is interesting to note that surface adsorption does not appear to be affected by the condition of the adsorbent, material aged for an hour being as effective as freshly precipitated hydroxide (Appleton and Leckie 1981).

The nature of the precipitant and its mode of addition also have an influence. Thus sodium bicarbonate, sodium carbonate and calcium hydroxide all produce differing results in terms of extent of precipitation, residual metal ion solubility, and character of precipitate. Furthermore, if the precipitant is added quickly to a localised part of the solution, locally high pH values develop (depending on both the metal concentration and the nature of the precipitant), resulting in the formation of species or precipitates that may be metastable with respect to the pH of the dispersed mixture. The equilibration of these products with a solution mixture may be very slow—quite often hours, and is some circumstances months—and may result in both pH shifts and modified solubilities of metal ions.

pH–solubility relationships

The practical expression of the preceding remarks can be appreciated by an examination of curves plotting pH against residual ion concentration. The plots of Figs. 2.8, 2.9, and 2.10 (Mattock 1977) were obtained by preparing solutions of simple metal salts and then adjusting these to increasing pH values with the chosen reagent, using vigorous mixing to simulate plant control conditions, followed by filtration through a Whatman No. 40 filter paper after elapse of the stated ageing period. The concentration of the metal in the filtrate was then determined by atomic absorption measurements. The curves show mean rather than individual values, for the reason that replicability is not high, so that too much emphasis must not be placed on particular determinations. The reason for this is that a number of variables affect

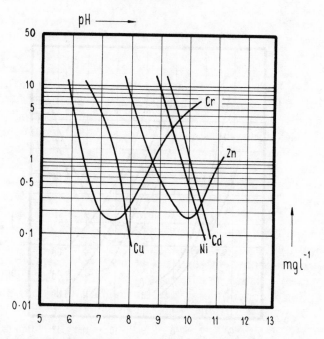

Fig. 2.8 — Residual soluble metal concentrations as a function of pH using NaOH as precipitant. Precipitate stood for 30 minutes before filtration.

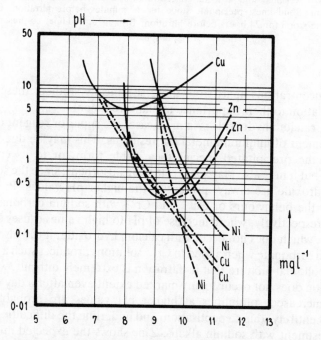

Fig. 2.9 — Residual soluble metal concentrations as a function of pH using Na_2CO_3 as precipitant. Solid lines: precipitate stood for 30 minutes before filtration. Broken lines: precipitate stood for 24 hours before filtration. Broken dotted lines: precipitate stood for 30 minutes before filtration, with polyelectrolyte present.

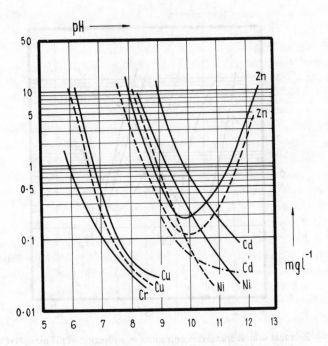

Fig. 2.10 — Residual soluble metal concentrations as a function of pH using Ca(OH)$_2$ as precipitant. Solid lines: precipitate stood for 30 minutes before filtration. Broken lines: precipitate stood for 24 hours before filtration. Broken dotted line: co-presence of excess phosphate.

residual concentrations apart from those specifically studied; these are discussed below in relation to their importance in plant treatment.

Clearly, calcium hydroxide is in most cases the superior reagent of the three for the achievement of minimum metal ion residuals. This may be due to its ability to equilibrate the precipitated species to a stable form, or to its colloid-breaking properties, but it may also play some role as an adsorbent, since the solid particles of calcium hydroxide are always present at the higher pH values. Particularly note-worthy are the behaviours of Cu^{2+} and Cr^{3+} with sodium carbonate and sodium hydroxide respectively, where increase of pH to high values causes an apparent re-dissolution, which does not occur with calcium hydroxide; it can also be mentioned that sodium carbonate when added to Cr^{3+} solutions produces such a fine precipitate or almost colloidal mixture that its filtration is extremely difficult, while settlement consolidation does not occur to any marked extent even after a day's standing. The apparent increased solubilities at higher pH values are probably (though not necessarily entirely) due to peptisation, and underline the difficulties that can occur in pH adjustment with sodium alkalis. Zinc shows the expected minimum with all three reagents. Lead (not described in the plots) constitutes an interesting special case. Hydrous lead oxide is relatively soluble, but lead carbonate is both insoluble and densely crystalline, settling well. Sodium carbonate is in fact fairly effective in lowering lead residuals to approximately 0.5 mg l^{-1}, using a precipitating pH of 7–9. The addition of an Fe(III) salt as co-precipitant, with polyelectrolytes, can also be beneficial (Hartinger 1973).

Ageing of the precipitates results in a lowering of the residual metal content, improving both settleability and filterability, demonstrating the time effect reviewed in the preceding sections. Conventional practice for the removal of precipitated hydrous metal oxides calls for settlement or flotation of the solids using anything between 1 and 4 hours' nominal residence time, usually in a continuous system, although the alternative of direct filtration after pH adjustment is sometimes mooted. If minimum residual metals' concentrations are required, neither procedure is entirely adequate, the first because settlement (or the alternative of flotation) of course does not remove all the precipitate and so some metal remains in the discharge, and the second because insufficient ageing is provided with most direct filtration systems. For best results, therefore, initial clarification needs to be followed by filtration to remove remaining traces of solids. (It should also be borne in mind that a nominal holding time does not correspond to the effective holding time for the majority of the fluid particles in an unstirred system such as a settlement tank. A useful rule of thumb is to multiply an experimental ageing time found for static conditions by a factor of 5–6 for use in a continuous system.)

Notwithstanding the indicative data given in Figs. 2.8–2.10, it must be stressed that many influences affect performance of a treatment plant, including not only the factors that are reviewed below, but also the design approach. Better results can be obtained where, for example, co-precipitation or co-adsorption can occur. Argaman and Weddle (1974), in their work on removal of heavy metals in physico-chemical treatment of sewage, used high pH (11.6) lime coagulation, settlement and recarbonation, obtaining residuals broadly those expected from Fig. 2.10, or slightly better. They were also better than those obtained by the use of ferric chloride to pH 8.

Minimum solubilities occur at different pH values with different metals, and these are not necessarily within the pH range of 6–9 (or 6.5–8.5) often specified in effluent discharge standards, implying that the common practice of pH adjustment of the metal-containing effluent discharges to approximately 8.5 may not be satisfactory for best results. Notably, nickel requires quite a high pH of at least 10, while the solubility minimum for zinc suggests an optimum value of no higher than 10. Stream separation for individual treatments may thus be called for within very strict discharge quality requirements, or at least pH control to alkaline levels, followed by thorough removal of the precipitate and final re-adjustment to neutrality if the discharge standards so demand.

Some of the metals less commonly found in waste waters can also be precipitated effectively by pH adjustment to give low residual soluble levels (although final carbon adsorption can be advantageously applied after removal of precipitated solids). Lime addition is effective with, for example, silver, beryllium, bismuth, cobalt, molybdenum, tin and titanium, although for mercury, antimony and vanadium co-precipitation with ferric salts is preferable (Hannah et al. 1977). Neither procedure is entirely satisfactory with selenium or thallium.

Polyelectrolyte flocculating agents can have a beneficial effect in promoting precipitative separation with the sodium alkalis where, for example, the effect with copper and chromium(III) is marked, the volume of precipitate also being reduced. (The use of polyelectrolytes is discussed further in Chapter 6.)

Although not deducible from Figs. 2.8–2.10, the volumes of precipitates formed from the different reagents, starting with the same quantities of metals in solution, are quite different. The view sometimes expressed that sodium alkalis produce smaller precipitate volumes than calcium hydroxide is often the reverse of fact, except for very high pH values when a large excess of lime reagent may be present. Calcium hydroxide usually gives denser, more readily settled precipitates that may be of no greater volume than those given by the sodium alkalis, which sometimes show a tendency to float. Unless flotation is the preferred method of solids' removal, this can be disconcerting, and the addition of polyelectrolytes does not always solve the problem.

Co-precipitants, such as phosphate, are sometimes useful, as is shown by the effect on cadmium residuals illustrated in Fig. 2.10. Certainly in our experience the use or adventitious presence of phosphates can aid in reducing heavy metal residuals in waste water streams: this can be useful for removing zinc, tin and lead to lower levels than are normally achievable by simple pH adjustment. (Note, however, that benefit does not always follow when lime is used, presumably from the competitive consequences of calcium phosphate precipitation.) Other types of anion, particularly those that can form basic salt precipitates in alkaline conditions, can be influential, and the co-presence of several different types of cation can also lower the residuals to levels less than those observable with a single precipitable metal. This can be advantageously employed when only relatively low concentrations of metals are present, by addition of a collector species, such as Fe^{3+} or Al^{3+}, which hydrolyses and presumably adsorbs the other metal ion hydrolysis products.

In addition to the foregoing features, others should be recognised. For example, the ionic strength of the solution can have a significant effect on apparent metal hydrous oxide solubilities, high strengths sometimes inhibiting full precipitation or causing peptisation, while at other times promoting precipitation, probably through salting out. With 'real' effluents (i.e. those found in plant operating conditions) the ionic strength is rarely predictable or constant, which means that laboratory data have to be applied with caution in predictive design studies.

The direction of approach to a given pH can be important. If the pH is lowered from an excess alkaline condition to a normal optimum precipitation pH the results may be different from those when alkali is added to an acid solution. In effluent streams where an amphoteric metal is present at high pH, and where it is desired to lower the pH to achieve a minimum solution residual, it may be necessary to reduce the pH so far as to dissolve insoluble metal compounds completely, and then raise it again to the optimum level. If this is not done the metallate anion (or equivalent species) may not re-equilibrate by structural modification in a reasonable period, and high apparent metal residuals may result.

Temperature plays a part. Both the degrees and the rates of hydrolysis and polymerisation increase with temperature, which promotes the condensation and ageing processes and often results not only in lower residual metal solubilities but also in more compact precipitates.

For further discussion and a presentation of more data on this important subject, the reader is also referred to the excellent book by Hartinger (1976).

The influence of complexing agents

Association between metal ions and coordinating ligands is a well-known phenomenon, and provides a competitive alternative reaction course to the hydrolysis process. (Many standard texts review this subject in detail, e.g. Dwyer and Mellor 1964, Basolo and Pearson 1967, and Martell 1971.) The ligands may be singly or multiply negatively charged ions which associate in a stepwise manner with the metal cations up to the limit imposed by the characteristic coordination number, N, e.g.

$$M^{z+} + L^- \rightleftharpoons (ML)^{(z-1)+}$$

$$(ML)^{(z-1)+} \quad + L^- \rightleftharpoons (ML_2)^{(z-2)+}$$

$$(ML)_{N-1}^{(z-N+1)+} + L^- \rightleftharpoons (ML_N)^{(z-N)+}$$

Coordination may proceed in this way to form anionic complexes. Neutral molecules may also associate, e.g.

$$M^{z+} + X \rightleftharpoons (MX)^{z+}$$

$$(MX)^{z+} + X \rightleftharpoons (MX_2)^{z+}$$

etc., giving rise solely to cationic complexes. A further complexing action can take place with chelating ligands where more than one atom in the ligand can associate with the metal ion, to form (mostly) anionic 'polydentate' complexes:

$$M^{z+} + L^{2-} \rightleftharpoons (ML)^{(z-2)+}$$

$$(ML)^{(z-2)+} + L^{2-} \rightleftharpoons (LML)^{(z-4)+}$$

The degree of association, as indicated by the association constants for the equilibria concerned, e.g.

$$K_{ML} = \frac{a_{ML}}{a_M a_L}$$

determines the extent to which any particular ligand will compete with the hydrolytic processes. Since the complexes are frequently highly soluble, the presence of coordinating ligands can inhibit maximum precipitation of metal hydrous oxides in pollution treatment. In a practical sense, any potentially complex-forming substance must be examined to determine whether it can be influential. Typical ligands that may be encountered in industrial waste waters include, for example: hydroxy organic acids such as gluconic, heptonic, tartaric, oxalic and citric; powerful complexing amine acids such as EDTA, EGTA and NTA; ammonia and amines; cyanide; peroxydisulphate; and various phosphates such as pyrophosphate and hexametaphosphate. An indication of the relative complexing powers of some common complexing agents is given in Table 2.4; but it must be appreciated that the mode and extent of complexation depends on the conditions, such as the pH and the relative

Table 2.4 — Relative complexing powers of some metal ion complexing agents.(Values are given as log K_1, where K_1 is the 1:1 association constant.)

Complexing agent	H⁺	Cd²⁺	Cu²⁺	Ni²⁺	Zn²⁺
Ammonia	9.24	2.55	4.04	2.72	2.21
Citrate	6.4	3.75(a)	5.90(a)	5.40(a)	4.98(a)
Monoethanolamine	9.5	2.8(c)	5.7(c)	2.98(c)	3.7(c)
Ethylenediamine	9.93	5.41	10.48	7.32	5.66
EDTA	10.17(c)	16.36(c)	18.7(c)	18.52(c)	16.44(c)
EGTA	9.4(c)	16.5(c)	17.57(c)	13.5(c)	12.6(c)
NTA	9.65(c)	9.78(c)	12.94(c)	11.5(c)	10.66(c)
Oxalate	4.3	3.89	6.23	5.16	4.87
Pyrophosphate	8.37	—	7.6(b)	5.94(c)	8.7
Quadrol (Tetrol)	8.84(d)	7.80(d)	9.75(d)	7.65(a)	6.09(d)
d-Tartrate†	4.4	—	—†	2.06(b)†	3.82†

†Meaningful comparisons are particularly difficult for tartrate complexes, in that these are mainly polynuclear.

Values refer to ionic strength $(I)=0$, temperature 25°C, except:
   (a) $I = 0.1$, 20°C
   (b) $I = 1.0$, 25°C
   (c) $I = 0.1$, 25°C
   (d) $I = 0.5$, 25°C.

concentrations of metals and ligands, and is governed by the stability constants applicable: there are often many of these, covering for example hydroxy and polynuclear species as well as simple ones. Log K_1, where K_1 is the first association constant, has been chosen for Table 2.4 as a relative measure of complexing ability, since in general the steric effects evident in higher complexes due to crowding of the coordination sphere of the metal ion are at a minimum for 1:1 complexation. This is not true of course with polydentate chelating compounds such as EDTA, EGTA and NTA. It must be stressed that direct comparison of values between ligands of different denticity is not valid in a strict sense because of the different numbers of metal ion coordination sites occupied by ligand donor atoms. However, the figures in Table 2.4 do give a qualitative idea of the relative strengths of the complexes formed between different metal complexes of the same ligand.

A semi-quantitative approach to define the competitive influence of complexing ligands with respect to hydroxyl ions has been given by Chaberek and Martell (1959), which is useful as an introduction to the calculation procedures. Another more generalized approach has been expressed by Ringbom (1963). This latter has been employed in a useful publication compiled by Kragten (1978), which correlates the known competitive equilibria for a range of 29 ligands on 45 metals, and presents the results in graphical form. From the graphs it is a simple matter to predict the extent of precipitation of metal hydrous oxides at any given pH in the presence of a complexing ligand. Once again, however, it must be stressed that these data must not be regarded as wholly quantitative: concentrations rather than activities have been used in the computations, and some of the literature data used may be suspect. For this reason the graphs should be used as a guide, and one should not be too surprised if in practice, where ionic backgrounds (for example) may be quite different and where there may be more than one complexing ligand present, the results are different from the predictions.

The nature of the association between metal ions and most complexing ligands is more complicated than in the special case of the OH^- ligand that has already been reviewed. The hydrated metal ion may associate in an electrostatic manner with the negatively charged ligand, and this is most likely to occur when the inner hydration sheath of the metal ion remains undisturbed. When this sheath is penetrated, with displacement of the water molecules, electron donation from the ligand to the metal ion can result in a firm association. Chelating agents particularly are capable of forming extremely stable complexes that are resistant to many forms of chemical attack, which often forms the basis of their industrial applications and hence their appearance in waste waters. (Comprehensive compilations of the stability of metal ion complexes are given by Martell and Smith (1974, 1977), Smith and Martell (1975, 1976), Perrin (1979), and Högfeldt (1982).)

The mechanism of complex formation is connected with the nature of the metal ion, and the kinetic characteristics of the complexation reaction are dependent on the lability of the water molecules associated with the metal ion where direct metal ion–ligand association occurs. (For reviews of this extensively studied subject see, e.g., Hunt 1963, Wilkins and Eigen 1965, Nancollas 1967, Taube 1970, and Petrucci 1971.) Alkali and alkaline earth metals (except Mg^{2+}) are subject to fast rates of water substitution by ligands, and the kinetics are dependent on the nature of the ligand. Many transition metal ions of environmental importance, however, have complex formation rate constants that are almost independent of the nature of the ligand, and which appear to be controlled by the rate of loss of water from the hydration sheath. A third group of ions, including Al^{3+} and Fe^{3+}, have relatively non-labile water co-ordination, and here the substitution is quite slow, and apparently governed by the rate of hydrolysis. The existence of these relative rate characteristics can sometimes be turned to advantage in waste water treatment to overcome complexing association.

Some experimental data can be presented to illustrate the practical orders of importance of the presence of complexing agents on the precipitative removal of metal ions by pH adjustment (P. Huddy, personal communication, 1974). A set of experiments was carried out on a solution containing initially $7.4 \, \text{mg} \, l^{-1}$ Ni and $93.7 \, \text{mg} \, l^{-1}$ Zn, having a starting pH of 3.5, into which were introduced $60 \, \text{mg} \, l^{-1}$ of various complexing agents. The pH was then adjusted by the addition either of 2.5% sodium hydroxide or 5% calcium hydroxide, the mixtures agitated for 5 minutes, and Dow A 23 anionic polyelectrolyte blended in to give a concentration of $2 \, \text{mg} \, l^{-1}$. The solutions were then aged for a further 60 minutes before being filtered through a No. 40 Whatman filter paper. In all cases the sodium hydroxide additions produced a gelatinous precipitate that was difficult to filter effectively in all cases, and there appeared to be some colloid formation. This did not occur with lime, supporting the remarks already made concerning the efficacies of the two reagents. Table 2.5 shows the quantitative results, from which it can be seen that calcium hydroxide is able to compete with the complexing agents to a far greater extent than is sodium hydroxide. Noteworthy is the obvious amphoteric effect between zinc and sodium hydroxide, which causes the zinc to stay in solution even though the complex may be broken. The reason for the superior performance of the lime could possibly be ascribed to the

Table 2.5 — The effects of various complexing agents on precipitation of a mixture of nickel and zinc by sodium hydroxide or lime

	Ni–NaOH pptn		Ni–Ca(OH)$_2$ pptn		Zn–NaOH pptn		Zn–Ca(OH)$_2$ pptn	
	pH 8.5	pH 10.5	pH 8.5	pH 10.5	pH 8.5	pH 10.5	pH 8.5	pH 10.5
	pH 3.5	Ni 7.4			Zn 93.7			
Test solution								
No additive	2.0	3.7	0.8	0.1	0.1	9.8	0.1	0.2
Sodium gluconate	3.7	6.3	1.2	0.4	2.4	92.0	0.5	0.2
Sodium citrate	2.7	7.2	4.7	0.6	4.7	93.0	0.3	0.4
Sodium tartrate	1.5	6.4	2.9	0.4	2.2	90.8	0.6	0.3
Sodium oxalate	4.3	7.3	7.1	1.0	9.2	92.3	0.3	0.3
Sodium heptonate	2.5	6.9	1.3	0.5	9.2	92.3	0.4	0.3
EDTA	2.3	7.2	2.0	0.7	10.8	92.4	5.4	4.8

All concentrations given are in mg l^{-1}. Concentration of complexing agent = 60 mg l^{-1}.

formation of insoluble calcium salts with tartrate and oxalate ligands, but this hardly applies to citrate, since calcium citrate has an appreciable solubility. With the EDTA species the possibility of competitive chelation by Ca^{2+} can be advanced; although the association constants for Zn^{2+} are relatively much stronger, a mass action effect is almost certainly occurring from the presence of excess calcium ions.

EDTA is a well-known powerful complexing agent for many metal ions, and its presence in solution inhibits strongly the precipitation of metal hydrous oxides. The tendency of EDTA–metal complexes to hydrolyse follows the sequence Ni<Co<Cd<Mn<Zn<Cu<Fe(III), but in all these cases it is difficult to precipitate the hydrous oxide unless a very high pH is used. The precipitative removal of metals from complexes by calcium hydroxide appears in some cases to be improved by first lowering the pH to a value, e.g. 1–2, where the complex dissociates and the free ligand acid is formed, and then raising the pH to a high value such as 12 (to gain maximum advantage from competitive mass action effects of OH^- and Ca^{2+}). For example, a typical commercial electroless copper-plating solution containing 23 g l^{-1} Na_2EDTA, 140 g l^{-1} potassium sodium tartrate tetrahydrate, and 170 ml l^{-1} 37% formaldehyde, diluted 100-fold (to simulate rinsing conditions on a process plant) to give a total copper concentration of 76 mg l^{-1}, on precipitation with lime, settlement and filtration, retained 31 mg l^{-1} (pH 11.7) and 33 mg l^{-1} (pH 12) residual copper. Prior adjustment to pH 2 before lime precipitation at pH 12 gave, however, a residual of only 1.5 mg l^{-1} copper. Wing and Rayford (1978) obtained essentially similar results by adding spent pickle liquor to an electroless copper rinse before addition of lime to a high pH, although here there may also have been a competitive complexing effect from the dissolved iron in the pickle liquor. It could be that the relative rates of complexation and of hydrolysis may be of relevance here: the formation of hydrolytic species through loss of protons from the hydrated metal ion, with subsequent polymerisation, may be faster than replacement of the hydrating water molecules by the complexing ligand. Rate of lime reagent addition is important, a fast one being preferable (which would usually be the case in plant pH control practice). Further support for this appears to be shown in experience in treating metal ion–cyanide–EDTA solutions, where a mixture of cyanide and EDTA metal complexes co-exist. Destructive chlorination of the cyanide at high pH releases the metal ions for hydrolytic precipitation, while free EDTA ions are also attacked by chlorine, reducing the level of metal–EDTA complexes. The result is a lower residual of metals in the filtrate.

Doane et al. (1977) have pointed out the benefits accruing from the addition of an excess of calcium, e.g. as lime, calcium chloride or calcium sulphate, at high pH (11.6–12.0) to improve precipitation of metal hydrous oxides from complexes. Using a molar ratio of at least 2.5:1 for $Ca^{2+}:Cu^{2+}$, residual soluble copper levels can be less than 2 mg l^{-1} for complexes with EDTA, NTA and HEDTA, although higher residuals were found by Wing with tartrate (13.5 mg l^{-1}), citrate (9 mg l^{-1}) and gluconate (23.9 mg l^{-1}), starting from initial copper levels of 50 mg l^{-1}. Lickskó and Takacs (1986) similarly recommend the use of Mg^{2+} at high pH to aid in the precipitation of heavy metals from complexes and colloidal association products.

Another approach used successfully by Wing et al. (1977) for the treatment of electroless copper rinses from printed circuit board manufacturing where strong complexing agents are present is by addition of ferrous sulphate after lowering of the pH to approximately 3 to dissociate the complex. Wing (1981) attributes the action of the Fe(II) to reducing Cu(II) to Cu(I), but this seems unlikely, and is more probably due to competitive complexing by the ferrous iron. The method, from our observations, appears to be effective when other metals are present, but excess Fe(II) is necessary, molar Fe(II):metal ratios of the order of 5:1 being required—more with more dilute metal concentrations. Of course, as with the other approaches so far reviewed, this does not remove the complexing agent, which is still liable to cause problems downstream if subsequent admixtures with other metals can occur.

Pyrophosphate is used in certain copper-plating solutions, and the complex present inhibits full removal of the copper by normal pH adjustment. However, advantage may be taken of the fact that the pyrophosphate ion, $P_2O_7^{2-}$, is subject to hydrolytic transformation, as discussed earlier. Thus it would appear that by lowering of the pH of a pyrophosphate–metal complex solution to approximately 2, not only will the complex dissociate but also hydrolysis will take place such that on subsequent elevation of the pH the metal ion will be fully precipitated as the hydrous oxide. The fact that pyrophosphate complexes with heavy metals dissociate to a substantial degree below a pH of 5, and simultaneously some precipitation of the insoluble metal compound takes place, is another line of approach. Thus adding ferrous or ferric salts to copper–pyrophosphate complex solutions and adjusting the pH to 4–5 causes precipitation of iron pyrophosphate, releasing the copper for subsequent alkali treatment with lime.

Another interesting group of ligands responsible for complexing includes gluconates, $CH_2OH_4CO_2^-$, and heptonates, $CH_2OH(CHOH)_4CO_2^-$ (reviewed by Sawyer 1964). Gluconic and heptonic acids in solution participate in equilibria with lactone ring compounds: for example, the gluconic acid–lactones' equilibria are

$$
\begin{array}{ccc}
\begin{array}{c}
CO_2H \\ | \\ HCOH \\ | \\ HOCH \\ | \\ HOCH \\ | \\ HCOH \\ | \\ CH_2OH
\end{array}
&
\begin{array}{c}
O = C \overline{} \\ | \quad\quad | \\ HCOH \quad | \\ | \quad\quad | \\ HOCH \quad O \\ | \quad\quad | \\ HCOH \quad | \\ | \quad\quad | \\ HC \overline{} \\ | \\ CH_2OH
\end{array}
&
\begin{array}{c}
O = C \overline{} \\ | \quad\quad | \\ HCOH \quad | \\ | \quad\quad | \\ HOCH \quad O \\ | \quad\quad | \\ HC \overline{} \\ | \\ HCOH \\ | \\ CH_2OH
\end{array}
\\
\text{gluconic acid} & \delta\text{-lactone} & \gamma\text{-lactone}
\end{array}
$$

The nature of the complexes formed with metals in solution is by no means certain, however: some of the high metal:complexant ratios from which stable solutions are formed suggest more than simple complex formation, and micelle structures may be responsible.

In precipitation of metal hydrous oxides from such complex solutions it would appear that advantage could be taken of the fact that as the pH is lowered the sequestering ability reduces and achievement of the acid–lactone equilibria is by no means instantaneous. At high pH values, gluconate anion formation will favour complexation, with a reduction in the lactones' levels, and conversely at low pH values a shift towards higher lactones' levels can be expected; the lactones will have less sequestering powers. If, therefore, a metal gluconate or heptonate solution is lowered in pH, allowed to achieve equilibrium (say 15–30 minutes), and the pH is then raised *rapidly*, the new equilibrium may not be adequately established before precipitation of the metal hydrous oxide. The relative quantities of complexant to metal are important: Van Haarst *et al.* (1974) found that provided the gluconate:metal ratio is less than the stoicheiometric 1:1, low residuals can be achieved by lime treatment to pH 9 (or sodium carbonate for cadmium). Addition of other non-toxic metal cations, e.g. Al^{3+}, can also be made to reduce this ratio in a given solution, thus to aid further the precipitation of the toxic metal(s). Ott (1982) has also shown the benefit of introducing a competitive ion through the use of Fe(III) with gluconate complexes for precipitation of heavy metals at pH 9–10. The amount of Fe(III) required depends on the gluconate concentration, and is less with NaOH than $Ca(OH)_2$ as the precipitant. Ott considers that since a reduction in the iron concentration as well as the gluconate concentration occurs, a sparingly soluble ferric gluconate species is formed.

The complexing action of the neutral ammonia molecule with metals is significant, association occurring at the residual negative charge associated with the nitrogen atom. Metal ammines are cationic, dissociating at low pH, and this can be used, for example, for the precipitation of silver from its ammine by addition of excess chloride; but they are less susceptible to calcium hydroxide treatment than are the anionic complexes already discussed, and it is not always possible to achieve a satisfactory degree of metal removal by simple pH adjustment of copper and nickel ammine solutions. Zinc ammines, however, can be treated relatively satisfactorily with calcium hydroxide, although at high starting concentrations of zinc somewhat higher residuals may be found. This is illustrated by the results in Table 2.6, where

Table 2.6 — Treatment of zinc ammine solutions (pH lowered to 2 with HCl before addition of $Ca(OH)_2$ to pH 10.0)

Total Zn, mg l^{-1}	NH$_3$, g l^{-1}	Residual Zn, mg l^{-1}
300	0	0.15
1000	10	72
600	10	7.0
50	10	0.4

the solutions were treated initially with a large excess of ammonia (1% w/w), acidified to pH 2, and then treated with calcium hydroxide to a pH of 10. The importance of the initial zinc concentration has been demonstrated by Rodenkirchen (1973), who started with a bright acid zinc-plating process solution containing NH$_4^+$ ions. From a dilution of the process solution to 10:1, giving a zinc concentration of 4000 mg l^{-1}, and adjustment to pH 7 with sodium hydroxide, the residual zinc in the filtrate was 429 mg l^{-1}. From a 1000:1 dilution (equivalent to a starting concentration 40 mg l^{-1} Zn), the filtrate of the treated solution contained 2.1–2.4 mg l^{-1} Zn.

Finally, it can be noted that the amine complexes formed by, for example, mono-, di- and triethanolamine with metals are, like the ammines, cationic in character, and behave rather differently from anionic complexes when subjected to alkali addition. These amines are somewhat less inhibitory to metal precipitation than, for example, EDTA and related materials, although ethylenediamine and Quadrol (a substituted ethylenediamine) are quite powerful complexing agents.

Complex formation is much more of a problem in waste water treatment practice than is often realised, particularly as a result of careless mixing of effluent discharges before attempted treatment. Inadequate prior stream separation of discharges before treatment can often result in perplexing difficulties in plant treatment when the possible contamination by complexing agents of metal streams is not appreciated. Even with the techniques mentioned above in this section, although it may be possible to remove metals to an appropriate degree, there is not necessarily any guarantee that the complexing species will be simultaneously fully removed. Thus although insoluble calcium compounds may be formed with the complexing anions, there may be a residual soluble fraction of significance after removal of metal-bearing precipitate; and this is obviously true with the ammine and amine complexes. This may not be serious if the total treatment plant design ensures that following precipitation treatment no recomplexing can occur with metals from other sources, but where the treated waste water passes to a sewage system already containing traces of heavy metal, the result may be interference with a normal primary sedimentation process, and also possibly the secondary biological treatment stage.

There are two ways of circumventing the problem, at least in principle (apart, of course, from eliminating all complexing agents, which is technologically difficult for many industrial processes). One good and obvious approach, as already mentioned in Chapter 1, is to minimise discharges by stream isolation and the use of concentrating techniques to recover the solution for re-use—although even then there will always be a purification 'blow-down' demanding treatment. Such concentration procedures are not always possible, and so the second alternative, that of chemical destruction of the complexing agent, must then be resorted to. This generally involves either oxidative or reductive attack, which are reviewed in Chapters 3 and 4.

PRACTICAL CONSIDERATIONS IN pH CONTROL PRACTICE

The various influences so far described must be borne in mind in the design of treatment systems. Additionally, however, there are certain practical considerations that relate to the ease of process control and associated operations, augmenting the discussion in Chapter 1.

Reaction time factors

In addition to reagent reaction lags there may be slow adjustment to pH equilibrium to which, for example, precipitating hydrous metal oxides may be subject, as analysed earlier. The characteristics of this demand that an efficient dispersion system be used for reagent introduction to minimise the local formation of meta-stable products or the entrapment of alkali within the precipitate in the vicinity of the reagent stream outlet. It means that solids should not be introduced directly, but that finely dispersed slurries or solutions be used. That this effect is not negligible is well borne out by practical experience, where drifts in pH, e.g. from desorption from precipitates, are not uncommon in post-reaction stages, e.g. in settlement tanks and at final outlet points. Where a precipitate is liable slowly to release alkali, a pH layering effect may be observed in a settlement tank, or the outlet pH from this tank may assume a consistently higher and perhaps undesirable value than that to which control is sought. This is unwelcome in, for example, the treatment of amphoteric metal species, or where peptisation may occur in high pH conditions. One way of minimising the problem is to use a dilute rather than a concentrated reagent solution; and an alkaline reagent solution having a lower pH also helps. A two stage reactor can be employed, but usually the nominal retention time should not be less than about 10 minutes overall, unless of course a higher discharge pH is not of particular importance.

Removal of precipitates
Some consideration has to be given in the design of the treatment system to the character of precipitate that is desired. Where settlement is proposed for post-reaction systems, as dense and as aggregated a floc as possible is to be sought, and this is usually achieved by use of calcium hydroxide, and as high a pH as possible. Flocculating agents such as polyelectrolytes (reviewed in Chapter 6), can be used, but with discretion and only after experimental trials, because not only may one type be superior to others with a particular precipitate, but also because the precipitating conditions (e.g. pH, background medium, reagent), are very important. Other aids, such as solids to provide adsorptive or density-increasing properties, e.g. diatomaceous earths, bentonite, cellulosic substances, are occasionally employed, but these give only marginal benefit. Their use is more relevant to the in-stream provision of filter aids where direct filtration after precipitation is to be effected.

One particularly advantageous approach has been described by Evans (1966), based on experiences in water treatment practice. The procedure applies to the treatment of acid solutions of heavy metals, and involves the preliminary admixing of a quantity of settled slurry with the incoming effluent *before* pH adjustment. This pre-conditioning results in a dense precipitate, the characteristics of which are retained when the effluent passes to the pH adjustment. The slurry that settles in the solids' removal stage may approach 10% w/v rather than the usual 2–4% w/v for single stage precipitation. It is important that the ratio of returned slurry to incoming metal content be kept relatively high; and furthermore, the incoming effluent must have the metal in the soluble condition (i.e. it must not be pre-precipitated). Evans postulates an ion exhange hypothesis to explain the action, where the incoming metal and hydrogen ions are exchanged in the primary reaction stage for calcium ions on the recirculating sludge, and hydrogen ions on the sludge are displaced by calcium ions from added calcium hydroxide in the secondary pH adjustment stage. Further studies on such precipitate densification have been described by Patel and Pearson (1977). Densification of aluminium hydroxide, a notoriously gelatinous material with high water content as normally precipitated, is particularly beneficial (D. Pearson, personal communication, 1980).

Increasing attention is nowadays being given to the alternative of flotation rather than settlement of precipitates. This is reviewed in Chapter 7, but mention can be made here of the particular significance in the treatment of hydrolysable metal ions. One criterion for effective flotation must be minimisation of the precipitate density, and sodium alkalis are usually superior to lime as precipitating reagents in this respect. However, flotation assistance is normally necessary, and the usual practice is to employ either very finely dispersed bubbles of air at the inlet to the flotation tank, or compressed air releasing at the inlet. Electrolytic generation of gas bubbles has also been used (see Chapter 5). The fine bubbles attach themselves to the floc particles and by entrainment lift the precipitates to the solution surface (see Chapter 7 for a discussion on dissolved air flotation).

Surfactants (such as sodium lauryl sulphate) may be employed in small quantities as 'collectors', these acting on the hydrolysis products. Hydrolysis data are useful in predicting the extent to which metal removal may be achieved by precipitate flotation, where the amount of collector used is unimportant except to maintaining the minimum necessary to achieve a stable foam at the solution surface. It is interesting that copper does not interact significantly with lauryl sulphate anion beyond electrostatic bonding, and so the sublate formed is soluble. For the removal of copper with sodium lauryl sulphate, therefore, the action is one of ion flotation, and stoicheiometric or larger quantities of the surfactant are needed. A further discussion is given in Chapter 6 on the different actions of surfactants.

pH ADJUSTMENT REAGENTS

The choice of a reagent for pH adjustment in a treatment plant depends on several things: cost, ease and safety of storge and handling, effectiveness (e.g. for removing heavy metals), buffer characteristics of the pH titration curve as they affect pH control, and availability, just to take the most important ones. Some of these aspects have been considered incidentally in preceding sections of this chapter, but it is worthwhile to provide a summary of the key properties of the more important reagents.

Alkalis

The three major reagents in use are hydrated calcium hydroxide, sodium carbonate, and sodium hydroxide, each having its own advantages and disadvantages.

Calcium hydroxide $Ca(OH)_2$ *(lime)*

This is prepared by hydration of quicklime, and for chemical treatment processes it should be in a fine powder state. It is not particularly deliquescent, but does absorb some carbon dioxide from air, thus changing its neutralisation characteristics.

The bulk density is variable, from a settled condition of about 560 kg m^{-3}, falling to 400 kg m^{-3} on aeration. (Aeration is necessary in bulk storage in silos, because the material tends to arch across conical-based containers.) It is normally used as a slurry, in strengths up to a usual maximum of 10% w/w, since above this it becomes difficult to pump and to distribute efficiently and rapidly in a reaction vessel. In practice lime usage is often as much as 15–25% inefficient, but it is still the cheapest of the three commonly used alkalis, the effective cost ratio in the UK being for many purposes approximately 1 ($Ca(OH)_2$):3 (Na_2CO_3):5 ($NaOH$).

Lime slurry can give a maximum pH at normal ambient temperatures of about

12.5. (Saturated lime slurry has a pH of 13 at 10°C, and 12.6 at 20°C, further decreasing as the temperature rises: the solubility decreases as the temperature rises.)

Sodium carbonate, Na₂CO₃ (soda)

Sodium carbonate, Na$_2$CO$_3$ (soda)

The anhydrous form made by the Solvay ammonia–soda process is the one most commonly employed. Both 'dense' (granular) and 'light' forms are commercially available. The granular quality has a bulk density of about 1000 kg m^{-3}, while the light has one of about 540 kg m^{-3}. This varies according to whether the material has settled or is in a free flowing condition, the figures given referring to the latter.

The solubility of sodium carbonate is limited (see Fig. 2.11), and it is usually best

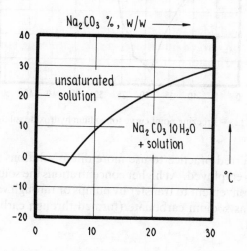

Fig. 2.11 — Solubility–temperature curve for sodium carbonate.

to use a concentration of about 5% if the ambient temperature may drop towards 0°C: crystallisation of sodium carbonate decahydrate in pipes, valves and dosing pumps can cause damage.

Sodium hydroxide, NaOH (caustic soda)

Sodium hydroxide, NaOH (caustic soda)

Sodium hydroxide is available commercially both as solid flakes and as a concentrated solution. The later is particularly convenient for use, because handling of the flakes is an unpleasant job, and presents the risk of causing skin burns and eye damage. Moreover, on dissolution it evolves substantial amounts of heat (the heat of solution being 43 kJ mol^{-1}), which can result in local overheating and solution spitting. On the other hand, concentrated solutions used for transportation of the commercial material (normally 46.6% w/w, specific gravity 1.5 at 20°C) have their own disadvantages. Quite apart from their aggressive nature the stronger ones (above 30% w/w) solidify at temperatures above 0°C, as shown in Fig. 2.12. For this

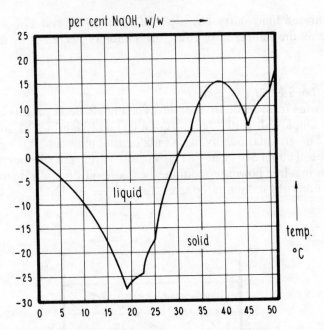

Fig. 2.12 — Freezing point curve for sodium hydroxide solutions.

reason it is the preferred practice to use more dilute solutions for reagent dosing: usually 2–10% w/v is employed. At higher concentrations the solution becomes quite viscous, and is less convenient to transfer by pumps or through valves. Moreover, at higher concentrations sodium carbonate (formed through carbon dioxide absorption) crystallises out.

Acids
The two most commonly employed acid reagents for pH adjustment are sulphuric and hydrochloric. With both, suitable protective clothing and goggles should be worn.

Sulphuric acid H_2SO_4
Sulphuric acid is normally shipped for industrial use either at 77% w/w strength (B.O.V., 'brown oil of vitriol') or 98% w/w (R.O.V., 'rectified oil of vitriol'). In both conditions it is a dense oily liquid, having specific gravities of 1.69 and 1.84 at 20°C, respectively. It is highly corrosive to metals at most strengths, although B.O.V. and R.O.V. do not attack mild steel.

The acid is generally employed for convenience in a more dilute form (up to 10% of the supply strength); the concentrated solutions are viscous, and not usually convenient for direct application in pH control. A major disadvantage with sulphuric acid arises from this, in that dilution evolves substantial quantities of heat, with a consequent risk not only of local overheating, but also explosive solution spitting. The rules are that concentrated sulphuric acid should *always* be added to water (not the other way round), that additions should be made slowly and uniformly, and with continuous blending, and that if the temperature does rise sharply, additions should be stopped immediately. With plastics dilution/storage tanks, particular care must be exercised to prevent temperature rises softening and thereby structurally weakening the material.

Hydrochloric acid, HCl

Various strengths of hydrochloric acid are available commercially, e.g. specific gravities 1.14 (27.6% w/w HCl), 1.16 (31.5% w/w HCl), and 1.18 (35.4% w/w HCl). All hydrochloric acid solutions are highly corrosive to most metals, and should be stored in non-metallic or lined steel vessels.

For dosing, although the concentrated acid can be used directly, it is normal to employ strengths up to 10% of the concentrated versions. Dilution presents no particular problem, but because the diluted acid is highly corrosive to stainless steel (as opposed to diluted sulphuric acid), mixers, tanks and dosing devices need to be non-metallic or of suitably lined steel.

REFERENCES

Amell, A. R. (1956) *J. Am. Chem. Soc.* **78** 6234.
Appleton, A. R., Jr., and Leckie, J. O. (1981) *Environ. Sci. Technol.* **15** 1383.
Argaman, Y., and Weddle, L. C. (1974) *A.I.Ch.E. Symp. Ser. No. 136* 400.
Baes, C. F., Jr., and Mesmer, R. F. (1976) *The Hydrolysis of Cations*, Wiley–Interscience.
Basolo, F., and Pearson, R. G. (1967) *Mechanisms of Inorganic Reactions*, 2nd edn, Wiley.
Bates, R. G. (1954) *Anal. Chem.* **26** 871.
Bates, R. G. (1973) *Determination of pH Theory and Practice*, Wiley.
Bates, R. G. (1981) *Crit. Rev. Anal. Chem.* **10** 247.
Bates, R. G. (1982) *Pure Appl. Chem.* **54** 229.
Bates, R. G., and Guggenheim, E. A. (1960) *Pure Appl. Chem.* **1** 163.
Bell, R. N. (1947) *Ind. Eng. Chem.* **39** 131.
Benjamin, M. M., and Leckie, J. O. (1981) *J. Colloid Interface Sci.* **79** 209.
Benjamin, M. M., Hayes, K. F., and Leckie, J. O. (1982) *J. Wat. Pollut. Control Fed.* **49** 2297.
BS 1647 (1984) *Pt. 1. Specification for pH Scale. Pt. 2. Specification for Reference Value Standard and Operational Standard Solutions*, British Standards Institution.
Burgess, J. (1978) *Metal Ions in Solution*, Ellis Horwood.
Chaberek, S., and Martell, A. E. (1959) *Organic Sequestering Agents*, Wiley.
Civit, E. M., Parm, M. A., and Lupin, H. M. (1982) *Wat. Res.* **16** 809.
Covington, A. K. (1981) *Anal. Chim. Acta* **127** 1.
Covington, A. K., Bates, R. G., and Durst, R. A. (1983) *Pure Appl. Chem.* **55** 1467.
Covington, A. K., Bates, R. G., and Durst, R. A. (1985) *Pure Appl. Chem.* **57** 531.
Department of Environment (1980) *Waste Management Paper No. 21. Pesticide Wastes*, Her Majesty's Stationery Office.
Doane, W. M., Wing, R. E., and Rayford, W. E. (1977) *Plat. Surf. Finish.* **64** 57.
Dwyer, F. P., and Mellor, D. P. (eds.) (1964) *Chelating Agents and Metal Chelates*, Academic Press.

Eden, G. E., and Truesdale, G. A. (1950) *J. Iron Steel Inst.* **164** 281.

Eisenman, G., (ed.) (1967) *Glass Electrodes for Hydrogen and Other Cations*, Marcel Dekker.

Evans, R. R. (1966) *Proc. 21st Ind. Waste Conf. Purdue Univ.* 511.

Gehm, H. W. (1944) *Sewage Wks. J.* **16** 104.

Griffiths, E. J. (1959) *Ind. Eng. Chem.* **51** 240.

Hannah, S. A., Johns, M., and Cohen, J. M. (1977) *J. Wat. Pollut. Control Fed.* **49** 2297.

Hartinger, L. (1973) *Metallobarfläche (München)* **27** 157.

Hartinger, L. (1976) *Taschenbuch der Abwasserhandlung für die Metallverarbeitende Industrie. Band I. Chemie.*, Carl Hanser.

Heinke, G. W., and Norman, J. D. (1969) *Proc. 24th Ind. Waste Conf. Pardue Univ.* 644.

Hoak, R. D., Lewis, C. L., Sindlinger, C. J., and Klein, B. (1947) *Ind. Eng. Chem.* **39** 131.

Hoerth, J., Schindewolf, U., and Zbinden, W. (1973) *Chem.-Ing. Technik* **45** 641.

Högfeldt, E. (1982) *IUPAC Chemical Data Series, No. 21. Stability Constants of Metal Ion Complexes. Part A. Inorganic Ligands*, IUPAC/Pergamon.

Hunt, J. P. (1963) *Metal Ions in Aqueous Solution*, Benjamin.

Ives, D. J. G., and Janz, G. J. (eds.) (1961) *Reference Electrodes*, Academic Press.

Jensen, M. B. (1958) *Acta Chem. Scand.* **12** 1657.

Jola, M. (1970) *Galvanotechnik (Saulgau)* **61** 1003.

Kennedy, M. V., Stojanovic, B. J., and Shuman, F. L., Jr. (1969) *Residue Rev.* **29** 89.

Kennedy, M. V., Stojanovic, B. J., and Shuman, F. L., Jr. (1972) *J. Agric. Food Chem.* **20** 341.

Kenyon, W. O., and Gray, H. L. (1936) *J. Am. Chem. Soc.* **8** 1922.

Korenowski, T. F., Penland, J. L., and Kitzert, C. J. (1977) *U.S. Patent Nos.* 4, 008, 162; 4, 045, 339.

Kragten, J. (1978) *Atlas of Metal–Ligand Equilibria in Aqueous Solution*, Ellis Horwood.

Krieble, U. K., and McNally, J. G. (1929) *J. Am. Chem. Soc.* **51** 3368.

Krieble, U. K., and Peiker, A. C. (1933) *J. Am. Chem. Soc.* **55** 2326.

Leckie, J. O., and James. R. O. (1974) *Aqueous Environmental Chemistry of Metals*, ed. Rubin, A. J., Ch. 1, Ann Arbor Science.

Leigh, G. M. (1969) *J. Wat. Pollut. Control Fed.* **41** R450.

Lickskó, J., and Takacs, I. (1986) *Wat. Sci. Technol.* **18** 19.

Marsh, J. D. F., and Martin, M. J. (1957) *J. Appl. Chem.* **7** 205.

Martell, A. E. (1971) *A.C.S. Monograph. Coordination Chemistry*, Vol. 1, Van Nostrand.

Martell, A. E., and Smith, R. M. (1974) *Critical Stability Constants*, Vol. 1, Amino Acids, Plenum Press.

Martell, A. E., and Smith, R. M. (1977) *Critical Stability Constants*, Vol. 3, Other Organic Ligands, Plenum Press.

Mattock, G. (1961) *pH Measurement and Titration*, Heywood.

Mattock, G. (1977) *Proc. Symp. Treatment of Wastewater from Chemical Industries*, Institute of Water Pollution Control/Institution of Chemical Engineers.

Mattock, G., and Band, D. M. (1967) *Glass Electrodes for Hydrogen and Other Cations,* ed. Eisenman, G., Ch. 2, Marcel Dekker.

Nancollas, G. H. (1967) *Interactions in Electrolyte Solutions,* Elsevier.

Nieweglowska, Z., and Bartkiewicz, B. (1980) *Physicochemical Methods for Water and Wastewater Treatment,* ed. Pawlowski, L., 221, Pergammon.

O'Melia, C. R., and Stumm, W. (1967) *J. Colloid Interface Sci.* **23** 437.

Ott, D. (1982) *Galvanotechnik (Saulgau)* **73** 453.

Patel, C., and Pearson, D. (1977) *Inst. Chem. Eng. Ser. No. 52* **2** 59.

Perrin, D. D. (1979) *IUPAC Chemical Data Series, No. 22. Stability Constants of Metal Ion Complexes. Part B. Organic Ligands,* IUPAC/Pergamon.

Petrucci, S. (ed.) (1971) *Ionic Interactions,* Vol. II, Academic Press.

Reidl, A. L. (1947) *Chem. Eng.* **34**(7) 100.

Resnick, J. D., Moore, W. A., and Ettinger, M. B. (1958) *Ind. Eng. Chem.* **50** 71.

Ricci, J. E. (1952) *Hydrogen Ion Concentration,* Princeton Univ. Press.

Ringbom, A. (1963) *Complexation in Analytical Chemistry,* Interscience.

Rodenkirchen, M. (1973) *Galvanotechnik (Saulgau)* **64** 782.

Sawyer, D. T. (1964) *Chem. Rev.* **64** 633.

Schwarzbach, E. (1974) *Wasser Luft u. Betrieb* **18** (1)26.

Shannon, J. E., and Lee, G. F. (1966) *Int. J. Air Wat. Pollut.* **10** 735.

Shinskey, F. G. (1973) *pH and pIon Control in Process and Waste Streams,* Wiley.

Sillén, L. G. (1959) *Quart. Rev.* **13** 146.

Siu, R. G. H. (1951) *Microbiol Decomposition of Cellulose with Special Reference to Cotton Textiles,* Reinhold.

Smith, R. M., and Martell, A. E. (1975) *Critical Stability Constants,* Vol. 2. Amines, Plenum Press.

Smith, R. M., and Martell, A. E. (1976) *Critical Stability Constants,* Vol. 4. Inorganic Complexes, Plenum Press.

Stumm, W., and Morgan, J. J. (1962) *J. Am. Water Works Assoc.* **54** 971.

Stumm, W., and Morgan, J. J. (1981) *Aquatic Chemistry,* 2nd edn, Wiley–Interscience.

Stumm, W., and O'Melia, C. R. (1968) *J. Am. Water Works Assoc.* **60** 514.

Taube, H. (1970) *Electron Transfer Reactions of Complex Ions in Solution* Academic Press.

Tomlinson, H. D., Thackston, E. L., Koon, J. H., and Krenzel, P. A. (1975) *J. Wat. Pollut. Control Fed.* **47** 562.

Van Haarst, W. F. M., Mulder, R. J., and Tervoort, J. L. J. (1974) *Belg.-Ned. Tijd. Oppervlakte.* **18** 264.

Watzel, R. (1942) *Die Chemie* **55** 356.

Wendt, T. M., and Kaplan, A. M. (1976) *J. Wat. Pollut. Control Fed.* **48** 660.

Wheatland, A. B., and Borne, B. J. (1962) *Water and Waste Treat.* **8** 560.

Wiegand, G. H., and Tremelling, M. (1972) *J. Org. Chem.* **37** 914.

Wilkins, R. G., and Eigen, M. (1965) *Advances in Chemistry Ser. 49. Mechanisms of Inorganic Reactions,* 53, American Chemical Society.

Wing, R. E. (1981) *Proc. Workshop Inst. Interconnecting and Packaging Circuits,* Anaheim, May 19–21.

Wing, R. E., and Rayford, W. E. (1978) *Metal Finish.* **76** 31.

Wing, R. E., Rayford, W. E., and Doane, W. M. (1977) *Plat. Surf. Finish.* **64** 39.

3

Chemical oxidation

THE SIGNIFICANCE OF CHEMICAL OXIDATION IN POLLUTION ABATEMENT

Oxyen demand by pollutants

Ultimately, the fate of all pollutant material discharged by man to the environment is dictated by the relative thermodynamic stabilities of such material compared to the stabilities of possible reaction products. In the aerobic aquatic ecosystem to which most material is consigned, the relative bond energies suggest that in the case of organic molecules the final products will be carbon dioxide, water, and oxyanions such as sulphate and phosphate. The ultimate fate of heavy metal ions is to be incorporated into sediments after precipitation as insoluble compounds such as hydroxides, hydrated oxides, oxides and sulphides. Unfortunately, many of these natural changes take place very slowly. The object of oxidative treatment processes is frequently to accelerate these changes by exploiting chemical and biochemical principles in order to surmount the kinetic restraints which are responsible for the slowness of some of the reactions.

The biochemical oxidation of organic matter is of course widely exploited in the familiar biologial treatment plants for domestic and industrial waste waters, where an adequate oxygen supply is maintained either by the use of devices such as large surface area trickling filters or by forced aeration as in activated sludge processes. Not all organic material is rapidly degraded by such means. There are many organic compounds, chiefly synthetic ones, that are not amenable to biological treatment either because they are toxic to the micro-organism population at the concentrations present or because of their metabolic inertness. It is in the treatment of organic material of this nature and in the treatment of certain inorganic pollutants that chemical oxidation processes are used to their greatest advantage. It must be remembered that in chemical oxidation processes it is not always possible, necessary or even particularly desirable that the organic content of the offending waste be removed entirely. Frequently all that is necessary is that toxic or biologically inert materials be modified sufficiently by oxidative degradation to enable them subsequently to be treated biologically, since for non-toxic biodegradable material biological treatment offers distinct economic advantages. Of course, there are occasions when simply modifying the organic content is inadequate, and the requirement may be to reduce the total organic load significantly before discharge.

THE NATURE OF OXIDATION–REDUCTION REACTIONS

General principles

The original defintion of oxidation as the chemical combination of oxygen with a substance to form an oxide was subsequently extended to include combination of other electronegative elements such as chlorine with substances, and the abstraction of hydrogen from substances. The definition is now generalised in terms of the removal of electrons from an element or compound. Reduction is conversely defined as the removal of oxygen or other electronegative elements, addition of hydrogen or electropositive elements or, more generally, as a gain of electrons. Since in chemical reactions electrons are merely transferred from one species to another, it follows that oxidation and reduction are complementary processes which are, in a real sense, inseparable. Often, for convenience, reactions are written as separate oxidation and reduction 'half-reactions', but it must always be remembered that this is a hypothetical exercise and that a chemical oxidation is impossible without a concomitant reduction.

The definition of oxidation proposed by Weber (1972) as a process in which the oxidation state of a *substance* is increased is misleading, since the notion of oxidation state is only applicable to individual atoms within molecules, and not to 'substances'. Nevertheless, the concept of oxidation state or oxidation number (the two terms are synonymous) is an extremely useful one, particularly when dealing with inorganic reactions. The oxidation number of an atom in a molecule or ion is assigned on the basis of a set of simple though somewhat arbitrary rules which may be found in almost any introductory text-book of inorganic chemistry (see, e.g., Mackay and Mackay 1981). Application of these rules leads for example to the assignment of $+3$ as the oxidation number of the nitrogen atom in nitrous acid and $+5$ as the oxidation number of nitrogen in nitric acid; in both these compounds, the oxidation number of oxygen is -2 and that of hydrogen is $+1$. Conversion of nitrous acid to nitric acid involves an increase in the oxidation number of nitrogen from $+3$ to $+5$ and is therefore classed as an oxidation. It must be emphasised that the oxidation numbers of $+3$ and $+5$ refer only to the nitrogen atoms in the molecules, not to the molecules themselves. It is also important to realise that the oxidation number of an atom has no direct physical significance regarding the actual charge distribution within the molecule or ion in which the atom is found. It is an entirely artificial device which is no more than a convenient accounting aid when dealing with the transfer of electrons from one species to another. In inorganic systems, where the assignment of oxidation numbers is usually straightforward and unambiguous, the concept is particularly useful, and greatly simplifies the stoicheiometric balancing of redox equations and calculations in analytical procedures involving oxidation–reduction reactions.

It is evident that in addition to being able to recognise qualitatively a particular substance as a potential oxidant or reductant, it is important to have some quantitative measure of its oxidising or reducing power under particular conditions. Such a measure is the standard electrode potential, $E^{\ominus}$. This is related to the standard Gibbs free energy change, $\Delta G^{\ominus}$, for the reaction by the equation

$$\Delta G^{\ominus} = -nE^{\ominus}F \tag{3.1}$$

where n is the number of electrons transferred and F is the Faraday constant.

A generalised oxidation–reduction reaction may be written as follows:

$$A_{ox} + B_{red} \overset{ne}{\rightleftharpoons} A_{red} + B_{ox} \tag{3.2}$$

where the subscripts denote the oxidised and reduced forms of A and B. It is convenient to consider this overall reaction in terms of the two half-reactions that notionally make it up:

$$A_{ox} + ne \rightarrow A_{red} \tag{3.3}$$

$$B_{ox} + ne \rightarrow B_{red} \tag{3.4}$$

For consistency, both reactions (3.3) and (3.4) are written as reductions, although, of course, in the overall redox reaction involving these half-reactions (equation 3.2) it is clear that one of them must be reversed.

Associated with half-reactions (3.3) and (3.4) are the standard electrode potentials $E_A^{\ominus}$ and $E_B^{\ominus}$ and the standard free energy changes $\Delta G_A^{\ominus}$ and $\Delta G_B^{\ominus}$.

For the overall reaction (3.2) it is clear that

$$\Delta G^{\ominus} = -nE^{\ominus}F = \Delta G_A^{\ominus} - \Delta G_B^{\ominus} = nF(E_A^{\ominus} - E_B^{\ominus})$$

hence

$$E^{\ominus} = E_A^{\ominus} - E_B^{\ominus}$$

Thus from tabulations of standard electrode potentials for half-reactions (Appendix 1) one can, in principle, calculate the standard electrode potential for any oxidation–reduction reaction,[a] and hence the value of $\Delta G^{\ominus}$ from equation (3.1) and the equilibrium constant for the reaction from the familiar equation

$$\Delta G^{\ominus} = -RT \log_e K \tag{3.5}$$

It must be remembered that the standard electrode potential $E^{\ominus}$ for a reaction or half-reaction implies standard state conditions (unit activity, temperature equal to

a Note that tabulated values of standard potentials, as in Appendix 1, refer to half-reactions written as reductions. If half-reactions are written as oxidations the sign of the potential is reversed and should not be referred to as an electrode potential: it is then an oxidation potential.

298 K) for both reactants and products. Rarely, if ever, in water treatment processes do conditions even approach the standard state and it is therefore necessary to consider the effect of non-standard conditions.

Non-standard conditions—the Nernst equation

The position of equilibrium and hence the electrode potential of a given redox half-reaction is, of course, a function of the concentration or, more correctly, the activity of all the species involved in the reaction. This relationship is expressed quantitatively in the familiar Nernst equation, which for a half-reaction written as a reduction:

$$ox + ne \rightleftharpoons red$$

It takes the form

$$E = E^{\ominus} + \frac{RT}{nF} \log_e \frac{a_{ox}}{a_{red}}$$

This equation is derived quite simply from the van't Hoff equation, which for the general reaction

$$aA + bB \rightarrow cC + dD$$

may be written as

$$\Delta G = \Delta G^{\ominus} + RT \log_e \frac{(a_C)^c (a_D)^d}{(a_A)^a (a_B)^b} \tag{3.6}$$

In the case of the above redox half-reaction this is simply

$$\Delta G = \Delta G^{\ominus} + RT \log_e \frac{a_{red}}{a_{ox}} \tag{3.7}$$

Reduction of one mole of oxidant to its reduced form requires the passage of nF coulombs of electricity against a potential difference of E volts, so the electrical work done by the system (at constant temperature and pressure) is nEF joules. This must, of course, be equal to the decrease in free energy of the system ($-\Delta G$), hence

$$\Delta G = -nEF \tag{3.8}$$

or, for standard state conditions,

$$\Delta G^{\ominus} = -nE^{\ominus}F \tag{3.9}$$

Substituting for ΔG and $\Delta G^{\ominus}$ in the van't Hoff equation (3.7) leads to

$$- nEF = - nE^{\circ}F + RT \log_e \frac{a_{red}}{a_{ox}}$$

or

$$E = E^{\ominus} + \frac{RT}{nF} \log_e \frac{a_{ox}}{a_{red}} \tag{3.10}$$

which is the familiar form of the Nernst equation. The more negative is ΔG for a reaction, the greater is the tendency for that reaction to occur, so it follows from equations (3.8) and (3.9) that the more positive is the value of E for a particular half-reaction, the greater the tendency for that half-reaction to occur *when it is written as a reduction.*

The great value of a table of standard electrode potentials in the field of waste water treatment is that it greatly simplifies the process of deciding whether or not a particular oxidation or reduction is thermodynamically feasible. A simple example illustrates this point. Suppose one wished to consider the possibility of recovering copper from an acid effluent stream by reaction with scrap iron. The relevant half-reactions and their associated standard potentials are as follows:

$$Cu^{2+} + 2e \rightleftharpoons Cu^0 \quad E^{\ominus} = + 0.34 \text{ V}$$

$$Fe^{2+} + 2e \rightleftharpoons Fe^0 \quad E^{\ominus} = - 0.44 \text{ V}$$

(Note that both reactions are written as reductions.) The simple rule to be used is that the half-reaction with the more positive value for $E^{\ominus}$ proceeds as written (since this will have the more favourable free energy change), and the other is reversed in the overall reaction, which will therefore be

$$Cu^{2+} + Fe^0 \rightarrow Fe^{2+} + Cu^0$$

The standard electrode potential for this overall reaction is simply the algebraic sum of the standard electrode potentials of the two half-reactions (remembering of course that in this case the second half-reaction is reversed):

$$E^{\ominus}_{overall} = + 0.34 - (- 0.44) = + 0.78 \text{ V}$$

This large positive value implies a large negative value for $\Delta G^{\ominus}$, and hence the reaction is thermodynamically favourable. Just how favourable one can easily calculate, since

$$-nE^{\ominus}F = \Delta G^{\ominus} = -RT \log_e K$$

i.e.

$$\log_e K = \frac{nE^{\ominus}F}{RT}$$

$$\log_{10} K = \frac{nE^{\ominus}F}{2.303\ RT} = \frac{2 \times 0.780 \times 96\ 500}{2.303 \times 8.314 \times 298} = 26.38$$

In other words, the equilibrium constant for the overal reaction is approximately 10^{26}, i.e., complete for all practical purposes. It is easy to see that if one considers reactions to be essentially complete if $K > 10^4$, then this implies a difference in electrode potential for the two half-reactions of approximately 240 mV for one-electron exchanges, or about 120 mV for two-electron processes. This somewhat rough and ready rule enables a rapid assessment of the thermodynamic feasibility of potential reactions to be made, although, of course, kinetic restraints may preclude certain thermodynamically feasible reactions. These are discussed in more detail later.

Oxidation state diagrams

A convenient graphical way of presenting redox potential data for elements such as manganese, which can exist in several oxidation states in aqueous solution, is by means of so-called oxidation state diagrams (Fig. 3.1). This obviates the need for extensive tabulations of data and allows one to identify immediately those oxidation states that are particularly prone to disporportionate and those that are particularly stable; and, when several elements are plotted on the same figure, to recognise quickly those overall reactions that are thermodynamically feasible and those that are not. This approach, which was originally suggested by Frost (1951), plots the free energy $(\Delta G_f^{\ominus})$ for the formation of M^{n+} from M against the oxidation state n. In order to be able to work with small numbers $\Delta G_f^{\ominus}$ is often divided by F, the Faraday constant, and this is the method adopted here. $\Delta G_f^{\ominus}/F$ is, of course, simply equal to $nE^{\ominus}$, since

$$\Delta G^{\ominus} = -\Delta G^{\ominus}_{(reduction)} = +nE^{\ominus}F$$

The quantity $nE^{\ominus}$ is referred to as the 'volt equivalent'. If one plots the volt equivalent $nE^{\ominus}$ against the oxidation number n as in Fig. 3.1, then the slope of a line joining any two oxidation states of the element is equal to the standard electrode potential for that redox couple. It follows, therefore, that any species represented by a 'convex' point, i.e., any species which is represented by a point lying above the line joining points representing adjacent species, will be unstable with respect to disproportionation to these species (e.g. $2Cu^+ \rightarrow Cu^0 + Cu^{2+}$).

It is apparent from an inspection of Fig. 3.1 that pH in general has a marked effect on the redox properties, higher oxidation states of transition metals being particularly stabilised in alkaline solution. It is therefore important to realise the limitations

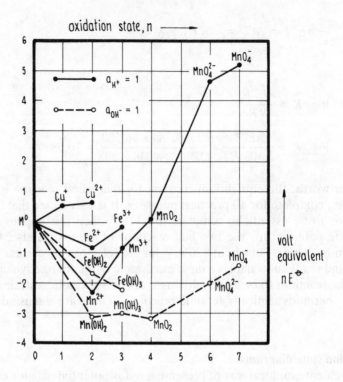

Fig. 3.1 — Oxidation state diagrams.

of such diagrams, and to recognise that diagrams drawn from standard electrode potential data (pH = 0) will not be the same for other pH conditions, nor will they necessarily be valid in the presence of precipitants or complexing ligands.

pH effects
So far we have considered only simple reactions in which single species exchange electrons. In many oxidation–reduction reactions occurring in aqueous solution, other molecules or ions may be involved. Water molecules, hydrogen and hydroxyl ions, complexing and precipitating reagents may have significant effects on the course of a reaction, and this is reflected in their effect on the electrode potentials of the various half-reactions in which such species are involved. A good example of this is afforded by the reactions, in acid solution, of the permanganate ion, for which the relevant half-reaction may be written

$$MnO_4^- + 8H^+ + 5e \rightleftharpoons Mn^{2+} + 4H_2O \qquad E^\ominus = +1.51 \text{ V}$$

The *standard* electrode potential $E^\ominus$ refers to conditions such that the activities of all species, including hydrogen ions, where involved in the reaction, are unity. For hydrogen ion activities other than unity, i.e. at pH values other than zero, the magnitude of the electrode potential and hence the oxidising power of permanganate will be different. The Nernst equation for the above half-reaction must be written as

$$E = E^\ominus + \frac{RT}{5F} \log_e \frac{(a_{MnO_4^-})(a_{H^+})^8}{(a_{Mn^{2+}})}$$

since H^+ ions are involved in the reaction. Simple arithmetic leads to the following conclusion:

At pH $= 0$ $E = E^\ominus = +1.51$ V

At pH $= 3$ $E = E^\ominus + (-0.28) = +1.23$ V

At pH $= 6$ $E = E^\ominus + (-0.56) = +0.95$ V

i.e. the oxidising power of MnO_4^- is strongly pH-dependent. However, it is important to note that in alkaline or neutral solutions, permanganate is reduced not to Mn^{2+}, but to MnO_2 by a three-electron step according to the equation

$$MnO_4^- + 4H^+ + 3e \rightleftharpoons MnO_2 + 2H_2O$$

for which $E^\ominus = +1.69$ V, or to the equation,

$$MnO_4^- + 2H_2O + 3e \rightleftharpoons MnO_2 + 4OH^-$$

for which $E^\ominus = +0.59$ V. (These two reactions are indistinguishable in aqueous solution, the difference in the numerical value of $E^\ominus$ being accounted for by terms in the Nernst equation involving the ionic product of water.) In acid solution this does not occur, since in the presence of excess hydrogen ions MnO_2 will be further reduced thus:

$$MnO_2 + 4H^+ + 2e \rightleftharpoons Mn^{2+} + 2H_2O$$

The nett effect is equivalent to the half-reaction for the MnO_4^-/Mn^{2+} couple previously discussed.

Similar calculations lead to the conclusion that oxygen is a better oxidising agent in acid solution than in alkaline solution, since for the half-reaction

$$\tfrac{1}{2}O_2 + 2H^+ + 2e \rightarrow H_2O$$

$E^\ominus = +1.23$ V (pH $= 0$), whereas at pH $= 14$

$$E = +1.23 + \frac{RT}{2F} \log_e 10^{-28}$$

$$= +0.41 \text{ V}$$

This result, suggesting as it does that oxygen will always be a better oxidising agent in acid solution than in alkaline solution, can be misleading if due thought is not given to the chemistry of the complete reaction. All chemical oxidations must involve two half-reactions, and the effect of pH on both must be considered. These effects are not always obvious. For example, contrary to what one might have expected from the above, it is known from practical experience that oxygen oxidises Fe^{2+} to Fe^{3+} much more readily and efficiently at a high pH than it does in acid solution, and yet, apparently, H^+ ions are not involved in the Fe^{3+}/Fe^{2+} couple:

$$Fe^{3+} + e \rightleftharpoons Fe^{2+} \qquad E^{\ominus} = +0.76 \text{ V}$$

However, in alkaline solution, the activities of Fe^{2+} and Fe^{3+} cannot approach anywhere near unity, and in fact, will be very small due to the insoluble nature of '$Fe(OH)_3$' and '$Fe(OH)_2$', for which the solubility products are approximately 10^{-36} and 10^{-14} respectively. At pH = 14, for example, when the activity of OH^- is unity, one can calculate the actual potential of the Fe^{III}/Fe^{II} couple from the Nernst equation as follows:

At pH = 14, $a_{OH^-} = 1$, and hence

$$a_{Fe^{3+}} = 10^{-36} \quad \text{and} \quad a_{Fe^{2+}} = 10^{-14}$$

since $\qquad K_{s.p.} \text{ (for } Fe(OH)_3) = 10^{36} a_{OH^-}^3 \times a_{Fe^{3+}}$

and $\qquad K_{s.p.} \text{ (for } Fe(OH)_2) = 10^{-14} = a_{OH^-}^2 \times a_{Fe^{2+}}$

$$E = E^{\ominus} + \frac{RT}{F} \log \frac{a_{Fe^{3+}}}{a_{Fe^{2+}}}$$

$$= +0.76 + 0.059 \log_{10} 10^{-22}$$

$$= -0.538 \text{ V}$$

Thus the oxidation of Fe^{2+} to Fe^{3+} by oxygen is thermodynamically much more favourable at pH = 14 than it is at pH = 0 ($E_{overall}$ at pH = 0 is $1.23 - 0.76 = +0.47$ V; $E_{overall}$ at pH = 14 is $0.41 + 0.54 = +0.95$ V).

It is generally true to say that metal ions which exist in more than one oxidation state will have the higher oxidation states stabilised preferentially in alkaline solution, since in most cases the hydroxide of the metal in its higher oxidation state is less soluble than the hydroxide of the lower state. The standard Winkler method (Department of Environment 1979, APHA/AWWA/WPCF 1985) for the determination of oxygen in water depends on the stabilisation of Mn^{3+} with respect to Mn^{2+}. In acid solution $E^{\ominus}$ for Mn^{3+}/Mn^{2+} is $+1.51$ V, and oxygen, for which $E^{\ominus}(O_2, H^+/H_2O)$ is only $+1.23$ V, cannot oxidise Mn^{2+}. However, in alkaline solutions the relevant potentials are $+0.15$ V for $Mn^{III}(OH)_3/Mn^{II}(OH)_2$ and $+0.414$ V for $O_2, H^+/H_2O$, and oxygen will oxidise $Mn(OH)_2$ to $Mn(OH)_3$.

Effect of precipitation and complex formation

The pH effects on the redox potentials of Fe^{3+}/Fe^{2+} and Mn^{3+}/Mn^{2+} referred to above are really only particular examples of the more general effects of complexing and precipitating agents. If any reagent preferentially removes one oxidation state of a redox couple relative to the other, then the electrode potential of the couple will be changed. Another example which does not involve hydrogen or hydroxyl ions is the stabilisation of the Cu^I state relative to Cu^{II} by the presence of Cl^- ions, with which Cu^I forms an insoluble precipitate. Consideration of the electrode potentials for the half-reactions

$$Cu^{2+} + e \rightleftharpoons Cu^+ \qquad E^\ominus = +0.15 \text{ V}$$

$$Cu^+ + e \rightleftharpoons Cu \qquad E^\ominus = +0.52 \text{ V}$$

or inspection of Fig. 3.1 indicates that Cu^+ should disproportionate in aqueous solution according to the equation:

$$2Cu^+ \rightleftharpoons Cu^{2+} + Cu$$

for which $E^\ominus_{\text{overall}} = +0.52 - 0.15 = +0.37 \text{ V}$.

This implies a value of K, the equilibrium constant for the reaction, of approximately 10^6, confirming that Cu^+ is, as observed, unstable in aqueous solution. However, if Cu^I can be preferentially removed from the system as an insoluble precipitate or, alternatively, as a complexed species, then the equilibrium will be displaced in favour of Cu^I. This occurs in the presence of Cl^- ions, with which Cu^I forms the insoluble CuCl, or in the presence of CN^-, when Cu^I is removed from the equilibrium as the soluble complex $[Cu(CN)_4]^{3-}$. These precipitations or complexing equilibria should be regarded as being in competition with the disproportion reaction. The extent to which Cu^I will be stabilised and $E^\ominus$ shifted will depend on the relative magnitudes of the various equilibium constants involved—indeed measurements of electrode potentials in the presence of complexing agents is one way in which equilibrium constants for metal ions have been measured (Rossotti and Rossotti 1961).

Even after taking into account the effects of complexing and/or precipitation, it is evident that the use of thermodynamic data alone has its limitations. It allows one to assess the thermodynamic feasibility of a particular reaction, offers guidance as to which reactions might be possible, and identifies those reactions which are apparently impossible. However, thermodynamic information such as $E^\ominus$ or $\Delta G^\ominus$ data supply may have no direct significance when one is considering the rates at which particular reactions occur in order to assess their practicability in terms of the temporal limitations of treatment processes. For example, consider the following two half-reactions:

$$O_2 + 4H^+ \rightarrow 2H_2O \qquad E^\ominus = +1.23 \text{ V}$$

$$CH_3CHO + 2H^+ + 2e \rightarrow C_2H_5OH \qquad E^\ominus = +0.2 \text{ V}$$
$$\text{acetaldehyde} \qquad\qquad \text{ethanol}$$

These $E^\ominus$ values suggest that air or oxygen should oxidise ethanol to acetaldehyde with large decrease in free energy. In practice, however, in the absence of catalysts, this reaction occurs at an immeasurably slow rate at room temperature. This is fortunate for those of us who enjoy an occasional gin, but would be distinctly disadvantageous if we wished to utilise ambient temperature air oxidation as a means of removing alcohols (or phenols, for which a similar situation obtains) from effluent streams. It is indeed often the case that the outcome of a particular reaction or a set of reactions is kinetically rather than thermodynamically controlled. In a series of competing reactions it may well be that the reaction with the most favourable free energy change contributes little to the overall reaction if other reactions, perhaps with considerably less favourable free energy changes, have lower activation energies and hence can proceed at faster rates. It is apparent that a firm grasp of the principles of the kinetics and mechanistic features of oxidation–reduction reactions is an essential prerequisite to any rationalisation of the observed chemistry.

OXIDATION–REDUCTION MECHANISMS

Two distinct types of oxidation–reduction reaction can be formally recognised. Firstly, there are those reactions in which the nett result is simply the transfer of electrons from the reductant to the oxidant, as for example in the reaction

$$Cu^+ + Fe^{3+} \rightleftharpoons Cu^{2+} + Fe^{2+}$$

Secondly, there are those reactions in which atom or group transfer occurs, but which are nevertheless regarded as oxidation–reduction reactions because the oxidation states of some of the atoms change concomitantly. An example of such a reaction is the oxidation of nitrite by hypochlorite

$$NO_2^- + OCl^- \rightarrow NO_3^- + Cl^-$$

in which an oxygen atom is transferred.

The following discussion reviews those aspects particularly relevant in waste water chemical treatment. For fuller analyses see, for example, Edwards (1964), Stewart (1964), Eigen and Wilkins (1965), Basolo and Pearson (1967), Benson (1968), McAuley and Hill (1969), Taube (1970), and Cannon (1980).

Electron transfer reactions

These may be subdivided into two categories. In so-called 'outer sphere' reactions, the coordination spheres of the oxidant and reductant remain intact and the transferring electron(s) must penetrate both. In the other category, the 'inner sphere' reactions, the oxidant and reductant are linked in an intermediate or activated complex by an atom or group which is common to both coordination spheres and which acts as a bridge across which the electron(s) transfer. The outer sphere mechanism is in principle open to all electron transfer reactions, whereas the inner sphere mechanism of course depends on the presence of a suitable bridging ligand; frequently both mechanisms may contribute to the overall reaction. It is worth pointing out at this stage that there is no evidence for electron transfer proceeding via monomolecular mechanisms; the lifetime of a solvated electron is extremely short in aqueous solution and any proposed mechanism involving loss of an electron by the reductant to the solvent before subsequent transfer to the oxidant is not feasible. All electron transfer mechanisms are therefore dimolecular, i.e. involve both participating species.

Outer sphere reactions

The simplest type of electron transfer reactions in aqueous solution are so-called electron 'exchange' reactions in which there is no nett chemical change, e.g.

$$Fe(H_2O)_6^{2+} + {}^*Fe(H_2O)_6^{3+} \rightarrow Fe(H_2O)_6^{3+} + {}^*Fe(H_2O)_6^{2+}$$

These reactions are studied by isotopic labelling of one of the exchanging species, and although there is no nett chemical reaction, they are nevertheless properly considered as oxidation–reduction reactions. Their direct relevance to water treatment processes where the object is to achieve chemical change is obviously somewhat limited. Electron transfer reactions in which a chemical change is observed are referred to as 'cross-reactions,' and the mechanism of such outer sphere cross-reactions can be considered to consist of three elementary successive steps:

(1) Collision of reactants with formation of a loosely associated 'collision complex' or 'precursor complex':

$$A + B \rightarrow [\{A\}\{B\}]$$

(2) Chemical activation of the precursor complex, with electron transfer and relaxation to the 'successor complex':

$$[\{A\}\{B\}] \rightarrow [\{A^-\}\{B^+\}]$$

(3) Dissociation of the successor complex to give separate products:

$$[\{A^-\}\{B^+\}] \rightarrow A^- + B^+$$

It is the activation of the precursor complex which requires the input of energy and which is hence the slowest or rate-determining step.

Inner sphere reactions

In reactions of this type, the coordination shells of the oxidant and reductant share a common ligand in an intermediate complex. Electron transfer occurs via this bridging ligand; concomitant atom transfer may also occur, but this is not always so. The mechanism of these reactions may be summarised as follows, where X^- represents an anionic bridging ligand (e.g. OH^- or Cl^-) and M^{m+} and N^{n+} represent the oxidant and reductant metal ions:

$$MX^{(m-1)+} + N^{n+} \rightarrow [M^{m+}...X^-...N^{n+}] \rightarrow [M^{(m-1)+}...X^-...N^{(n+1)+}] \rightarrow products$$

The bridging ligand X may be a simple inorganic species as above, or it may be one of a wide variety of organic ligands. The essential requirement of a bridging species is, of course, the availability on the molecule (or ion) of two lone pairs of electrons which enable simultaneous bonding of the bridge species to both the oxidant and reductant centres.

Before leaving the subject of inner and outer sphere mechanisms, it must be pointed out that in many cases the possibility of either mechanism exists. In principle, all redox reactions involving transition metal ions could proceed via an outer sphere path provided sufficient activation energy were available. The extent to which any particular mechanism contributes to the overall reaction is of course reflected in the rate expressions, which frequently may be quite complex for apparently simple reactions. For example, the exchange of an electron between Fe^{2+} and Fe^{3+} in the presence of Cl^- ions is described by the following rate expression:

$$Fe^{*II} + Fe^{III} \overset{Cl^-}{\rightleftharpoons} Fe^{*III} + Fe^{II}$$

$$\frac{\partial [Fe^{II}]}{\partial t} = [Fe^{*II}][Fe^{III}](k_1 + k_2/[H^+] + k_3[Cl^-])$$

The three terms including the rate constants k_1, k_2 and k_3 respectively correspond to three different pathways, all of which contribute to the overall reactions to an extent that depends on conditions such as pH and Cl^- concentration. The first term, $k_1[Fe^*(II)][Fe(III)]$, corresponds to a direct outer sphere electron transfer. The second and third terms correspond to inner sphere electron transfers involving hydroxide and chloride bridges respectively.

Two electron transfers: complementary and non-complementary reactions

So far, we have considered redox reactions in which a single electron is transferred. The same principles, however, apply to two-electron transfers, whether these be exchange reactions or cross-reactions. Reactions in which the oxidant and reductant change their oxidation states by the same number of electron equivalents are called *complementary* reactions, e.g.

$$Zn + Cu^{2+} \rightarrow Zn^{2+} + Cu$$

or

$$Fe^{3+} + Cu^+ \rightarrow Cu^{2+} + Fe^{2+}$$

When the oxidant and reductant differ in the number of electrons gained or lost, then the reactions are referred to as *non-complementary*. An example of a non-complementary reaction is the oxidation of Fe(II) by Cr(VI), in which the oxidation state of iron changes by one electron equivalent ($Fe^{II} \rightarrow Fe^{III}$), whereas a three electron change occurs at chromium ($Cr^{VI} \rightarrow Cr^{III}$). Many oxidations involving permanganate also fall into this category. Reactions such as these generally proceed by more complicated mechanisms than do complementary reactions, since trimolecular or higher order steps are highly improbable. Since only dimolecular steps are considered likely, then the existence is implied of intermediates which are unstable oxidation states of either oxidant or reductant. Further examples are the reactions of bisulphite, HSO_3^-, with Fe(III) and Cr(VI), which are considered in Chapter 4.

Free radical reactions

Free radical mechanisms
A free radical may be defined as any atom or molecule which has one or more unpaired electrons in s or p orbitals. Transition metal ions with unpaired electrons in d orbitals are not usually considered to be radicals and are excluded from the above definition. Free radicals are usually highly reactive species and indeed most of them are too unstable to be isolated, e.g. $OH\cdot$, $Cl\cdot$, $H\cdot$, $SO_4\cdot$, $NO_3\cdot$. On the other hand, there are several stable isolable molecules having unpaired electrons in s or p orbitals which are, therefore, quite properly classed as radicals. Examples of these are NO_2, NO, ClO_2, and of course, O_2, which has two unpaired electrons and is therefore a diradical. The relative stability of these molecules is readily understood in terms of molecular orbital theory or alternatively on the basis of Pauling's (1960) three-electron bond theory. The unstable radicals are usually produced by homolytic bond dissociation induced either thermally or photochemically. Chemical reaction can also lead to free radicals as, for example, in the case of the Fenton reaction, involving catalysed oxidation of ferrous ion by hydrogen peroxide, for which the first step may be written

$$H_2O_2 + Fe^{2+} \rightarrow Fe^{3+} + OH^- + OH\cdot$$

In general, reactions involving free radicals proceed via complex mechanisms with several steps. Many of these steps may be competitive and frequently no one step can be identified as being uniquely rate-determining: variable stoicheiometry is often observed.

Chain reactions
Certain free radical reactions proceed via mechanisms in which a series of intermediate steps constitute a cyclic process within which most of the products are formed, but in which no *nett* gain or loss of radicals occurs. Such steps are referred to as 'chain propagation'. Typically, a chain reaction can be considered to consist of three distinct stages:

(1) initiation, in which radicals are produced;

(2) propagation; and
(3) termination, in which the radicals are destroyed.

An example of such a reaction is the Fe(II)-catalysed decomposition of hydrogen peroxide, the mechanism of which was first investigated by Haber and Weiss (1934) and Weiss (1935), and which has been the subject of numerous other investigations since then. Here the three stages are:

Initiation

$$Fe^{II} + H_2O_2 \rightarrow Fe^{III} + OH\cdot + OH^-$$ (3.11)

Propagation

$$OH\cdot + H_2O_2 \rightarrow HO_2\cdot + H_2O$$ (3.12)

$$HO_2\cdot + H_2O_2 \rightarrow O_2 + H_2O + OH\cdot$$ (3.13)

or, more likely,

$$OH\cdot + H_2O_2 \rightarrow HO_2\cdot + H_2O$$ (3.14)

$$HO_2\cdot \rightleftharpoons H^+ + O_2^-$$ (3.15)

$$O_2^- + Fe^{III} \rightarrow Fe^{II} + O_2$$ (3.16)

$$Fe^{II} + H_2O_2 \rightarrow Fe^{III} + OH\cdot + OH^-$$ (3.17)

Termination

$$Fe^{II} + OH\cdot \rightarrow Fe^{III} + OH^-$$ (3.18)

Again, the variable stoicheiometry results as a consequence of competition from reaction (3.18) and reactions (3.14)–(3.17), exess Fe(II) favouring chain termination via (3.18) and excess hydrogen peroxide favouring propagation via reactions (3.14)–(3.17).

The usefulness of Fe(II) as a catalyst for peroxide oxidation of organic molecules was first discovered by Fenton (1894), and the Fe(II)/H$_2$O$_2$ reagent is still referred to as Fenton's reagent (see Fenton and Jones 1900). The extremely aggressive OH· radicals produced by the interaction of Fe(II) with H$_2$O$_2$ will attack many otherwise resistant organic species, including benzene, which is oxidised either to phenol or diphenyl:

COMMON OXIDANTS AND THEIR APPLICATIONS

Oxygen

General properties

Dioxygen (O_2) gas comprises approximately 20% of the atmosphere and is, therefore, prima facie, an excellent source of oxidising power. It is self-evident that oxygen, if effective as an oxidant, is the reagent of choice on the grounds of convenience, cheapness, and lack of health hazard. Unfortunately, two important factors offset the apparent advantages. Firstly, the solubility of oxygen in water is quite low, being approximately 10^{-3} mol l^{-1} at one atmosphere pressure. Since the partial pressure of oxygen in air is only about 0.2 atmosphere, it follows from Henry's law that an air-saturated aqueous solution contains only about 2×10^{-4} mol l^{-1}. In biological treatment processes in which air (or oxygen) is used almost exclusively, most of the process design effort has been directed towards improvements in the efficiency of gas transfer to the aqueous phase.

In non-biological chemical processes, a second factor militates against the widespread use of oxygen as a general oxidant (although it is used in some instances). Oxygen is not as good an oxidising agent as might be supposed from a consideration of its redox potential. For the reaction

$$O_2 + 4H^+ + 4e \rightarrow 2H_2O$$

$E^\ominus = +1.23$ V, which suggests that oxygen should be a powerful oxidant, comparable with chlorine ($E^\ominus_{Cl_2/Cl^-} = 1.36$ V), bromine ($E^\ominus_{Br_2/Br^-} = 1.07$ V) or dichromate ($E^\ominus_{Cr_2O_7^{2-}/Cr^{3+}} = 1.33$ V). However, reactions with molecular oxygen at ambient temperatures are generally observed to proceed very slowly both in the gas phase and in homogeneous solution. Part of the reason for this slowness lies in the fact that in order to reduce completely the oxygen molecule and hence realise its full oxidising power, the addition of four electrons is required. The synchronous transfer of four electrons is clearly highly improbable, since reducing agents in general supply only one or two electrons per molecule, and so the required four-electron reduction of oxygen to water is constrained to take place via separate two-electron or even one-electron steps. The standard potential for the first two-electron step

$$O_2 + 2e + 2H^+ \rightarrow H_2O_2$$

is only $+0.68$ V, and for the first one-electron step

$$O_2 + e \rightarrow O_2^-$$

$E^\ominus$ is -0.32 V and the formation of superoxide from oxygen is thermodynamically unfavourable (see Fig. 3.2). Oxygen is therefore a much poorer oxidant, if con-

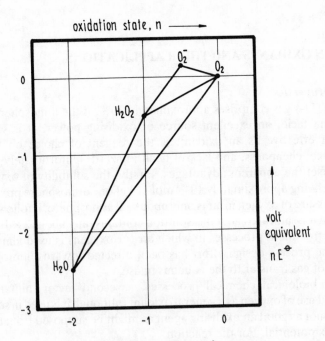

Fig. 3.2 — Oxidation state diagram for oxygen.

strained in a mechanism involving one- or two-electron steps, than the overall electrode potential might suggest. It is particularly noteworthy that the biochemical (enzymatic) reduction of oxygen to water can take place extremely rapidly. Biological systems have circumvented the need for multistage reduction by using enzymes in which several electron donor centres are present in the same molecule and which can provide all four electrons required.

There are also quantum mechanical restrictions (Taube 1965) on the rate of reaction of dioxygen with organic molecules, the details of which are beyond the scope of this present text, but which are related to the fact that whereas most organic molecules have no unpaired electrons, the lowest energy state of the oxygen molecule has two unpaired electrons. These quantum mechanical restrictions do not apply to the oxidation of transition metal ions such as Fe^{2+} and Mn^{2+}, which contain unpaired electrons; such oxidations can therefore be expected to occur more rapidly than those of organic species. The problem of unfavourable thermodynamics referred to above is, however, still present.

Oxidation of metal ions by molecular oxygen

Both divalent Fe^{2+} and Mn^{2+} are frequently encountered in ground waters of low oxygen content, in acid mine drainings, and also in the effluents associated with some metal treatment processes. The practicability of air oxidation followed by precipitation of the trivalent forms has been well established as a treatment process (see Stumm and Lee 1961, Weber 1972). The oxidation reaction is strongly pH dependent, occurring rapidly in neutral or alkaline solution, but much more slowly in acid. This pH dependence is not surprising since if we assume that the first stage of the oxidation is a one-electron transfer to give Fe^{3+} and a superoxide radical ion

$$Fe^{2+} + O_2 \rightleftharpoons Fe^{3+} + O_2^-$$

then consideration of the relevant redox potentials indicates that in acid solution the overall free energy change is decidedly unfavourable, with the equilibrium lying well to the left-hand side of the equation; superoxide will in fact reduce Fe^{3+} to Fe^{2+}. In weakly acid or alkaline solution, however, the overall thermodynamics become more favourable because of the effect on the Fe^{3+}/Fe^{2+} redox potentials (as discussed earlier). An intial two-electron transfer to give hydrogen peroxide

$$2Fe^{2+} + 2H^+ + O_2 \rightarrow 2Fe^{3+} + H_2O_2$$

is thermodynamically much more favourable but since this, if occurring directly, would involve a highly unlikely polymolecular collision complex, the implication is that prior formation of an (unstable) intermediate $Fe^{2+}.O_2$ would need to occur followed by reaction with a second Fe^{2+} ion and electron transfer.

In weakly acid or neutral solutions, the overall mechanism proposed by Haber and Weiss (1934) and Weiss (1935) involves the generation of free radicals in the first rate-determining step, followed by a sequence of reactions thus:

$$Fe^{II}OH^+ + O_2 \rightarrow Fe^{III}OH^{2+} + O_2^-\cdot$$

$$O_2^- + H^+ \rightleftharpoons HO_2\cdot \ (pK = 4.8)$$

$$Fe^{2+} + HO_2\cdot \rightarrow Fe^{3+} + HO_2^-$$

$$HO_2^- + H^+ \rightleftharpoons H_2O_2 \ (pK = 11.8)$$

$$Fe^{2+} + H_2O_2 \rightleftharpoons Fe^{III}OH^{2+} + OH\cdot$$

$$Fe^{2+} + OH\cdot \rightarrow Fe^{III}OH^{2+}$$

(For the sake of clarity and brevity, water molecules in the coordination spheres of the metal ions have been omitted.)

Under conditions where extensive hydrolysis of the iron species can occur, Stumm and Lee (1961) have demonstrated that the rate equation shows a higher order dependence on H^+ (or OH^-) concentration. The rate expression obtained by them,

$$\frac{-\partial[Fe^{II}]}{\partial t} = k[Fe^{II}][OH^-]^2 . p_{O_2}$$

is indicative of hydrolysed species being the reactive entities. In addition to the marked effect of pH on the reaction rate, catalysis by Cu^{2+} ions occurs. This can be explained in terms of an enhanced production of O_2^- via

$$Cu^{2+} + Fe^{2+} \rightleftharpoons Fe^{3+} + Cu^+$$

$$Cu^+ + O_2 \rightarrow Cu^{2+} + O_2^-$$

the resultant superoxide radical entering the Weiss type chain mechanism we have just described. The fact that the equilibrium between Cu^{2+} and Fe^{2+} lies very much to the left-hand side of the equation is not necessarily evidence against this catalysis mechanism, since provided equilibium is re-established rapidly a significant turn-over of reagents can occur with a very low standing concentration of Cu^+.

In practical terms, the marked sensitivity of the rate of oxidation to pH changes is of great importance. The time for 90% reaction at 20°C and 0.21 atmosphere partial pressure of oxygen is reduced from 43 minutes at pH 6.9 to only 8 minutes at pH 7.2 (Weber 1972). In the oxidation of manganese(II) by atmospheric oxygen, the rate is again very pH-dependent and also very much slower than the iron(II) oxidation—so much slower in fact that air oxidation as a means of removing manganese(II) can only be effected at high pH. Weber points out that even using pure oxygen rather than air as oxidant, the time for removal of 90% of Mn(II) is 80 minutes at pH 9.3, reducing to 50 minutes at pH 9.5. It is of significance in relation to the practical application of air oxidation that the products of the reaction, which are the hydrated oxides or hydroxides of higher oxidation states of manganese and iron, have powerful adsorptive capacities (Posselt et al. 1968, Fair et al. 1968). The composition of the product of the oxidation of Mn^{2+} has been observed to have a variable stoicheiometry ranging from $MnO_{1.3}$ to $MnO_{1.9}$ rather than the theoretical MnO_2. This is a reflection of the adsorption of Mn^{2+} ions on manganese dioxide produced in the first stage of the proposed mechanism:

$$Mn^{2+} + O_2 \rightarrow MnO_2$$

$$MnO_2 + Mn^{2+} \rightarrow Mn^{2+} \cdot MnO_2 \quad \text{(adsorption)}$$

$$Mn^{2+} \cdot MnO_2 + O_2 \rightarrow 2MnO_2$$

The implication of this is of course that complete oxidation of Mn^{2+} is not required for its removal, since a significant amount of Mn^{2+} will be adsorbed on the manganese dioxide and removed in subsequent sedimentation or filtration stages of the process.

Direct oxygenation of organic substances

Because of the kinetic restraints previously referred to, the effective oxidation of organic material by molecular oxygen, unless carried out enzymatically in biological treatment processes, requires either high temperatures, i.e. combustion, or the use of catalysts if the oxidation is to proceed at an acceptable rate. Where direct oxygenation is applied, as in the Zimmerman process (see Teletzke 1964), it is to concentrated solutions, e.g. sewage sludge and pulping black liquor effluents: temperatures of 250°–300°C and pressures up to 250 atmospheres are used. Rarely is combustion an economic alternative in waste water treatment, and even in the treatment of industrial waste gases energy-conserving catalytic processes are frequently sought.

In the absence of catalysts, the reactions of organic molecules with molecular oxygen (so-called autoxidation) generally proceed slowly and in a manner characteristic of radical chain processes, frequently with long induction periods. The induction period, which may be shortened or even eliminated entirely by the presence of catalysts, is a consequence of the requirement for, but lack of, availability of radicals to initiate the chain reaction. The mechanism of many of the uncatalysed autoxidations has been discussed in terms of the following simple steps (Stirling 1965, Stern 1971, Kochi 1973), although it must be remembered that the reactions and products are frequently more complex due to side reactions and chain branching:

$$
\left.
\begin{array}{l}
RH \rightarrow R\cdot + H\cdot \\
RH + In\cdot \rightarrow\; + InH
\end{array}
\right\} \quad \text{Initiation (In is an unspecified initiator radical)}
$$

or

$$
\left.
\begin{array}{l}
R\cdot + O_2 \rightarrow ROO\cdot \\
ROO\cdot + RH \rightarrow ROOH + R\cdot
\end{array}
\right\} \quad \text{Propagation}
$$

$$
\left.
\begin{array}{l}
R\cdot \rightarrow R - N \\
ROO\cdot + R\cdot \rightarrow R - O - O - R \\
2ROO\cdot \rightarrow ROOR + O_2
\end{array}
\right\} \quad \text{Termination}
$$

At temperatures in excess of 100° thermal decomposition of the hydroperoxide formed in the propagation steps can lead to autocatalysis by producing further initiators (Stirling 1965), for example

$$ROOH \rightarrow RO\cdot + OH\cdot$$

The efficiency of the chain reaction (chain length), of course, depends on the relative rates of the propagation and termination reactions. Subsances such as alcohols and phenols can act as effective autoxidation inhibitors because of the rapidity with which potential chain propagating radicals are removed by reaction with such substances (Twigg 1962). This is not to say that phenols and similar hydroxylic compounds cannot be oxidised by molecular oxygen, but the mechanism is not a chain process of this type and catalysts are required to sustain the production of radicals.

It is interesting to note that the most effective catalysts both for chain and non-chain oxidations are usually those involving transition metals which can undergo one-electron oxidation or reduction, e.g. Fe^{3+}/Fe^{2+}, Mn^{3+}/Mn^{2+}, Co^{3+}/Co^{2+}, Cu^{2+}/Cu^+. Initiation of reactions can then occur via production of radicals thus:

$$M^{n+} + O_2 \rightarrow M^{(n+1)+} + O_2 \cdot^- \text{ (or } (M^{(n+1)+}O_2\cdot^-)$$

Generally speaking, free radical attack on organic molecules is indiscriminate as regards functional groups, the site of attack usually being simply dictated by the thermodynamic stability of the bonds. Thus $C-H$ bonds are susceptible to homolytic attack, particularly if they are activated (weakened) by an adjacent functional group. Alcohols, aldehydes, ethers and ketones are all attacked at the $C-H$ bond nearest the oxygen atom. Ketones are less susceptible to attack than the first three classes of molecule because the target $C-H$ group is less activated, since it is one carbon atom further away:

$$
\begin{array}{cc}
\textcircled{H} & \textcircled{H} \\
| & | \\
R-C-H & R-C=O \\
|, & \\
R &
\end{array}
$$

$$
\begin{array}{ccc}
\textcircled{H} \quad \textcircled{H} & & \textcircled{H} \quad \textcircled{H} \\
| \qquad | & & | \qquad | \\
R-C-O-C-R & \text{but} & R-C-C-C-R \\
| \qquad | & & | \quad \parallel \quad | \\
\textcircled{H} \quad \textcircled{H} & & \textcircled{H} \quad O \quad \textcircled{H}
\end{array}
$$

(The circled H atoms are those most susceptible to attack.)

In the case of aromatic hydroxylic compounds such as phenols the $O-H$ bond is frequently attacked. The products of extended free radical attack on organic substrates are invariably complex and may include both degraded and polymerised material. Rarely in waste water treatment processes is complete oxidation achieved, although significant reductions in C.O.D. are readily realised in many cases.

Anions

Apart from certain metal ions, the reactions for which have already been discussed, various anions are oxidisable by air, including sulphite and nitrite. The reactions of the former are discussed in Chapter 4.

Nitrite is readily oxidised by air, as can be anticipated from the standard potentials for the redox reactions:

(acid medium) $NO_3^- + 3H^+ + 2e \rightleftharpoons HNO_2 + H_2O \quad E^\ominus = +0.94$ V

(alkaline medium) $NO_3^- + H_2O + 2e \rightleftharpoons NO_2^- + 2OH^- \quad E^\ominus = +0.01$ V

The mechanism proceeds by oxidation of nitric oxide formed by decomposition of the acid:

$$4HNO_2 \rightarrow N_2O_4 + 2NO + 2H_2O$$
$$2NO + O_2 \rightarrow N_2O_4$$
$$N_2O_4 + H_2O \rightarrow HNO_2 + HNO_3$$

The various sulphur oxy-anions such as $S_2O_3^{2-}$, $S_3O_6^{2-}$, and $S_4O_6^{2-}$ are all oxidisable by air in alkaline medium, the rates increasing with increase in temperature. The kinetics of the decomposition reactions have been studied by Rolla and Chakrabarti (1982), who noted that the rates are extremely slow in acid conditions.

Thiosulphate, $S_2O_3^{2-}$, is oxidised to sulphate, SO_4^{2-}, by dissolved oxygen in alkaline solutions, the reaction having the order characteristics

$$K \propto [S_2O_3^{2-}] \, [OH^-]^{1.1} \, [pO_2]^{1.6}$$

The trithionate ion, $S_3O_6^{2-}$, reacts with water in the pH range 5.5–12 to give $S_2O_3^{2-}$ and SO_4^{2-}, the reaction apparently being uninfluenced by either $[OH^-]$ or the level of dissolved oxygen. Tetrathionate, $S_4O_6^{2-}$, decomposes to $S_2O_3^{2-}$ and $S_3O_6^{2-}$ at pH values greater than 10, with first-order reaction characteristics with respect to both $[S_4O_6^{2-}]$ and $[OH^-]$.

The oxidation of sulphides and related sulphur-containing species is discussed later (p. 115).

Catalytically assisted oxygenation

Organic substances

In the absence of catalysts, the oxidation of organic material present in effluent streams by air or oxygen only takes place at significant rates at high temperatures. Borkowski (1967), for instance, has shown that temperatures in excess of 1000°C are required for the complete oxidation of phenols in evaporated effluent, whereas in the presence of a copper oxide catalyst oxidation to carbon dioxide and water occurs at temperatures in the range 300–500°C, with a consequent energy saving. Because of the large amounts of energy required to evaporate aqueous solutions (in excess of 2000 kJ l^{-1}) it is unlikely that vapour phase oxidations will in general be economically viable except for effluents containing high concentrations of organic material, where the heat of reaction can provide a significant contribution to the energy requirements of the process. Because of this consideration, aqueous phase oxidative processes have been sought. Moses and Smith (1954) described a process in which waste water containing a variety of organic substances was passed over catalysts consisting of mixtures of transition metal oxides in a packed bed reactor. Operating conditions varied from 125°C and 27 atmospheres for relatively easily oxidised substances such as methanol and formaldeyde to 300°C and 136 atmospheres for more intractable compounds like acetic acid and benzene. Hamilton *et al.* (1969) used an aqueous suspension of manganese dioxide as catalyst in order to oxidise phenol. At 25°C the oxidation rate was very slow, and even at 100°C the quoted rate (1 mg phenol per g catalyst per hour) was probably too slow to be of practical use. Increased partial pressures of oxygen and higher temperatures were said to increase the rate of oxidation but details are omitted. In a more exhaustive study of phenol oxidation, Katzer *et al.* (1976) achieved 99% conversion of phenol to carbon dioxide and water within 9 minutes using a supported copper(II) oxide catalyst at a temperature of 200°C, and an oxygen partial pressure of 9.5 atmospheres. The same authors concluded that the claims for rapid catalytic oxidation of organic material at ambient temperatures and atmospheric pressures are probably unfounded, and that the reported reductions in C.O.D. are more likely to be the result of adsorption of organic material on the catalyst particles. Certainly the high catalyst loadings (up to 100 g catalyst per litre of effluent) and the fact that the transition metal oxides, of which many of these catalysts are composed, are excellent adsorbents is not inconsistent with this view. More recently, however, Robinson (1978) and Cuthbertson and Robinson (1980) have claimed that a manganese-based catalyst (Robinson 1976) achieves significant oxidation of a variety of organic substances at ambient temperature and atmospheric pressure. The effectivenes of the catalyst is dependent on pH and catalyst concentration, and is initiated by the addition of either sodium hypochlorite or hydrogen peroxide, both of which are potential sources of free radicals. Typical performance data are illustrated in Fig. 3.3 and Table 3.1.

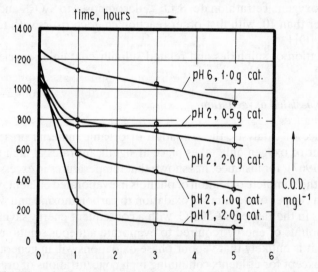

Fig. 3.3 — The effect of pH on the catalytic oxidation of phenol: variation in C.O.D. after addition of 4 ml 1.9×10^{-3} M NaOCl to 50 ml 5.3×10^{-3} M phenol (after Robinson 1978).

Table 3.1 — C.O.D. and colour removal by catalytic oxidation (data from Robinson 1978)

Effluent composition	Catalyst loading, g l^{-1}	NaOCl added, g l^{-1}	H$_2$SO$_4$ added, g l^{-1}	Retention time, hours	C.O.D. removal, %	Colour removal, %
Vat dyes + disperse dyes and unspecified organics	10	0.08	0.15	2	67	99
Chicken-processing waste	10	0.04	0.06	2	91	—

Another means of promoting oxidation by air has been used by Röhrer (1977), who introduced the additional energy required by passing an electrical discharge through the air, used along with metal oxides as catalysts. Röhrer claims that a rapid induction (of the order of minutes) of the oxidation reaction is thereby achieved, and quotes efficient oxidation of alcohols and benzoic esters at room temperatures. He also describes the treatment of a complex textile effluent, where the C.O.D. was reduced from 1620 mg l^{-1} to 11 mg l^{-1}, with colour removal from bright red (890 mg l^{-1} Pt) to 'colourless' (15 mg l^{-1} Pt) and elimination of turbidity, by operation at the effluent discharge temperature of 35°C. Röhrer nevertheless acknolwedges that tests have to be carried out to determine the appropriate catalyst in any given case.

A photolytic technique for increasing the rate of oxygenation of sewage has been examined by Acker and Rosenthal (1977) using methylene blue as a photosensitizer, and sunlight as the energy source. These workers noted a decrease in the C.O.D. of a recirculation oxidation pond treating municipal sewage of approximately two-thirds, with reductions in turbidity and suspended solids' levels by more than 50%. In yet another approach, Chen (1973) used ultrasonic agitation in conjunction with Raney nickel to promote oxidation of secondary sewage effluent, and observed reductions in C.O.D. of 78% within 2 hours.

It must be said in conclusion, however, that notwithstanding all the interest that has been shown in catalytically assisted oxygenation, no completely satisfactory general technique has yet been evolved for the treatment of most organic wastes at ambient temperatures.

Oxidation of sulphides

Sulphides arise in industrial waste waters from a variety of sources, including, for example, from certain pulp and paper processes, fellmongering, the coking industry, and, particularly, petrochemical operations. Treatment practices for the petroleum and petrochemical industries have been described in the *API Manual* (API 1969), by Beychok (1967), and for tannery effluents by Aloy *et al.* (1976).

The chemical oxidation of sulphides by oxygen is a complex process, and has not been fully elucidated. Chen and Morris (1972) concluded from the experimental evidence that in slightly acid to neutral and alkaline solutions the mechanism is

$$HS^- + O_2 \rightarrow HS\cdot + O_2^{\overline{}}$$
$$HS\cdot\ + O_2 \rightarrow HO_2 + S$$
$$HS\cdot\ + O_2^{\overline{}} \rightarrow S + HO_2^-$$
$$HS^- + (x-1)S \rightarrow H^+ + \frac{x}{2} S_2^{2-} \quad (x \text{ can be } 2.5)$$

The reaction rate is very slow at pH values below 6, where H_2S is the predominant species, but increases sharply as the pH increases from 6 to a maximum value at pH 8. It then decreases to a minimum at pH 9, increasing again to a second maximum at pH 11, and then decreases again in more alkaline conditions. Chen and Morris used starting concentrations of 0.5–2.0×10^{-4} M sulphide and 1.6–8.0×10^{-4} M oxygen, and found that with increasing sulphide concentrations the first peak shifts from pH 8 to pH 7, attributing this to the presence of polysulphides as intermediates. The characteristic induction period of the reaction, during which the concentrations of the reactants remain unchanged, ranges from approximately 6 hours at pH values near 7 to 0.2–1 hour around pH 9 and 11. The oxidation rate can be increased by raising the temperature to 85°–95°C and operating at elevated pressures (e.g. 6 atmospheres), and this has been used for sulphide oxidation in petroleum industry effluents: in these conditions the rate appears to be first order (API 1969).

That the oxidation of sulphides by oxygen can be catalytically assisted by transition metal ions has been known for some time—see, e.g., Krebs (1929) and Meunier and Kapp (1931). Such catalysis can indeed cause problems, as in the atmospheric oxidation of iron pyrites in mines, leading to acid mine wastes. However, the reactions can be turned to useful account, as was demonstrated by Bailey and Humphreys (1967), who found that the addition of relatively small amounts of manganese salts (50–100 mg l^{-1} Mn^{2+}) to even large concentrations of sulphide in tannery waste waters catalysed the oxidation sufficiently to make the reaction practical. Potassium permanganate and nickel sulphate are also suitable, Martin and Rubin (1977) claiming that as little as 1 mg l^{-1} Mn is effective with permanganate. About 90% of the sulphide can be oxidised in a period of 1–6 hours (depending on the oxygenation rate and reaction conditions). Eye and Clement (1972) have also studied the application to tannery wastes and confirmed the previous results; the procedure is now relatively commonly used and has been used in other contexts such as waste waters from cellulose-sponge manufacture (Paschke *et al.* 1977), textile manufacture (Sidwick and Barnard 1981) and in the oxidation of petroleum refinery effluents containing sulphides and other sulphur-containing species.

In simple terms, the reaction broadly proceeds as

$$HS^- + O_2 \overset{Mn^{2+}}{\underset{\text{catalyst}}{\rightarrow}} S_2O_3^{2-}, SO_3^{2-}, SO_4^{2-}$$

but it is undoubtedly complex. There is, for example, a pH effect, the extent of oxidation over a given period falling from a maximum at pH 7–8 to a lower plateau at pH 10–11 (Paschke *et al.* 1977). In our experience with fellmongering wastes, operation at high pH, which is convenient because of their alkaline nature, can give residual levels as low as 5–10 mg l^{-1}, but if the pH of the treated sample is then lowered the values can rise again to as much as 50–100 mg l^{-1}. The reason for this is not clear, but one possible simple explanation is that part of the treatment actually involves formation of MnS, which will tend to re-dissolve as the pH is lowered. In practical terms it means that the liquid must be clarified thoroughly before pH adjustments can be permitted.

Another catalytic procedure, used for oxidising hydrogen sulphide gas wastes with oxygen in air, is the Stretford process, developed by North-West Gas, Manchester (UK). The hydrogen sulphide is absorbed in an aqueous solution of sodium vanadate, 2,7-anthraquinone disulphonic acid (ADA) and a buffering agent. The solution also soon contains sodium bicarbonate, thiosulphate and sulphate, as side reaction products:

$$H_2S + Na_2CO_3 \rightarrow NaHS + NaHCO_3$$

$$NaHS + NaHCO_3 + 2NaVO_3 \rightarrow Na_2CO_3 + Na_2V_2O_5 + S + H_2O$$

$$2,7\text{-ADA}(O_2) + Na_2V_2O_5 \rightarrow 2,7\text{-ADA} + 2NaVO_3$$

$$2,7\text{-ADA} + O_2(\text{air}) \rightarrow 2,7\text{-ADA}(O_2)$$

The optimum operating temperature is 35–37°C; above 40°C side reactions lead to an increased formation of thiosulphate and sulphate, and in general it is better that these should not constitute more than about 2% of the H_2S oxidation. The Stretford process was originally developed for the cleaning of gasworks, coke oven and town gases, but it has been successfully applied to the final clean-up of tail gases from other sulphur recovery processes, and is in use in the SASOL petrochemical complex in South Africa.

Oxidation of hydrazine
Hydrazine appears as a contaminant in, for example, waste water discharges associated with rocket propellants, and from boiler water blowdowns where the compound is used in the boiler water circuit. It appears that catalytically assisted aeration effects oxidation of hydrazine (Ebara-Infilco Co. 1981): in one example quoted, 200 mg l^{-1} N_2H_4 at a pH of 10.5 after aeration for 2 hours in the presence of copper sulphate (which can be present adsorbed on to an ion exchange resin) was lowered to a concentration of 2.2 mg l^{-1}.

Ozone
General properties
Ozone, O_3, is a decidedly unstable allotrope of oxygen, the enthalpy of formation from O_2 being approximately $+142\,kJ\,mol^{-1}$. Concentrations of up to 10% of ozone can be prepared by the action of a high-voltage, silent electric discharge on dry air or oxygen in commercially available ozonisers. It is also possible to produce ozone by

the action of ultraviolet radiation on oxygen. In both methods oxygen atoms are produced by homolytic dissociation of O_2; these subsequently combine with oxygen molecules to give the triatomic ozone molecule. The pure gas, which has a perceptible blue colour, condenses at $-112°C$ to give a deep blue diamagnetic liquid that is dangerously explosive. Ozone is an extremely powerful oxidising agent, the standard electrode potential for the reaction

$$O_3 + 2H^+ + 2e \rightarrow O_2 + H_2O$$

being $+2.07$ V, a value exceeded only by fluorine, oxygen difluoride and some radicals.

Reaction characteristics

The gas-phase decomposition of ozone to dioxygen occurs slowly at room temperature, the kinetics being consistent with a deceptively simple mechanism:

$$O_3 \rightleftharpoons O_2 + O$$
$$O + O_3 \rightarrow 2O_2$$

Other diluent gases, such as oxygen or nitrogen, each have different effects on the rate of the reactions, and therefore should be included in any reaction scheme as mediators (Benson and Axworthy, 1959). In aqueous solution the kinetic behaviour of ozone with respect to decomposition is much more rapid and complex (Peleg 1976) and is strongly pH-dependent (Hoigné and Bader 1976, Stackelln and Hoigné 1982). Peleg suggests the following steps as contributing to the overall reaction scheme:

$$O_3 + H_2O \rightarrow O_2 + 2OH·$$
$$O_3 + OH· \rightarrow O_2 + HO_2·$$
$$O_3 + HO_2· \rightarrow 2O_2 + OH·$$
$$OH· + OH· \rightarrow H_2O_2$$
$$OH· + OH· \rightarrow H_2O + \tfrac{1}{2}O_2$$
$$HO_2· + HO_2· \rightarrow H_2O_2 + O_2$$

At high pH there are also further steps involving the ozonide ion O_3^- and the oxide radical O^-. Gurol and Singer (1982) consider that the decomposition of ozone can be expressed by

$$\frac{-\partial[O_3]}{\partial t} = k_0[OH^-]^{0.55}[O_3]^2$$

although Roth and Sullivan (1983) describe the kinetics as being essentially first order over the pH range 0.45–10.2, as do Teramoto *et al.* (1981) (see also Sullivan and Roth 1980).

In the context of the use of ozone in water and waste water treatment a number of important points emerge in relation to the above schemes. Firstly, there is the production of highly reactive radical species, particularly the hydroxyl radical, which is extremely aggressive and potently destructive of organic material. Secondly, Norrish and Wayne (1965) have proposed that hydroxyl radicals are produced during the flash photolysis of mixtures of ozone and water. If this is so, then we have perhaps a ready explanation for the enhanced effect of ozone/ultraviolet treatment (Mauk and Prengle 1976), in that the first reaction in the above scheme will be photocatalysed and lead to an increase in the number of reactive radicals present. It is evident that the readiness of ozone to enter into side reactions makes mechanism evaluation difficult.

The kinetics of the reactions of ozone with several classes of organic compounds are, according to the work of Hoigné and Bader (1983), first order with respect to ozone and solute concentrations of the compound. This is in conflict with the conclusion of Hewes and Davison (1973), who considered that the rate is relatively independent of dissolved ozone concentration. However, it has to be recognised that a problem in studying ozone reactions is the difficulty in eliminating secondary processes, which can lead to confusion in assessment of mechanisms for specific oxidations.

Hoigné *et al.* (1985) have also studied the reaction rates of ozone with various inorganic species, including compounds of sulphur (e.g. H_2S, H_2SO_4, $HOCH_2SO_3H$), chlorine (e.g. Cl^-, $HOCl$, NH_2Cl, ClO_2), bromine (e.g. Br^-, $HOBr$), nitrogen (e.g. NH_3, NH_2OH, N_2O, HNO_3) and oxygen (e.g. H_2O_2). They found second-order rate constants, most reactions exhibiting an increase of rate constant with increasing pH. This appears to be due to many compounds being protected when the reaction site becomes protonated, as for example with $OCl^- \rightarrow$ $HOCl$, and $NH_3 \rightarrow NH_4^+$. Therefore at pH values below the pK of the protonation reaction the rate constants generally decrease by a factor of 10 per unit pH decrement.

Ozone generation and use

Ozone generation equipment for plant use has been available for several years, and has been applied for disinfection purposes as an alternative to chlorine. A supply of dry air or oxygen is passed through a high-voltage electric field, giving a gaseous product that is injected directly into the water. The transfer of ozone into the solution phase is of course governed by its solubility in that phase, which is increased by increasing the partial pressure of ozone in the gas and by lowering the solution temperature. This implies that deep level injection at the lowest possible temperature will be favourable for efficient gas transfer and hence utilisation, while the smaller the bubbles the larger the surface area for transfer.

Ozone has a short life time (approximately 25 minutes in pure water at room temperatures), and, as has already been indicated, exhibits a strong tendency to decompose by entering into side reactions—for example, high pH conditions reduce the half-life time dramatically. Although ozone is more soluble than oxygen in water by an order of magnitude, the ozone produced by electrogeneration is not concentrated, being typically 1–2% using air, and the resulting solution concentration is only of the order of 10 mg 1^{-1} at 10°C and 1 atmosphere pressure. All of this means that highly efficient dispersion is essential, and that optimum reaction conditions must be provided to minimise side effects.

Ozone generators are expensive in capital terms, and power consumption is of the order of 20–25 W per g ozone produced. Because of this, ozonation is preferably applied as a tertiary process after preliminary treatments have reduced the level of material to be treated to the practical minimum. Ozone is toxic as well as corrosive, and no risk of venting unreacted material should be allowed. Gas recycling can be employed, or excess ozone can be decomposed, e.g. by passing the discharge gas after reaction through activated carbon, by heating, or by catalytic action.

Ozone in waste water treatment

A substantial interest in the use of ozone for waste water treatment has developed over the past 20 years, as is demonstrated by the number of papers that have appeared on the subject (not wholly reflected by the extent of plant applications, although these are growing). Much of the attention has been directed to fundamental studies in attempts to elucidate reaction mechanisms, but agreement is still not apparent on many aspects.

Evidence for the involvement of free radical species in the attack of ozone solutions on organic material lies in the marked similarity between these reactions and the reactions of Fenton's reagent (a known source of OH radicals) with the same molecules. For example, phenol on ozonation yields hydroquinone, o-quinone and catechol as intermediate degradation products (see Eisenhauer 1968, 1971, Gould and Weber 1978, Reutskii et al. 1981, Roth et al. 1982, Singer and Gurol 1983, and Gurol and Singer 1983 for studies on phenol ozonation); treatment of phenol with Fenton's reagent yields similar products (Stein and Weiss 1951). There are many other examples of degradation of organic compounds by ozone that have been studied (Hoigné and Bader 1976, 1983).

The ultimate products of the ozonation of phenols are of course carbon dioxide and water. Because of economic and temporal constraints it is not usually possible to effect the complete oxidation of phenol (although chemically this is feasible). Nevertheless, oxidation to inoffensive and more readily biodegradable molecules such as glyoxal (OHC.CHO), glyoxylic acid (OHC.COOH) and oxalic acid (HOOC.COOH) is frequently possible without unduly long retention times (Eisenhauer 1968, 1971, Hillis 1977). Eisenhauer concludes that oxidation of phenol to intermediate products is economically feasible with ozone:phenol mole ratios of 5:1, a point confirmed by Doré et al.'s (1978) observation that the aromatic ring opens at this ratio. Hillis (1977), in an investigation of the reaction of ozone with a variety of phenolic substances, established that the rate of phenol removal was pH-dependent, with higher pH favouring a more rapid reaction (Table 3.2). However, Li and Kuo (1980) recommend the use of a neutral pH, because of the greater decomposition rate of ozone in high pH conditions. It is noteworthy that whereas chlorinated phenols react more slowly, xylols and then cresols react more rapidly than phenol itself (Bauch et al. 1970); chlorobenzene is relatively slow to react.

Legube et al. (1981) studied the reactivity of various aromatic compounds with ozone, and also the biodegradability of the reaction products. Opening of the aromatic ring leads to the formation of aliphatic acids and aldehydes resistant to

Table 3.2 — Effect of pH on rate of phenol removal by ozone. Initial phenol concentration 30 mg l^{-1} (data from Hillis 1977)

pH	Time to reduce phenol concentration to:		
	5 mg l^{-1}, minutes	1 mg l^{-1}, minutes	<0.1 mg l^{-1}, minutes
4	4	5.7	ca. 8
7	3.6	5.4	8
8.5	2.5	4.7	8
10	1.2	2.8	8

ozone; and if post-chlorination is carried out after insufficient ozonation to open the aromatic rings, toxic species can result.

The practical application of ozonation to phenolic contaminants has been reviewed by Nebel *et al.* (1976), who cite fibreboard manufacturing, resin manufacturing, paper mill, coke plant and petroleum refinery effluents. Other substances which may be present in waste waters and which have been treated with ozone include pesticides (Buescher *et al.* 1964), photographic colour-processing wastes (Bober and Dagon 1975), NTA (Games and Staubach 1980), amines (Elmghari-Tabib *et al.* 1982), and dyestuffs (Snider and Porter 1974, Nebel and Stuber 1976, Netzer 1976, Erndt and Kurbiel 1980). Erndt and Kurbiel, for example, found in a laboratory investigation that 10–14 mg l^{-1} ozone applied for 5–40 minutes resulted in a 70–90% colour removal from dyes of the azo, dispersed and indigo groups, but that vat dyes were more resistant. A considerable reduction in the contact times required for particular treatment levels may be achieved by using multistage rather than single-stage processes. (The industrial use of ozone to decolorise dyestuff effluents has been reviewed by Beszedits 1980.) Ozone also decolorises kraft mill waste waters, but is not so effective in lowering B.O.D. and T.O.C. levels (Ng *et al.* 1978).

Cyanides, both free (Zeevalkink *et al.* 1980) and complexed to metal ions (Khandelwal *et al.* 1959, Novak and Sukes 1981), are rapidly oxidised by ozone in alkaline solution, although cyanate is decomposed more slowly (Teramoto *et al.* 1981). Some metal ions catalyse the cyanide reaction, Khandelwal *et al.* (1959) showing that the rate of oxidation of cyanide is almost doubled in the presence of Cu(II) ions. Cyanide complexed to iron as either ferrocyanide or ferricyanide is, however, much more resistant to oxidation. Interestingly, metal–EDTA complexes appear to be more readily broken down (by an order of magnitude) than EDTA alone (Shambaugh and Melnyk 1978).

Sulphides and mercaptans are oxidised by ozone, and this is utilised in the treatment of petroleum refinery effluents (API 1969), and in the treatment of gases at sewage pumping stations and filter press houses (Anderson and Greaves 1983). Various water authorities in the UK use wet scrubbing with ozone for odour control: these gases are extracted and passed through a scrubbing tower using an ozonated alkaline adsorption medium (pH 10–11). Not all organic materials are oxidised and the treatment can be unsuccessful (Haan 1983), but it appears to be more effective than using hydrogen peroxide.

For the purpose of comparison of the ease with which various substances are oxidatively degraded, Prengle *et al*. (1975) have used a 'refractory index' (RFI) scale in which the higher the value of the RFI for a particular substance, the more difficult it is to oxidise. RFI values are determined experimentally under a set of standard conditions as follows:

(1) ozone is used as an oxidant
(2) temperature 25–27°C
(3) initial concentration of reactant 50–100 mg l^{-1}
(4) efficient mixing.

The refractory index is then calculated from the equation

$$\text{RFI} = B_c^0 . t_{\frac{1}{2}} / A^0$$

where $t_{\frac{1}{2}}$ is the time in hours for 50% conversion of the reactant as measured by T.O.C. (total organic carbon) for organic compounds, or by an appropriate method for inorganic species, B_c^0 is the cumulative ozone concentration pumped into the liquid phase from $t = 0$ to $t = t_{\frac{1}{2}}$ (mg l^{-1}), and A^0 is the initial concentration (mg l^{-1} of reactant). Table 3.3 gives RFI values for a variety of refractory materials commonly encountered in industrial waste waters (Ventron Technology 1977). These values, obtained using ozone as an oxidant, may not be directly applicable for all other oxidants, although the relative refractory nature of substance should be unchanged for air, oxygen, or hydrogen peroxide as oxidants, since in most cases reactions involving these oxidants proceed via similar free radical mechanisms.

The great advantage of ozone as a treatment is, of course, that the products of its reduction (O_2 or eventually H_2O) do not add to the pollution load, and that excess ozone eventually decomposes spontaneously or, if required, may be rapidly removed by catalytic decomposition on an active carbon filter. As well as being a powerful oxidant, ozone has potent bactericidal properties, and has an advantage over chlorine as a potable water supply disinfectant in that characteristic tastes and odours imparted as a result of the chlorination of phenolic impurities are absent, although it does not retain the residual disinfection properties over a period of time as does chlorine. Unfortunately, most of the advantages of ozone are to some extent vitiated by the relatively high cost of ozone generation equipment and inefficiencies associated with the low solubility of the gas in water. It is almost invariably economic factors rather than chemical considerations which dictate the feasibility or otherwise of ozone treatment. For this reason, the most widespread use of ozone at present is in the treatment of potable water supplies or in tertiary waste water treatment, where the concentrations of impurities is low and consequently the ozone generating requirement is less. Ultraviolet radiation-enhanced ozone treatment (see below) may, however, be indicated for particularly refractory substances (Peyton *et al*. 1982) and there may also be advantages in the conjoint use of ozone with hydrogen peroxide for these difficult applications.

Table 3.3 — RFI values for various compounds

Compound	RFI value	Qualitative scale
Potassium cyanide (KCN)	0.41	
Phenol	0.44	
Chloroform	0.53	Slightly resistant
Complexed Cd-cyanide	0.96	(RFI < 1)
Pentachlorophenol	1.6	
Ammonium ion	8	
Glycine	19.7	Resistant
Palmitic acid (NH_4^+ salt)	27.3	(RFI = 1–100)
Nitrosodimethylamine	31	
Dichlorobutane	56	
Methanol	88	
Glycerol	112	
o-dichlorobenzene	113	Highly resistant
Polychlorinated diphenyls		(RFI = 100–1000)
Ethanol	245	
Ferricyanide	270	
DDT	297	
Acetic acid	>1000	Very highly resistant
Malathion	>1000	(RFI > 1000)

For general reviews on the application of ozone and its technology, see Selm (1959), Evans (1972), Heist (1974), Murphy and Orr (1975), Peleg (1976), Rice and Browning (1981), and Rice and Netzer (1982).

Ozone – UV processes

A modification to conventional ozonolysis is by the simultaneous use of ultraviolet radiation (Prengle *et al.* 1975, Garrison *et al.* 1975a, b). This apparently effects considerable enhancement of the rates of oxidation of a variety of refractory substances (Prengle and Mauk 1978), almost certainly as a direct consequence of the increased production of reactive free radicals: it is claimed, for example, that marked improvements in the oxidation of such substances as ferricyanides and acetic acid can be achieved, although for the former excessive amounts of ozone are needed (Prober *et al.* 1977). PCB residues can also be broken down (Arisman *et al.* 1980a, b), as can various pesticides (Kearney *et al.* 1984).

Chlorine and sodium hypochlorite
General properties

Chlorine, Cl_2, is a greenish-yellow, extremely reactive gas which is both toxic and highly irritant to mucous membranes. It is a powerful oxidising agent and, like oxygen, can support combustion, thus presenting a potential fire risk. Because of its relatively high boiling point ($-34°C$) and ease of liquefaction, chlorine is normally supplied in liquid form in steel cylinders that can be arranged to deliver either gas or liquid. Although dry chlorine does not significantly attack steel cylinders or pipework, in the presence of moisture it is extremely corrosive. Some of the relevant properties of the element are given in Table 3.4.

Table 3.4 — Properties of chlorine

Boiling point	$-34.15°C$
Freezing point	$-102.4°C$
Critical pressure	78 atmospheres
Critical temperature	144°C
Solubility in water	7.3 g l^{-1} at 20°C and 1 atmosphere
$E^{\ominus}\ (\frac{1}{2}Cl_2 + e \rightarrow Cl^-)$	$+1.36$ V

Sodium hypochlorite is a more acceptable form in which to store and dose chlorine, because of the risks attendant upon use and storage of the gas. It is supplied normally as a solution containing 14–15% active chlorine (s.g. approximately 1.28), which can be stored in concrete, rubber-lined steel or unplasticised pvc, or UV-stabilised polyethylene or polypropylene tanks. During storage, the undiluted solution deposits sodium chloride, and account must be taken of this in equipment usage. The solution does tend to degrade after a time (into sodium chloride, sodium chlorate and oxygen), a process accelerated by light, heat and most metals (which should not be allowed to come in contact with the solution).

An alternative that has sometimes been used is a slurry from bleaching powder, effectively $Ca(OCl)_2$ plus $Ca(OH)_2$. The presence of the lime provides additional benefit in some circumstances from its flocculating action, and the available chlorine concentration normally present in commercial supplies of the solid (36–38%), coupled with the relatively low cost, makes the material attractive where the extra solids introduced do not adversely affect plant operation or economics.

In terms of cost-effectiveness, chlorine, either as the gaseous element or in the form of hypochlorites, is probably the most generally useful chemical oxidant in the physico-chemical treatment of water and waste water. In general, one can recognise chlorine as having two useful functions, firstly as an oxidant, and secondly as a disinfectant of water supplies and effluent discharges to destroy or at least inactivate pathogenic organisms. A disadvantage of chlorine which has significant implications in either of the above roles is the fact that it will react readily with most organic molecules to give chlorinated products (chlorophenols, for example) which impart unpleasant tastes and odours to potable water supplies, and the presence of which in discharged waste waters is environmentally undesirable.

In aqueous solution, chlorine undergoes hydrolytic disproportionation to give chloride and hypochlorite, in which species the oxidation number of chlorine is -1 and $+1$ respectively:

$$Cl_2O + H_2O \rightleftharpoons HOCl + Cl^- + H^+$$

The equilibrium constant for this reaction is approximately 4×10^{-4} (Connick and Chia 1959), which means that under the conditions normally obtaining in chlorine treatment processes (pH > 6) there is very little free elemental chlorine present in solution at equilibrium. Additionally, since hypochlorous acid is a weak acid (with a pK of approximately 7.5) it dissociates accordingly:

$$HOCl \rightleftharpoons H^+ + OCl^-$$

It is apparent therefore that the relative equilibrium concentrations of the three active chlorine species Cl_2, $HOCl$ and OCl^- will be strongly pH-dependent, and hence it is to be expected that the properties of chlorine in aqueous solution will also be pH-dependent to an extent dictated by the relative reactivities of the three species towards potential substrates. This is amply illustrated in succeeding sections.

In terms of oxidising equivalents it is noteworthy that one atom of chlorine in $HOCl$ or OCl^- is equivalent to two atoms of chlorine in the elemental form. This is because in each case two electrons are required to reduce the species to chloride ions, since the oxidation state of chlorine is zero in the element but $+1$ in hypochlorous acid or hypochlorites. Hence

$$Cl_2 + 2e \rightarrow 2Cl^-$$

$$HOCl + 2e \rightarrow Cl^- + OH^-$$

$$OCl^- + H_2O + 2e \rightarrow Cl^- + 2OH^-$$

Specific applications
Cyanide detoxification
Both free cyanide and most metal cyanide complexes react rapidly with chlorine or hypochlorous acid to give cyanogen chloride according to the equations

$$Cl_2 + CN^- \rightarrow ClCN + Cl^-$$
$$HOCl + CN^- \rightarrow ClCN + OH^- \tag{3.19}$$

Cyanogen chloride is lachrymatory and toxic, probably being at least as toxic as HCN. It also exerts a significant vapour pressure over its aqueous solution, thus rendering it a hazardous species. In acid solution (pH 2) cyanogen chloride is relatively stable, but under alkaline conditions hydrolysis occurs to give the very much less toxic cyanate ion according to the overall equation

$$ClCN + 2OH^- \rightarrow NCO^- + Cl^- + H_2O \tag{3.20}$$

Price *et al.* (1947) showed that this reaction exhibits second-order kinetics, with

$$\frac{-\partial [ClCN]}{\partial t} = k[ClCN][OH^-]$$

The rate constant k was measured as being approximately $600\,l\,mol^{-1}\,min^{-1}$ at 25°C. More extensive work by Eden and Wheatland (1950) indicated values for the rate constant ranging from $80\,l\,mol^{-1}\,min^{-1}$ at 0°C to $530\,l\,mol^{-1}\,min^{-1}$ at 25°C. Complexation of CN^- with metals such as copper and zinc does not appear to affect the reaction rates significantly; the more stable $Ni(CN)_4^{2-}$ complex is oxidised only slowly, with extra chlorine consumption due to the co-formation of black nickelic trioxide, Ni_2O_3, while $Fe(CN)_6^{4-}$ is only oxidised to $Fe(CN)_6^{3-}$ at room temperatures. Some decomposition does occur, however, in the presence of a large excess of available chlorine $(1-2\,g\,l^{-1})$ and above 80°C. It has also been claimed (Baden and Webster 1972) that silver ions catalyse destructive oxidation of $Fe(CN)_6^{3-}$ by hypochlorite in alkaline medium at temperatures between 25°C and 68°C.

In solutions of adequate buffer capacity such that the pH remains essentially constant throughout the course of the reaction, pseudo first-order kinetics are observed:

$$\frac{-\partial [ClCN]}{\partial t} = k'[ClCN]$$

the value of k' of course depending on the pH at which the reaction takes place. In practice it is found that in order to achieve rapid hydrolysis of cyanide to cyanate and to minimise the problems associated with the undesirable intermediate cyanogen chloride, the reaction must be carried out at pH 10 or above. Problems arise in solutions more acid than pH 10, due not only to the slowness of reaction (3.20), with consequent emission of cyanogen chloride, but also because of the possible evolution of HCN if insufficient chlorine is present. HCN is a weak acid ($pK = 9.5$) and exerts a significant vapour pressure of HCN gas over solutions containing cyanide ion even at pH 10. Since the hydrolysis of cyanogen chloride (which is the rate-determining step in the conversion of cyanide to cyanate) exhibits pseudo first-order kinetics at high pH, it is apparent that the reaction time to achieve any particular cyanide or cyanogen chloride residual concentration will depend on the initial cyanide concentration as well as on the pH of the solution. The time required to reduce the initial concentration by 50%, the so-called 'half-life' time, is given by the following expression, the general form of which is valid for first-order or pseudo first-order reactions (see Chapter 1):

$$t_{\frac{1}{2}} = \log_e 2/k[OH^-]$$

At the optimum operating conditions (pH = 11) this equation simplifies to

$$t_{\frac{1}{2}} = 0.693/k[10^{-3}] = 693/k$$

assuming for simplicity here that $[OH^-]$ at pH 11 = 10^{-3}. Recourse to values of k (e.g. those of Eden and Wheatland 1950) in principle enables half-life times to be calculated for particular reaction temperatures, and hence minimum contact times in order to achieve any desired residual concentration of cyanogen chloride. It is apparent, for example, that after seven half-life times the cyanogen chloride concentration will have fallen to less than 1% (100×0.5^7) of the original, and that after ten half-life times the residual concentration will be less than 0.1% (100×0.5^{10}) of the original. There are, however, complicating features of this particular sequence of reactions which, under certain circumstances, may invalidate calculations of half-life times or at best render them only semi-quantitative. According to equations (3.19) two chlorine atom equivalents are required to oxidise cyanide to cyanogen chloride. The subsequent hydrolysis of cyanogen chloride to cyanate (equation 3.20) does not, in itself, consume further chlorine. However, chlorine concentrations in excess of that required by the stoicheiometry of equations (3.19) have two extremely important effects. Firstly, it was shown by Eden and Wheatland (1950) and subsequently confirmed by Bailey and Bishop (1973) that excess chlorine exerts a catalytic effect on the hydrolysis of cyanogen chloride at pH 10–11. The exact mechanism of this catalysis is not certain, but Bailey and Bishop suggest that free elemental chlorine is the catalytically active species, on the basis of a comparison between the catalytic efficiency of freshly prepared and aged sodium hypochlorite solutions. It is difficult, however, to reconcile this with the kinetic data of Eigen and Kustin (1962), which indicate that the hydrolysis of chlorine to give hypochlorite and chloride takes place extremely rapidly. Whatever the origin of the catalytic effect, it is certainly real and of extreme importance in terms of treatment practice. Rate increases for the catalysed reaction of up to ten times the uncatalysed reaction have been observed (Eden and Wheatland 1950), typical results being shown in Fig. 3.4.

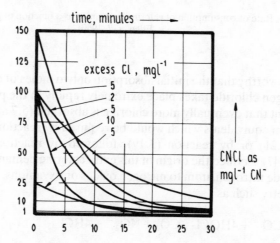

Fig. 3.4 — The effect of excess available chlorine, Cl, on the hydrolysis of cyanogen chloride, CNCl, at different starting concentrations of CNCl. Temperature = 5°C, pH = 11.0.

Secondly, in addition to the purely catalytic effect on cyanogen chloride hydrolysis, chlorine or hypochlorite will react with both cyanogen chloride and cyanate according to the equations below at rates which are strongly pH-dependent:

$$2ClCN + 3HOCl + H_2O \rightarrow N_2 + 2CO_2 + 5HCl \qquad (3.21)$$

$$2NCO^- + 3HOCl \rightarrow N_2 + 2CO_2 + H_2O + HCl + 2Cl^- \qquad (3.22)$$

These reactions, which consume a further three equivalents of chlorine per mole of cyanide, take place extremely slowly at pH 11 but (unlike the catalysed hydrolysis) take place rapidly at pH 6.5 (Eden *et al.* 1950, Zabban and Helwick 1980). Fig. 3.5 shows the rate of consumption of available chlorine as a function of pH.

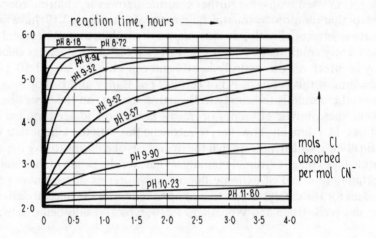

Fig. 3.5 — Rate of consumption of chlorine by cyanide as a function of pH (reproduced from Eden *et al.* 1950).

It is noteworthy that the initial absorption of two moles of chlorine by cyanide to form cyanogen chloride takes place extremely rapidly in the pH range studied. It is also apparent that eventually more chlorine is absorbed (5.7 atom equivalents) than the five atom equivalents which would be required by reactions (3.19), followed by reaction (3.21) or by reaction (3.19), followed by reactions (3.20) and (3.22). Dobson (1947) ascribed the origin of this extra chlorine demand to further oxidation of the cyanide nitrogen atom to nitrogen oxides or oxyanions. One might envisage a stoicheiometry such as

$$NCO^- + 4HOCl \rightarrow CO_2 + NO_3^- + 4HCl$$

Eden *et al.* (1950) have, in fact, shown that the 0.75 atom equivalent excess of chlorine over that required for the overall oxidation of cyanide nitrogen to elemental nitrogen closely corresponds to the amount of nitrate produced. In their experiments, approximately 12% of the original cyanide nitrogen was converted to nitrate. This figure is fairly typical of those achieved in current treatment practice, although considerable variations do occur. In view of the fact that the optimum conditions for the hydrolysis of cyanogen chloride to cyanate (high pH) are incompatible with those for the subsequent cyanate oxidation (much lower pH), then it is apparent that the effective removal of cyanide beyond cyanate requires a two-step process. The first stage is conducted at high pH in the presence of excess chlorine, which exerts a catalytic effect as previously discussed, and under these conditions rapid conversion to cyanate occurs without significant release of cyanogen chloride. The second stage is carried out in the pH range 6–8 and completes the oxidation to nitrogen and carbon dioxide together with small amounts of nitrate. (It must be pointed out, however, that this calls for care in plant engineering for continuous treatment operation, to ensure that full breakdown of cyanide and cyanogen chloride has been achieved before acidification is carried out.)

In acid solution (pH < 2.5) cyanate is rapidly hydrolysed to carbon dioxide and ammonium ions without the need of hypochlorite or chlorine, according to the equation

$$NCO^- + 2H^+ + H_2O \rightarrow CO_2 + NH_4^+$$

In many situations this route to cyanate removal may appear to offer an attractive alternative (in cost terms) to further hypochlorite treatment, particularly if waste acid streams are available from other processes which can be utilised for pH adjustment: it is reviewed in detail in Chapter 2 (p. 62 *et seq.*). There is of course the same safety risk from premature pH reduction as with additional chlorination, but additionally there is the disadvantage that in this method the cyanate is being converted not to nitrogen gas but to the less desirable ammonium ion. This may be acceptable in some circumstances but if it is subsequently necessary to remove the ammonia nitrogen, by breakpoint chlorination for example, then the exercise is clearly not as advantageous as might at first be supposed. Certainly in the co-presence of heavy metal ions the ammonia will exert an undesirable complexing action.

Breakpoint chlorination of ammonia

The so-called 'breakpoint chlorination' process for the removal of ammonia nitrogen from water has been practised for many years, and has been studied for treatment of domestic waste waters (see, e.g., Atkins *et al.* 1973). In waste waters ammonia originates from the enzymatic breakdown of urea, proteins and other nitrogen-containing substances, or from industrial sources, where it may be complexed to metal ions. In aqueous solution chlorine, as hypochlorous acid, reacts with ammonia to form a series of chloramines:

$$NH_3 + HOCl \rightarrow NH_2Cl + H_2O \tag{3.23}$$

$$NH_2Cl + HOCl \rightarrow NHCl_2 + H_2O \tag{3.24}$$

$$NHCl_2 + HOCl \rightarrow NCl_3 + H_2O \tag{3.25}$$

The addition of increasing amounts of chlorine or hypochlorites to waste water followed by measurements of the total residual chlorine concentration after a fixed period of time (say 30 minutes) results in a dose–response curve similar to the idealised curve shown in Fig. 3.6, which is usually referred to as a 'breakpoint curve'. A typical experimental breakpoint curve is shown in Fig. 3.7 (adapted from the data of Pressley *et al.* 1972). In the following discussion, 'total free chlorine' means the sum of the concentrations $[Cl_2] + [HOCl] + [OCl^-]$ and 'total residual chlorine' means the total available reactive chlorine, being numerically equal to the sum of the total free chlorine plus the concentration of chloramines present.

The significance of breakpoint curves such as those illustrated in Figs. 3.6 and 3.7 lies in the fact that at the breakpoint where the residual chlorine concentration is at a minimum, the ammonia nitrogen concentration tends to zero, or at least to very low values (< 0.1 mg l^{-1}). Beyond the breakpoint, the free available chlorine increases linearly with added reagent as expected. In order to be able to optimise treatment processes it is necessary to understand the reasons for the shape of breakpoint curves of this type. Referring to Fig. 3.6, in the region A–B an immediate chlorine demand

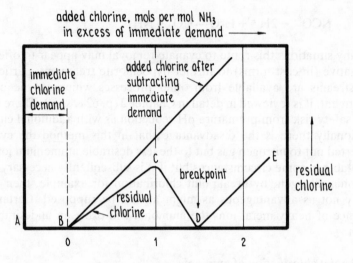

Fig. 3.6 — Idealised breakpoint chlorination curve.

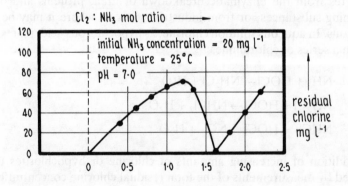

Fig. 3.7 — Experimental breakpoint chlorination curve.

is exerted due to the presence in the waste water of readily oxidised substances such as sulphides or metal ions in low oxidation states (e.g. Fe^{2+}, Mn^{2+}). The measured residual chlorine concentration thus remains at a very low level until all the easily oxidisable material has reacted. In the region B–C chloramines are produced by the consecutive (and competitive) reactions (3.23) and (3.24) above. At point C and beyond, significant amounts of dichloramine start to be formed, since the theoretical $1:1$ molar ratio of Cl_2 (or HOCl) : NH_3 required for the stoicheiometric production of monochloramine is now exceeded. The presence of two electronegative chlorine atoms in the dichloramine molecule activates it towards nucleophilic attack by a water molecule, leading to formation of the reactive intermediate [NOH]:

$$NHCl_2 + H_2O \rightarrow [NOH] + 2Cl^- + 2H^+ \tag{3.26}$$

Rapid attack on both mono- and dichloramines is then possible according to the equations

$$NH_2Cl + [NOH] \rightarrow N_2 + H_2O + H^+ + Cl^- \tag{3.27}$$

and

$$NHCl_2 + [NOH] \rightarrow N_2 + HOCl + H^+ + Cl^- \tag{3.28}$$

This results in a decrease in total residual chlorine as it is reduced to chloride and the ammonia nitrogen is oxidised to molecular nitrogen. The nett effect of reactions (3.23), (3.24), and (3.26) followed by (3.27) and (3.28) is given by the following equation, in which the $3:2$ stoicheiometry accounts for the theoretical breakpoint position in Fig. 3.6:

$$3HOCl + 2NH_3 \rightarrow N_2 + 3H_2O + 3H^+ + Cl^- \tag{3.29}$$

Because of the pH dependence of the rates of reactions (3.23) to (3.28) the detailed shape of experimental breakpoint curves is also pH-dependent, and because of a competing, pH-dependent reaction leading to the 'over-oxidation' of ammonia to nitrate, the breakpoint in experimental curves usually occurs at $Cl_2:NH_3$ ratios in excess of $3:2$ (see Fig. 3.7).

The rate of reaction (3.23) in particular is important in deciding the overall efficiency of the conversion of ammonia nitrogen to molecular nitrogen. This reaction exhibits second-order kinetics (see Fair *et al.* 1968) with a rate law as follows:

$$\frac{\partial [NH_2Cl]}{\partial t} = k_2 \, [HOCl] \, [NH_3]$$

Since it is the neutral molecular species which are predominantly involved in the rate-determining step rather than ionic NH_4^+ or OCl^-, then it is to be expected that a graph of rate vs. pH should show a maximum with rate decreasing at pH values below the optimum due to $NH_3 \rightarrow NH_4^+$, and above the optimum due to $HOCl \rightarrow OCl^-$.

This is illustrated in Fig. 3.8, in which the relative concentrations of HOCl and NH_3 are plotted as a function of pH. The curves were calculated from the equations

$$pH = pK + \log \frac{[NH_3]}{[NH_4^+]} \qquad (pK_{NH_4^+} = 9.25)$$

and

$$pH = pK_{HOCl} + \log \frac{[OCl^-]}{[HOCl]} \qquad (pK_{HOCl} = 7.5)$$

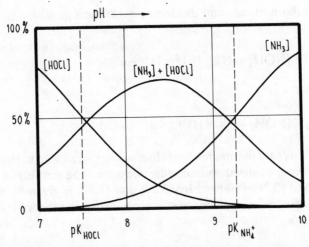

Fig. 3.8 — pH-dependence of NH_3 and HOCl concentrations in breakpoint chlorination.

It is readily apparent either from simple calculation or from an inspection of Fig. 3.8 that the pH at which the product $[NH_3][HOCl]$ is a maximum is numerically equal to the mean value of the two pK values. It is therefore to be expected that the maximum rate for reaction (3.23) would occur at this pH (8.3–8.4), and this is observed to be so (see Fair *et al.* 1968). However, in terms of operating practice, this is not necessarily the pH of choice, since there are other complicating factors to be considered. At high pH ammonia nitrogen can be further oxidised to nitrate; the overall equation may be written as

$$NH_3 + 4OCl^- \rightarrow H^+ + NO_3^- + H_2O + 4Cl^- \qquad (3.30)$$

Compared to the desired reaction (3.29), reaction (3.30) consumes considerably more chlorine, since the 3:2 stoicheiometry required for nitrogen production is increased to 4:1 when nitrate is the product. In addition to the obvious economic disadvantage of this, the presence of nitrate ions, in the context of pollution control, is not desirable. Pressley *et al.* (1972) have shown that the amount of nitrate in breakpoint chlorination decreases from about 10% at pH 8 to about 1.5% at pH 5. In view of this, it is desirable to operate breakpoint chlorination processes for nitrogen removal at a pH lower than the calculated optimum for the maximum rate of reaction (3.23). In practice (Pressley *et al.* 1972) it has been found that in the pH range 6.5–7.5 more than 95% of the ammonia present is oxidised to nitrogen gas within a few minutes, from initial concentrations of ammonia nitrogen in the range of 15–20 mg 1^{-1}. Because of these considerations, and in view of the fact that the addition of chlorine to water and waste water produces a marked drop in pH due to the hydrogen ions released in the disproportionation reaction and the ensuing oxidations, it is necessary to provide adequate pH control in breakpoint chlorination processes. The advent of relatively inexpensive microcomputer systems has led to their incorporation in many control systems, and the strategy for their implementation in breakpoint chlorination processes has been described by Fertik (1978). The main problem encountered in fully automated procedures lies in the reliability (or lack of it) of the continuous free ammonia and free chlorine analysers. Fertik and Sharpe (1980) have discussed in detail the problems involved in the optimisation of the control strategy, and have pointed out the necessity for a flexible approach involving both full feed forward–feedback control and manual intervention. However, it has been shown (Eilbeck and Mattock 1982, Eilbeck 1984) that simple redox control of breakpoint chlorination of both ammonia and metal ammine complexes is feasible. This is discussed in more detail later in this chapter (p. 163 *et seq.*).

With regard to the general problems associated with the presence of organic substances in waste waters which are subjected to breakpoint chlorination, it is important to realise that most organic molecules react with chlorine much more slowly than does ammonia. Thus with careful control of the chlorination process and subsequent removal of excess chlorine in dechlorination stages it is possible to remove ammonia selectively without extensive chlorination of organic materials such as amino acids and proteins. However, the special problems associated with the formation of chlorinated phenolic material are such that breakpoint chlorination is not an appropriate treatment when phenolic and related substances are present.

Chlorination of ammine complexes
Ammonia is frequently present in the form of metal ammine complexes. The mechanism of the important reaction (3.23) in waste water streams emanating from the surface treatment (including electronics components' production) industries is likely to involve interaction of the non-bonded pair of electrons on the ammonia nitrogen atom with the chlorine atom of HOCl prior to electron transfer, as

$$
\begin{array}{ccc}
\text{H} & \text{H} & \text{H} \\
\diagdown & \diagdown & \diagdown \\
\text{H} - \text{N:} \ \text{Cl} - \text{O} - \text{H} \rightarrow & \text{H} - \text{N}^+ - \text{Cl} + \text{OH}^- \rightarrow & \text{N} - \text{Cl} + \text{H}_2\text{O} \\
\diagup & \diagup & \diagup \\
\text{H} & \text{H} & \text{H}
\end{array}
$$

Since in metal ammine complexes (e.g. $Cu(NH_3)_4^{2+}$) the important, previously non-bonded electron pairs of the ammonia molecules are now intimately involved in bonding to the metal ion, they are considerably less accessible to the attacking chlorine species, and it might be expected that the rate of reaction of complexed ammonia would be very much less than that of free ammonia. Although this may indeed be the case, since metal ammine complexes have stability constants which in general are rather low compared to those of metal chelate complexes such as $Cu(EDTA)^{2-}$ (see Martell and Sillén 1954), then under the conditions usually encountered in breakpoint chlorination processes, significant amounts of free, uncomplexed ammonia will be in equilibrium with metal complexed ammonia. This free ammonia will of course react normally, and provided the metal ammine complexes are sufficiently labile, dissociation of these complexes will occur as the equilibrium is disturbed by removal of the free ammonia: thus eventual oxidation of the total ammonia present can occur. The extent to which the reaction occurs at a slower rate than in the absence of metal ions will depend on the stability constants of the complexes (which control the free ammonia concentration) and also on the kinetic lability of the complexes, i.e. whether or not the rate of dissociation of metal complexed ammonia can match the rate of removal of free ammonia. If not, then the dissociation of the complex becomes rate-limiting. For many metal ions this is not a problem: for example, solutions containing copper ions and ammonium peroxydisulphate can be successfully treated by breakpoint chlorination at pH 6–7 under redox control. The chlorine demand in this case is governed to some extent by the peroxydisulphate concentration, because in stronger solutions the peroxydisulphate more significantly tends to oxidise chloride formed in the reaction back to hypochlorite. Rapid oxidation of ammonia to nitrogen occurs, releasing the copper, which may then be precipitated as copper hydrous oxide by conventional pH adjustment techniques.

Chlorination of EDTA and other complexing agents

A common problem in the treatment of waste waters from metal finishing, photographic and other processes where strong complexing agents such as EDTA, NTA and similar metals are used is the extreme difficulty of removing heavy metals to an acceptable level. Extractive methods for metals' removal have been reviewed in Chapter 2. These, however effective for metals, do not break down the complexing species, which remain to pose potential problems in discharged waste waters. It is thus useful to examine the possibilities of oxidation by chlorine.

Claims have been made in the patent literature (Bober *et al*. 1973) for the alkaline chlorination of EDTA, and unpublished work (Grainger and Mattock 1973) has confirmed that reaction does occur to some extent at pH 6. However, significantly, it was found that although the EDTA (or its sodium salt) is susceptible to breakdown, heavy metal complexes appear to be resistant, so in order to make the process possible it is necessary to dissociate the complexes to release the EDTA. One method is by lowering the pH to 1–2. It was found (Grainger and Mattock 1973) that a chlorine : EDTA weight ratio of at least 8 : 1 is needed, and that the reaction is slow, requiring at least 3 hours for significant breakdown. It is possible that chlorine itself acts partly as catalytic agent, in that more is required for reaction than is consumed, and that below minimum quantities virtually no reaction occurs. A treatment technique for heavy metal complexes is thus conceivable, by lowering the pH to 1–2, chlorination of the EDTA, and subsequent addition of alkali to precipitate the hydrous metal oxides. However, it appears that chlorination is not only slow but also incomplete at low pH values—a pH in the region of 6 being towards optimum—so not all of the heavy metal is released by such low pH treatment. In similar manner, heavy metal tartrate and citrate complexes exert some chlorine demand, indicating some breakdown, but the demand is also variable with pH.

Amines can be oxidised by chlorination, and the kinetics of the decomposition of ethanolamine has been studied by Autelo *et al.* (1981). These workers found that the reaction at high pH occurs in two stages. In the first, *N*-chloroethanolamine is formed, at a rate proportional to $[ClO^-]$, [amine] and $[NaOH]^{-1}$. In the second, the decomposition rate is proportional to [*N*-chloroethanolamine] and [NaOH]. The reaction is however extremely slow in the pH range 6–8.

Oxidation of sulphides
Sulphides and polysulphides are readily oxidised by chlorine, resulting in the formation of elemental sulphur (sometimes colloidal) or sulphates and acid in the presence of excess reagent, and the procedure is useful when relatively small amounts of sulphide are to be treated. For larger quantities, however, catalysed oxidation by air (discussed earlier in this chapter) is generally to be preferred on economic grounds.

Oxidation of organic materials
Although chlorine has been used for disinfection for many years, it is often ineffective or can produce undesirable end-products with industrial organic wastes, as from the formation of stable chloro-organic compounds, e.g. chlorophenols from phenols. For this reason it is not usually favoured as a reagent for tertiary sewage treatment. It means that if chlorine is considered as a potential oxidant the full nature of the organic species present in the waste water must be clearly identified to verify that no problems can be caused.

There is evidence that some insecticides can be oxidised, although not necessarily completely. Low concentrations of parathion can be destroyed to 95%–100% by chlorination (Gomaa and Faust 1971a), although contact times of $1-1\frac{1}{2}$ hours are necessary: the oxidation of both parathion and paraoxon are faster in alkaline than in acid conditions. However, even here a degradation product may be *m*-nitrophenol, which itself may then be chlorinated to give a stable and undesirable contaminant. Lindane is unaffected by chlorination, but aldrin is attacked (Buescher *et al.* 1964). Calcium hypochlorite has apparently shown some promise in reducing levels of chlorinated hydrocarbons (Kennedy *et al.* 1972), although the data are sketchy.

Hypochlorite has some application in colour removal from certain dyes, but is not generally effective.

UV chlorinolysis
The simultaneous use of ultraviolet radiation and chlorine has been shown to enhance the rate of oxidation of ammonia nitrogen (Kimura *et al.* 1980). This occurs, however, at the expense of an increased concentration of nitrate ion in the treated effluent, and since conventional breakpoint chlorination reactions proceed at an acceptably rapid rate, it is difficult to see any advantage in UV chlorinolysis over conventional treatment of ammonia.

It can also be mentioned here that the chlorinative destruction of ferrocyanide has been claimed (Tokyo Electric Co. 1972) by simultaneous irradiation with

360–440 nm light at pH 11.5–12.5. It has long been known that ferrocyanide slowly decomposes in natural waters, and it has been assumed that this is at least partially due to the influence of sunlight.

Ferrate ion

The ferrate ion, FeO_4^{2-}, described by Wood (1958), is a strong oxidant, having a standard potential of $+2.2$ V in acid medium. It is prepared by the oxidation of ferric nitrate with 10% sodium hypochlorite; the insoluble salt K_2FeO_4 is then precipitated as a purple-black solid by the addition of saturated potassium hydroxide.

It is claimed that ferrate kills a wide variety of microorganisms , and Carr *et al.* (1981) have studied its attack on NTA and various amines. More work would clearly be worthwhile to establish its relative value as a waste water treatment reagent, e.g. to destroy metal ion complexes.

Chlorine dioxide
General properties

Chlorine dioxide, ClO_2, is a yellow, explosive gas (boiling point ca. 10°C) that is very soluble in water: at 4°C water will dissolve approximately twenty times its own volume of ClO_2. The gas is spontaneously explosive at partial pressures in excess of 70 mm Hg, but its solutions in water are safe and reasonably stable provided they are not stronger than 1–2 g l^{-1}. Solutions containing up to 8 g l^{-1} decompose only slowly to give a mixture of HCl and $HClO_3$, but in strong light or in alkaline solution the decomposition is more rapid. Chlorine dioxide is a strong oxidising agent, as is evidenced by the standard electrode potential of the couple

$$ClO_2 + 4H^+ + 5e \rightarrow Cl^- + 2H_2O \qquad E^\ominus = +1.50 \text{ V}$$

which is approximately the same as that for $MnO_4^- \rightarrow Mn^{2+}$. The reaction utilises five electrons per molecule of ClO_2, and hence on a mole for mole basis ClO_2 has $2\frac{1}{2}$ times the oxidising power of chlorine or hypochlorite, an important consideration when calculating relative dose rates. In alkaline solution, a one-electron change may occur:

$$ClO_2 + e \rightarrow ClO_2^- \qquad E^\ominus = +1.16 \text{ V}$$

Besides its properties as an oxidant, chlorine dioxide is also a potent disinfectant with pronounced bactericidal (Longley *et al.* 1980) and viricidal activity (Aieta *et al.* 1980), for which it is superior to chlorine.

Generation of chlorine dioxide

Because of the hazardous nature of the material chlorine dioxide is generated at the site of application, usually as a dilute aqueous solution and preferably at concentration of less than 1 g l^{-1}. There are many methods of synthesis (see Masschelein 1979), the choice usually being dictated by the quantity and purity required. The main methods of industrial scale production fall into two main categories: methods based on the reduction of chlorates in acid solution, and methods in which oxidation or disproportionation of chlorites occurs, again under acidic conditions.

Production from chlorates

In strongly acidic solution (e.g. 7–9 M HCl) the following reaction occurs:

$$2ClO_3^- + 4HCl \rightarrow 2ClO_2 + Cl_2 + 2H_2O + 2Cl^-$$

which should yield a mixture of chlorine dioxide and chlorine in the proportion of 2:1. In practice, however, the proportion of chlorine is higher due to the influence of competing side reactions such as

$$ClO_3^- + 6HCl \rightarrow 3Cl_2 + 3H_2O + Cl^-$$

There are many industrial variations of this process, for example, the R-2 and SVP processes (see Masschelein 1979), in which sulphuric acid and sodium chloride are used instead of hydrochloric acid. In this case, the analogous reactions are as follows:

$$2NaClO_3 + 2NaCl + 2H_2SO_4 \rightarrow 2ClO_2 + Cl_2 + 2Na_2SO_4 + 2H_2O$$

and, competing,

$$NaClO_3 + 5NaCl + 3H_2SO_4 \rightarrow 3Cl_2 + 3Na_2SO_4 + 3H_2O$$

Operating conditions are optimised to suppress the latter reaction as much as possible, since this produces only chlorine.

The addition of reducing agents in the industrial synthesis of chlorine dioxide from chlorates increases the yield of chlorine dioxide and lessens the amount of chlorine produced compared to the simple acidification processes. Sulphur dioxide is the most frequently used reducing agent for this purpose. The main reaction is as follows:

$$2NaClO_3 + H_2SO_4 + SO_2 \rightarrow 2ClO_2 + 2NaHSO_4$$

Industrial processes using chlorates as starting materials require relatively complex and expensive equipment, and the high capital costs involved are only justified for the production of very large quantities of chlorine dioxide such as are used in pulp bleaching, for example. For the majority of water treatment applications, the smaller quantities of chlorine dioxide required are more usually and cheaply prepared from oxidation or disproportionation of sodium chlorite.

Production from chlorites

Sodium chlorite reacts readily with chlorine to give chlorine dioxide according to the equation:

$$Cl_2 + 2NaClO_2 \rightarrow 2ClO_2 + 2NaCl$$

In practice the reaction occurs rapidly when aqueous solutions of the two reagents are mixed, giving an aqueous solution of chlorine dioxide. Typically, commercial sodium chlorite stock solution (26% w/w) is metered into chlorine water containing 500–1000 mg l^{-1} chlorine. In order to ensure complete reaction of the chlorite, a slight excess of chlorine is used. An alternative approach to the preparation of aqueous solutions of chlorine dioxide which obviates the need for a supply of chlorine is simply to mix solutions of sodium chlorite and hydrochloric acid. Under acid conditions, chlorite ions disproportionate to give chlorine dioxide and chloride according to the equation

$$5NaClO_2 + 4HCl \rightarrow 4ClO_2 + 5NaCl + 2H_2O$$

Some chlorine is usually formed from the competing reaction

$$4NaClO_2 + 4HCl \rightarrow 3ClO_2 + \frac{1}{2}Cl_2 + 4NaCl + 2H_2O$$

In practice, solutions of 26% w/w sodium chlorite and commercial 28% hydrochloric acid are mixed in carefully controlled conditions (the reaction is vigorous) and the resultant solution immediately diluted and used. Concentrations of chlorine dioxide in the diluted solution should not be allowed to exceed 2000 mg l^{-1}. Sodium chlorite itself has to be handled with care because of the readiness with which it will support combustion, although both the solid and the solution are stable if stored properly.

Specific applications
The main use of chlorine dioxide in water treatment processes is in the disinfection and removal of tastes and odours from water which is destined to become a potable supply, where it has been shown to be a more effective bactericide and viricide than chlorine. It has particular advantages in the disinfection of waters containing significant concentrations of ammonia nitrogen, since unlike chlorine it does not react with ammonia to give chloramines.

Chlorine dioxide is particularly effective against tastes and odours which arise from the presence of phenolic impurities. Unlike the undesirable and obnoxious chlorophenols which are produced on chlorination of phenol-containing waters, the main products of the reaction of chlorine dioxide with phenols are quinones and ultimately carboxylic acids, which do not present taste or odour problems. It is important however to ensure that sufficient chlorine dioxide is used for the oxidation. Inadequate treatment results in the formation of quinones, which may be chlorinated and subsequently lead to the formation of chlorinated phenolic derivatives with the attendant taste and odour problems. With chlorine dioxide:phenol mole ratios of between 3:1 and 4:1, only carboxylic acids are produced within 30 minutes (Masschelein 1979). For the treatment of raw water supplies dose rates of 0.15–0.25 mg l^{-1} are usually adequate unless significant pollution is present. The rate of reaction with phenol increases as the pH rises to 7, and Wajon et al. (1982) have suggested a mechanism involving

$$PhO^- + ClO_2 \rightarrow PhO\cdot + ClO_2$$

as being rate-determining.

Chlorine dioxide oxidises both diquat and paraquat, the former more quickly than the latter, and in this respect is superior to both chlorine (which is ineffective in acid conditions) and $KMnO_4$, the reaction rate for which also increases in alkaline conditions (Gomaa and Faust 1971b).

It is worth noting the observations of Rav-Acha and Blitz (1985) with polyaromatic hydrocarbons. Whereas chlorine may add, substitute or oxidise, chlorine dioxide acts mainly as a pure oxidant and one-electron acceptor. Resistance to oxidation by chlorine dioxide does not therefore necessarily produce chlorinated products, which can be the case with chlorine.

Hydrogen peroxide
General properties
Hydrogen peroxide is a clear, colourless, water-like liquid which is both a powerful oxidant and a somewhat less powerful reductant. It can for example reduce permanganate to Mn^{2+}, and chromate to Cr^{3+} (at pH < 8.5). The relevant electrode potentials for hydrogen peroxide acting as an oxidising agent and reducing agent are as follows:

$$H_2O_2 + 2e + 2H^+ \rightarrow 2H_2O \qquad E^\ominus = +1.77 \text{ V}$$
$$O_2 + 2e + 2H^+ \rightarrow H_2O_2 \qquad E^\ominus = +0.69 \text{ V}$$

Consideration of these potentials or inspection of Fig. 3.2 indicates that H_2O_2 should be thermodynamically unstable with respect to disporportionation to water and oxyen, thus:

$$2H_2O_2 \rightarrow 2H_2O + O_2$$

This reaction occurs only slowly in the absence of catalytic impurities (chiefly transition metals or their compounds) but proceeds rapidly, even explosively, in the presence of suitable catalysts. The reaction is strongly exothermic ($\Delta H^\ominus = -98$ kJ mol^{-1}), and at concentrations of hydrogen peroxide of 65% or more in aqueous solution, the heat liberated is sufficient to vaporise the solution completely. It is this reaction which has resulted in the somewhat undeserved reputation of hydrogen peroxide as an extremely hazardous material. Admittedly, solutions of high concentration do present a significant fire risk if allowed to come into contact with certain organic materials, but with normal precautions to avoid the ingress of contaminants, stabilised solutions of hydrogen peroxide can be stored and handled quite safely. The commercially available material is supplied in aqueous solutions of various concentrations up to about 70%, and is usually diluted on site before use.

Many types of stabiliser have been employed for commercial supplies of hydrogen peroxide, including some materials, such as EDTA, that can be undesirable in waste water treatment, although phosphates and tin salts are the most common. It is advisable to check on the nature of the stabiliser before a commercial product is employed, to determine whether it is acceptable for the application intended. Hydrogen peroxide is normally supplied in very slightly acid condition (pH 5), because above pH 6 the decomposition rate is significant, increasing significantly with rise in pH.

The material of construction for storage vessels, valves, pumps and pipework needs to be selected with care. Thus although passivated aluminium (Cu content $\leqslant 0.002\%$) can be used, anodised aluminium should not. Passivated stainless steel is satisfactory provided the finish is of high quality, to ensure that passivation is effective, for example at welded joints. Plastics materials such as ptfe, polyethylene and unplasticised pvc are suitable, but mineral fillers and plasticisers should not be present; and polyethylene, if used, should contain antioxidants and UV stabilisers (such as carbon black), and the H_2O_2 concentration should not exceed 50%.

Some of the physical properties of hydrogen peroxide solutions are given in Table 3.5.

Table 3.5 — Properties of aqueous hydrogen peroxide

Concentration, wt. %	35	50	70
Specific gravity, g ml^{-1} at 20°	1.133	1.196	1.288
Freezing point, °C	− 33	− 52	− 40
Boiling point, °C	108	114	126
Viscosity, cp. at 20°C	1.11	1.17	1.27
Available oxygen, wt. %	16.5	23.5	32.9

Areas of application

In waste water treatment one can recognise three distinct categories of hydrogen peroxide usage:

(1) *As an oxygen source.* The disproportionation of hydrogen peroxide under the influence of inorganic catalysts or by enzymes (such as catalase) which are naturally present in micro-organisms can provide a convenient ancillary source of dissolved oxygen in biological treatment plants (Cole *et al.* 1974). This is particularly valuable in situations where acute overload has occurred, perhaps due to accidental spillage or seepage of high B.O.D. industrial wastes, or from seasonal variations in domestic waste waters due, for example, to a large tourist influx. The short-term use of hydrogen peroxide as a supplementary oxygen source under these circumstances may enable considerable capital savings to be made on the biological treatment plants, since the capacity of smaller plants, which may be adequate for most of the time, can be temporarily uprated as conditions demand. A limiting factor is, however, the cost of the reagent.

(2) *As a reductant.* Hydrogen peroxide can be used as an alternative reductant to the more commonly employed sulphur dioxide, as for example in the removal of residual chlorine subsequent to breakpoint chlorination processes. This aspect of the chemistry and applications of hydrogen peroxide is discussed in more detail in Chapter 4.

(3) *As an oxidant.* Hydrogen peroxide is a strong oxidant, as evidenced by its high redox potential of + 1.77 V for the two-electron reduction to water. From the known redox potentials (Stein and Weiss 1951) for the separate one-electron reductions

$$H_2O_2 + e \rightarrow OH\cdot + OH^- \quad E^\ominus = +0.8 \text{ V}$$

and

$$OH\cdot + e + H^+ \rightarrow H_2O \quad E^\ominus = +2.74 \text{ V}$$

it is clear that most of the free energy drop occurs after the peroxy $O - O$ bond is broken when the resultant hydroxyl radical is reduced. The powerful oxidising ability and extreme kinetic reactivity of the hydroxyl radical is well established and has been referred to earlier in this chapter. The rate-determining step therefore in reactions where hydrogen peroxide is acting as an oxidant is the breaking or activation of the $O - O$ bond. In this context it is noteworthy that this bond in hydrogen peroxide is much weaker than the bond in the oxygen molecule because of the two extra electrons in the antibonding orbitals. It is therefore to be expected, and is observed, that hydrogen peroxide is a better reagent than oxygen in oxidation reactions. $O - O$ bond breaking in hydrogen peroxide may occur by homolytic fission either thermally or photochemically (UV) induced to give two hydroxyl radicals, or it may be brought about as the result of activation by metal ions, as in Fenton's reagent:

$$Fe^{2+} + H_2O_2 \rightarrow (FeOH)^{2+} + OH\cdot$$

The ensuing radical reactions involving organic substrates have been discussed in the sections on catalytic oxidations and ozone reactions.

Specific applications
Cyanides
Hydrogen peroxide acts as an atom-transfer oxidant in its reaction with free cyanide or labile cyanide complexes. The mechanism involves electrophilic attack by the peroxide oxygen atom on CN^-, leading to the formation of cyanate:

$$CN^- + H_2O_2 \rightarrow OCN^- + H_2O$$

This reaction is less exothermic than the reaction of cyanide with chlorine, and can therefore be useful in the decomposition of highly concentrated (up to 50 g l^{-1}) cyanide wastes. Other advantages of hydrogen peroxide over chlorine in the treatment of high-level cyanide wastes are the absence of problems associated with the intermediate production of cyanogen chloride in chlorinolysis, and the fact that hydrogen peroxide treatment does not add to the total dissolved solids. Disadvantages, apart from the extra cost of hydrogen peroxide, arise in the treatment of more dilute cyanide wastes (< 5 g l^{-1}) because of the slowness of the reaction. This necessitates the addition of catalysts such as Cu^{2+} in order to avoid inordinately long reaction times (up to three hours for low cyanide levels). One of the problems associated with the long reaction times needed for many uncatalysed hydrogen peroxide oxidations, particularly of organic material, is the loss of reagent that occurs via the disproportionation reaction of hydrogen peroxide giving oxygen and water. This leads to the requirement of excessive amounts of hydrogen peroxide, with the economic disadvantages that this entails.

Oxidation of organic material
Hydrogen peroxide will oxidise many organic substances, particularly where there are unsaturated carbon sites at which attack can take place. However, attack is often enhanced by the addition of a catalyst such as ferrous ion, the well-known Fenton's reagent. The mechanism of oxidation here involves free radical attack and hydrogen atom abstraction. The products of such reactions are similar to those observed in

reactions in which the same material is oxidised using oxygen or ozone, and which proceed via similar mechanisms. In the case of Fenton's reagent, radicals are generated by reactions such as

$$Fe^{2+} + H_2O_2 \rightarrow FeOH^{2+} + \cdot OH$$

and

$$FeOH^{2+} + H_2O_2 \rightarrow Fe^{2+} + HO\cdot_2 + H_2O$$

For phenol oxidation, hydrogen peroxide in the presence of an iron catalyst (usually ferrous sulphate, although ferric iron can also be used) generally provides a less expensive chemical treatment (Sims 1981) than processes involving the other effective oxidants for phenol (ozone, chlorine dioxide, or permanganate). (Chlorine is not an appropriate oxidant for phenols due to the production of undesirable chlorophenols.) Sims has reported the optimum operating conditions for phenol levels of approximately 100 mg l^{-1} to be pH 4 (below 3 and above 5 the reaction loses effectiveness) with a hydrogen peroxide:phenol molar ratio of 3:1 and an iron concentration of 10 mg l^{-1}. With effluents of high oxygen demand or elevated phenol levels, operating experience suggests increased H_2O_2:phenol ratios (up to 8 : 1) and increased iron concentrations may be necessary. Substituted phenols are also degraded rapidly by Fenton's reagent, as in indicated in Table 3.6. Keating *et al.* (1977) have also reported on application to phenolic waste water problems.

Table 3.6 — Oxidation of phenols by Fenton's reagent (data from Sims 1983)

	H_2O_2 : phenol molar ratio								
	2 : 1			4 : 1			6 : 1		
	Residence time, minutes								
	5	15	30	5	15	30	5	15	30
Species	Phenol concentrations, mg l^{-1}								
Phenol	—	—	—	16	5	2	9	2	<1
p-Chlorophenol	8	6	4	4	4	3	5	<1	<1
p-Nitrophenol	—	—	—	24	7	<1	19	3	<1
p-Phenolsulphonic acid	45	<1	<1	16	<1	<1	<1	<1	<1

Initial concentration of phenol=100 mg l^{-1}; initial pH=4; Fe^{2+} concentration=20 mg l^{-1}.

The practical advantages of Fenton's reagent in attacking phenolic species has also been shown in the reduction in permanganate value achieved in rinses from a tin-plating process for steel containing phenolsulphonic acid from 60 mg l^{-1} to approximately 10 mg l^{-1} (M. J. V. Wayman, personal communication, 1984). The presence of ferrous iron in the effluent obviated need for iron addition; the H_2O_2 : phenolsulphonic acid concentration (weight) ratio was 1.6 : 1.

The simultaneous removal of phenol and chromium from paint-stripping waste waters using iron-catalysed hydrogen peroxide has been described by Barnes *et al.* (1981), using a H_2O_2:phenol ratio of only 3:1, and subsequent lime precipitation with clarification. Sims (1983) refers to the same type of application, claiming C.O.D. reductions from 25 000 mg l^{-1} down to 3300 mg l^{-1}. Schmitt (1982) also found that the presence of Cr(VI) in the paint waste seems to assist the oxidation of phenol by the hydrogen peroxide, being reduced to the trivalent condition at the same time.

Eisenhauer (1965) investigated the oxidative degradation of biologically 'hard' alkyl benzene sulphonates using Fenton's reagent. Two notable features of the reaction were, firstly, the relatively large mount of reagent required (H_2O_2:Fe^{2+}:ABS molar ratios of 9:6:1). This is, however, not surprising in view of the high molecular weight of ABS and the multiple sites available for radical attack. The second point of practical importance is that although approximately 80% of the ABS was removed in the first ten minutes it required a further 15–20 hours to achieve 99% removal (see Fig. 3.9). This behaviour is again not unexpected, since even in the

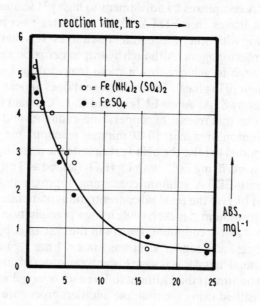

Fig. 3.9 — The oxidation of alkylbenzene sulphonate, ABS, by Fenton's reagent (from Eisenhauer 1965). Starting concentration of ABS = 50 mg l^{-1}.

absence of substrate, decomposition of Fenton's reagent occurs rapidly and hence there are only significant amounts of reagent present in the initial stages. The subsequent slow reaction in which the ABS concentration drops from approximately 4 mg l^{-1} to < 1 mg l^{-1} may simply be due to the catalysed autoxidation, the oxidant in this case being dissolved oxygen resulting from the peroxide decomposition. If this is the case then successive additions of Fenton's reagent (at, say, five-minute intervals) would be expected to reduce the residual ABS concentration rapidly. Eisenhauer (1965) in fact notes that with a second addition of Fenton's reagent after 30 minutes, the ABS concentration was reduced to less than 1 mg l^{-1} in under two hours. Similar improvements have also been noted (Eilbeck 1983) in the treatment of phenol using successive additions of Fenton's reagent. Reaction time to achieve 1 mg l^{-1} residual phenol is reduced from 30 minutes (see Table 3.6) to approximately 10 minutes if the reagent is added incrementally.

Fenton's reagent, although more vigorous in its action than hydrogen peroxide alone, is not powerful enough to break down fully the more stable organic species, although it can be partially effective. For example, Kinoshita and Sumada (1972) found that 10 mg l^{-1} H_2O_2 with 30 mg l^{-1} ferrous sulphate at pH 3.0 reduced a polychlorinated diphenyl concentration of 100 mg l^{-1} by 42% over a 24 hour period. It is therefore worthwhile in any specific case to examine the performance of the reagent in relation to requirements.

Oxidation of metal–EDTA complexes

The discussion in Chapter 2 indicated that the precipitative removal of heavy metals from their EDTA complexes by adjustment to high pH values is not fully effective, and that at best leaves free EDTA which may later re-complex with metals on subsequent mixing with other waste water discharges. It is therefore clearly better to destroy the complexing agent. Although hydrogen peroxide alone does not appear to be powerful enough to achieve this at room temperature (although there may be some effect at high pH values), Fenton's reagent does appear to offer some benefit. Work carried out by S. A. Abou El Teen and C. L. Graham (personal communication, 1975) suggests that release of copper, zinc, cadmium and nickel ions does occur on addition of Fenton's reagent: 10–20 minutes reaction time was allowed, and lime slurry was then added to raise the pH to precipitate the metal hydrous oxides. Typical quantities used were 70 mg Fe^{2+} with 1 g H_2O_2 (added as 3 ml 3% H_2O_2 solution) to one litre of metal–EDTA solution containing approximately 50 mg metal. At a precipitation pH of 11.0 the residual concentrations of metals increased in the series zinc ≈ copper < cadmium < nickel, while using a precipitation pH of 12.0 the series is copper < nickel < zinc < cadmium. Except in the case of copper precipitated at pH 12.0, where the residual copper level was low (< 1 mg l^{-1}) even without Fenton's reagent, the residual metal levels were lower by an order of magnitude by application of the oxidative treatment than without it. Even so, removal was not complete. More work may be justified using the multiple addition procedure mentioned earlier for phenol destruction by Fenton's reagent.

Oxidation of sulphur-containing compounds

Hydrogen peroxide reacts with most sulphur-containing compounds, including H_2S and inorganic sulphides, mercaptans and sulphite. In acid solutions sulphides are oxidised to sulphur by a 1:1 molar reaction (see Hofmann 1977):

$$H_2O_2 + H_2S \rightarrow S\downarrow + 2H_2O$$

Above pH 8.5 polysulphides and eventually sulphate are formed, with a higher reagent demand:

$$4H_2O_2 + S^{2-} \xrightarrow[10]{pH} SO_4^{2-} + 4H_2O$$

Various applications of hydrogen peroxide for oxidising sulphides have been found (Sims 1983). For example, sour water at an oil refinery containing 150 mg l^{-1} sulphide at a pH of 7 was oxidised in 24 hours to 1 mg l^{-1} sulphide concentration using 1.5:1 H$_2$O$_2$ to sulphide molar ratio, and in 15 minutes at a 2 : 1 ratio. Some sulphur was formed, giving milkiness, but by using a pH of 10 for the treatment, with a molar H$_2$O$_2$:sulphide ratio of 4:1, a clear solution was obtained after 15 minutes.

Sims also quotes other examples of sulphide oxidation. One involved treatment of 65 mg l^{-1} sulphide in an effluent from cellophane manufacture. Using a 1.5:1 molar ratio of H$_2$O$_2$:S^{2-} at a pH of 7.5, the sulphide level was reduced to 13 mg l^{-1} after 1 hour, and 5 mg l^{-1} after 3 hours. Benefit was however gained by introducing 2 mg l^{-1} Fe^{2+}, when the sulphide level was reduced to 0.1 mg l^{-1} in 2 hours. In another application—a tar distillery steam distillate containing 500 mg l^{-1} sulphide, 1200 mg l^{-1} phenols, 1000 mg l^{-1} ammonia, and 500 mg l^{-1} petroleum extractables —hydrogen peroxide selectively oxidised the sulphide completely when used at a molar ratio of 3.1 H$_2$O$_2$:1 S^{2-} at a pH of 8.5–9 in a continuous flow reactor having one hour retention. Tannery waste waters have also been treated with hydrogen peroxide to reduce sulphide levels down to less than 0.5 mg l^{-1} (O'Neill *et al.* 1978).

As with chlorine, the use of hydrogen peroxide as a sulphide oxidant must be viewed in the light of operating economics, as against catalytically induced aeration: it would appear to be most useful when employed to remove residual sulphide after preliminary aeration where this is practicable.

Other miscellaneous oxidations

Junkermann and Schwab (1978) have demonstrated the stoicheiometric oxidation of formaldehyde by hydrogen peroxide in alkaline conditions (pH > 10). Formaldehyde in concentrations up to 10% was oxidised, giving residuals of less than 0.01% with reaction times of 2–30 minutes; the reaction is exothermic.

Hydrogen peroxide can be used as an alternative to hypochlorite in the oxidation of materials such as sulphites and nitrites, with the advantage of not adding to the dissolved solid concentrations. In weakly buffered solutions, a starting pH of 3.5 may be used, but in more strongly buffered solutions, a lower pH is desirable (Rodenkirchen 1983). The mechanism of nitrite oxidation probably involves formation of peroxynitrous acid:

$$HNO_2 + H^+ \rightleftharpoons NO^+ + H_2O$$

$$H_2O_2 + NO^+ \rightarrow HO.ONO + H^+ \rightarrow NO_3^- + 2H^+$$

Hydrogen peroxide oxidises hydrazine readily, even at high pH, although in the presence of methanol the reaction is inhibited (Eden *et al*. 1951). Methanol + hydrazine + hydrogen peroxide mixtures can be present in waste water discharges where rocket fuels are used, so the inhibition can be significant. However, Eden *et al*. found that the presence of 1 mg l^{-1} Cu^{2+} catalyses oxidation, leaving some formaldehyde as well as excess methanol.

UV irradiation with hydrogen peroxide
Since it is the production of ·OH radicals which is responsible for the effectiveness of Fenton's reagent, it might be expected that a similar enhancement of the oxidising power of hydrogen peroxide could be achieved with ultraviolet radiation. Photolytic cleavage of the O − O bond in hydrogen peroxide leading to formation of hydroxyl radicals is known to take place readily, and the technique of simultaneous hydrogen peroxide addition/UV irradiation has proved effective in a number of cases. For example, Andrews (1980) in a US Government report demonstrated the feasibility of treating explosive-contaminated waste water containing TNT (2,4,6-trinitrotoluene) and RDX (1,3,5-trinitro-1,3,5-triazacyclohexane) by H$_2$O$_2$/UV photolysis, and concluded that this process is more economical than UV ozonation or active carbon adsorption. Equipment suitable for H$_2$O$_2$/UV photolysis has been described by Konbek (1977), who claims that most refractory organic substances (apart from fluorinated hydrocarbons) can be oxidised to CO$_2$ + H$_2$O. For reasons of economy, it is better to apply the process after conventional primary treatments have been carried out.

Peroxymonosulphuric acid
In general, peroxyacids and their salts are powerful oxidising agents which react rapidly with many organic and inorganic substances. In particular, peroxymonosulphuric (permonosulphuric) acid, H$_2$SO$_5$, also known as Caro's acid, is a much more powerful oxidant than hydrogen peroxide, and will for example rapidly oxidise even very low concentrations of cyanide and thiocyanate. The acid as 60–70% concentration is prepared either via the electrolysis of sulphuric acid or by the action of concentrated hydrogen peroxide on either sulphuric acid or chlorosulphuric acid:

$$H_2O_2 + H_2SO_4 \rightarrow H_2SO_5 + H_2O$$

$$H_2O_2 + HSO_3Cl \rightarrow H_2SO_5 + HCl$$

It can also be prepared to give approximately 15% concentration by reacting sulphuric acid with ammonium peroxydisulphate, but this product is of little value for waste water treatment purposes owing to the undesirable co-presence of ammonium ions. Salts of the acid such as KHSO$_5$ can only be obtained in an impure state admixed with K$_2$SO$_4$ and KHSO$_4$.

Aqueous solutions of either the free acid or its salts decompose readily, giving oxygen or hydrogen peroxide and sulphate, and care must be observed in handling the product when used at the higher concentrations. The reason for the increased rate of oxidations with Caro's acid compared to the equivalent oxidations with hydrogen peroxide can be appreciated if we consider the mechanism of the cyanide oxidation for the two reagents:

(With an excess of H_2SO_5 there is a further breakdown of cyanide from cyanate to carbon dioxide and nitrogen:

$$2CN^- + 5H_2SO_5 + 2OH^- \rightarrow 2CO_2 + N_2 + 5H_2SO_4 + H_2O.)$$

If the reaction is recognised as being a nucleophilic attack of the cyanide carbanion on the electrophilic peroxide oxygen atom (Edwards 1964) it is immediately apparent that since HSO_4^- is a much better leaving group than OH^- then peroxymonosulphuric acid will react much more readily than hydrogen peroxide. It will, for example, break down virtually completely phenol, o-cresol and m-cresol over a period of 5–30 minutes (the time depending on the concentration). Fairly large excesses are needed—of the order of 100 H_2SO_5:1 phenol—at a pH of 9–10.

The main disadvantage of inorganic peroxyacids is their relative cost; also, of course, due to their enhanced activity, extra safety precautions must be adopted. Furthermore, in the presence of chlorides, chlorine or hypochlorite may be formed, and this may then take over the oxidising action: which is of course tantamount to using chlorine or hypochlorite in the first place, but at a higher cost.

Potassium permanganate
General properties
Potassium permanganate, $KMnO_4$, is a stable, almost black, crystalline solid that is moderately soluble in water (6.4 g per 100 ml at 20°C) giving solutions that have a characteristic intense purple colour. The permananate ion, MnO_4^-, is a powerful oxidising agent and will rapidly oxidise Fe^{2+}, Mn^{2+}, sulphides, cyanides, phenols, and many other organic substances. The products of the reduction of permanganate by such substances depend on the pH at which the reaction takes place (see earlier, p. 98, for a discussion). In strongly acidic solution permanganate behaves as a five-electron oxidant giving Mn^{2+}, whereas in the pH ranges commonly encountered in water treatment processes (pH 3–11) it behaves as a three-electron oxidant with the formation of manganese dioxide, MnO_2. Inspection of Fig. 3.1 indicates that permanganate should oxidise Mn^{2+} to give MnO_2; this in fact happens, so that in conditions of excess permanganate MnO_2 will be formed even in acid solutions. The green manganate ion, MnO^{2-4}, which disproportionates in acid or neutral solution, is more stable in strongly alkaline solution (see Fig. 3.1), and consequently, under such conditions (pH > 12) permanganate may act as a one-electron oxidant at least in the initial stages of the reaction. The manganese(V) state (hypomanganate MnO_4^{3-}) only exists in extremely concentrated alkali (> 10 M KOH) and is not detectable in the usual reactions of permanganate occurring under less extreme conditions, although its transient existence as a short-lived reactive intermediate in the oxidation of cyanide by permanganate has been postulated (Steward and Van der Linden 1960). The involvement of Mn(V) in permanganate oxidations of some organic molecules has also been suggested (Stewart 1964). This is not unreasonable, since such molecules behave as two-electron reductants, and the initial formation of Mn(V) (MnO_4^{3-}) followed by its immediate disproportionation to Mn(IV)(MnO_2) and Mn(VI) (MnO_4^{2-}) is perfectly feasible and not at variance with the observed kinetics. The disproportionation reaction may be represented thus:

$$2MnO_4^{3-} + 2H_2O \rightarrow MnO_2 + MnO_4^{2-} + 4OH^-$$

Specific applications
Cyanide oxidation
In the pH range 12–14 permanganate oxidises cyanide to cyanate as represented by the following nett equation:

$$2MnO_4^- + CN^- + 2OH^- \rightarrow NCO^- + 2MnO_4^{2-} + H_2O$$

The reaction is described by the following rate equation (Stewart and Van der Linden 1960, Freund 1960):

$$\frac{-\partial[MnO_4^-]}{\partial t} = k[MnO_4^-][CN^-]$$

which is in accordance with the mechanism

$$MnO_4^- + CN^- \rightarrow [O_3Mn-O-CN]^{2-} \rightarrow MnO_3^- + NCO^-$$

$$2OH^- + MnO_3^- + MnO_4^- \rightarrow 2MnO_4^{2-} + H_2O$$

It has been observed by Posselt (1972) that Ca^{2+} ions change the stoicheiometry of the above reaction of cyanide with permanganate. In saturated calcium hydroxide (pH 12.4) permanganate behaves not as a one-electron oxidant as above, but as a three-electron oxidant, resulting in the formation of MnO_2 as the product:

$$MnO_4^- + 3CN^- + H_2O \rightarrow 2MnO_2 + 3NCO^- + 2OH^-$$

Since at this pH any manganate (MnO_4^{2-}) formed would be expected to disproportionate to give MnO_2 and MnO_4^-, the Ca^{2+} ions may simply be acting as a catalyst for this disproportionation, i.e. the effect has kinetic rather than thermodynamic origins. Posselt suggests that the reason for the observed stoicheiometry change is the removal by Ca^{2+} of an 'active' form of Mn(IV) from some so-called 'quasi-equilibrium'. However, an alternative explanation is that the Ca^{2+} ion acts as a bridging ion in the activated complex of the disproportionation reaction, thus facilitating electron transfer between manganese centres which would otherwise be constrained to proceed via an outer sphere mechanism:

$$\left[\begin{array}{c} O \\ O \end{array} \diagdown Mn \diagup \begin{array}{c} O \\ O \end{array} \diagdown Ca^{2+} \diagup \begin{array}{c} O \\ O \end{array} \diagdown Mn \diagup \begin{array}{c} O \\ O \end{array} \right]^{2-}$$

Support for this explanation is provided by studies of the electron exchange between MnO_4^- and MnO_4^{2-}, where it has been shown that a similar accelerating effect on the exchange rate is observed in the presence of alkali metal ions, and that it is due to the formation of a bridged complex of this type (Sheppard and Wahl 1957, Gjertsen and Wahl 1959). Whatever the reason for the change in the observed stoicheiometry, it is potentially of great practical significance in view of the widespread use of $Ca(OH)_2$ in water treatment processes. Although the reaction of permanganate with cyanide is rapid, particularly in the presence of Cu^{2+} (which has a marked catalytic influence), permanganate is not generally used for the removal of cyanide from waste waters because it is a relatively expensive reagent. Indicative experimental data have been provided by Kieszkowski and Krajewski (1968), from which it is apparent that permanganate offers no advantage over hypochlorite.

Oxidation of Mn^{2+} and other metal ions

Mn^{2+} may be present in waste waters originating from a number of industrial processes including metal treatment (of manganese steels) and ink and dye manufacture. It also occurs together with Fe^{2+} in some ground waters and mine drainage waters. Unlike Fe^{2+}, which is readily oxidised by aeration at or near neutral pH, Mn^{2+} is only oxidised rapidly by air at high pH. Potassium permanganate will, however, rapidly oxidise Mn^{2+} to insoluble MnO_2 in solutions more alkaline than pH 5 according to the equation (Ladbury and Cullis 1958)

$$3Mn^{2+} + 2MnO_4^- + 2H_2O \rightarrow 5MnO_2 + 4H^+$$

and this reaction has been successfully used for Mn^{2+} removal (see Patterson 1975). Fe^{2+} is often present together with Mn^{2+}, and this too is rapidly oxidised by permanganate. Depending on the amount of Fe^{2+} present, economic considerations may suggest pre-aeration to oxidise Fe^{2+} prior to the addition of potassium permanganate. The marked sorptive capacities of manganese dioxide and hydrated ferric oxides towards divalent metal ions, which have been mentioned earlier and in Chapter 2, frequently result in a lower permanganate requirement than that suggested by the formal stoicheiometry of the oxidation reactions.

Oxidation of organic material

In preparative organic chemistry potassium permanganate is probably the most useful, powerful oxidant available. It reacts with a variety of organic substances (see

Table 3.7 — Oxidations of organic molecules by permanganate (from Weber 1972)

$-CH=CH-$	$\rightarrow$	$\overset{\displaystyle OH\ \ OH}{\underset{\displaystyle\ }{-CH-CH-}}$
$R-CHO$	$\rightarrow$	$R-COOH$
$HCHO, HCOOH$	$\rightarrow$	$CO_2 + H_2O$
$R-CH_2OH$	$\rightarrow$	$R-COOH$
R_2CHOH	$\rightarrow$	$R_2C=O$
$R-NH, R_2NH, R_3N$	$\rightarrow$	CH_3COOH, NH_3, and other similar compounds
$R-SH$	$\rightarrow$	$R-SO_3H$
$R-S-R$	$\rightarrow$	$R-SO_2-R$
$R-S-S-R$	$\rightarrow$	$2R-SO_3H$
$R-SO-R$	$\rightarrow$	$R-SO_2-R$

(Benzene rings with NH_2, RNH, OH substituents) $\rightarrow$ (Complex)CO_2, NH_3, $RCOOH$, etc.

(Benzene rings with CH_3, CH_2OH, CHO substituents) $\rightarrow$ (Benzene ring with $COOH$)

(Thiophene) $\rightarrow$ SO_4^{2-}, CO_2, H_2O

Table 3.7) by mechanisms which in general result in the nett transfer of an oxygen atom from the permanganate ion to the organic substrate.

In general, the rates of reaction are very pH-dependent, usually occurring much more rapidly in alkaline solution than at neutral pH (Stewart 1964). Parathion and paraoxon are, for example, more effectively oxidised by permanganate in alkaline medium (Gomaa and Faust (1971b), although the reverse is true for *p*-nitrophenol. Although the oxidation of organic material by permanganate rarely results in the complete degradation of the molecule, it is often the case that the products of the oxidation of obnoxious or odorous compounds are considerably less unpleasant, and it is this feature of the reactions that has led to the use of permanganate for odour abatement in potable water supplies and in waste water. In this respect it is superior to chlorine when phenolic materials are present.

Aldehydes

Aldehydes (RCHO) or their bisulphite addition compounds react with cyanide to give cyanhydrins by a mechanism involving nucleophilic attack of the cyanide ion on the active carbon of the aldehyde group:

$$R - \overset{\delta+}{\underset{\underset{O}{\parallel}}{C}} \,\, CN^- \rightleftharpoons R - \underset{\underset{O^-}{|}}{\overset{\overset{H}{|}}{C}} - CN \overset{H_2O}{\rightleftharpoons} R - \underset{\underset{OH}{|}}{\overset{\overset{H}{|}}{C}} - CN + OH^-$$

Formaldehyde in particular has been proposed as a reagent for the detoxification of cyanide waste waters, since the cyanhydrin formed in the above reaction (glycolonitrile in this instance) is hydrolysed in alkaline solution to give first glycolamide and subsequently glycolic acid. Both of these products are readily biodegradable and non-toxic, although they do, of course, impose an additional B.O.D.:

$$H - \underset{\underset{OH}{|}}{\overset{\overset{H}{|}}{C}} - CN \overset{H_2O/OH^-}{\longrightarrow} H - \underset{\underset{OH}{|}}{\overset{\overset{H}{|}}{C}} - \underset{\underset{NH_2}{\parallel}}{\overset{\overset{O}{/}}{C}} \overset{H_2O/H^-}{\longrightarrow} CH_2OH \cdot COOH + NH_3$$

glycolonitrile   glycolamide   glycolic acid

There are distinct disadvantages in employing this series of reactions for the treatment of cyanide waste waters. The main problems arise from the relatively long reaction times and/or high operating temperatures, and the difficulties which are encountered in the presence of those metal ions that form stable complexes with cyanide. This last problem arises from the fact that the cyanhydrin-forming reaction is significantly reversible, and the removal of free cyanide from the system by complexation with metal ions drives the equilibrium in the opposite direction to that desired. Also, the nucleophilic character of cyanide is considerably reduced on complexing, and hence complexed cyanide is a poorer participant than free cyanide in the above reaction.

In an attempt to overcome some of the difficulties encountered when using formaldehyde alone, du Pont (USA) introduced the 'Kastone' process in which formaldehyde and hydrogen peroxide are used simultaneously. Details of the process have been described by Lawes *et al.* (1973), who point out that the overall efficiency of the process is very dependent both on the pH (optimum 10–11) and on the molar ratios of $[HCHO]:[CN^-]:[H_2O_2]$ (optimum 0.9:1:1.1). Excess formaldehyde reduces the rate of cyanide removal. This effect of excess formaldehyde is not unexpected, since the improved rate of removal of cyanide using a mixture of formaldehyde and hydrogen peroxide over the rate observed with either reagent alone is almost certainly due to the generation of reactive intermediates (such as free acyl radicals and peroxyformic acid) in the reactions between formaldehyde and hydrogen peroxide. This latter reaction is known to proceed via a free radical chain mechanism. Peroxyformic acid, for example, would be expected to react readily both with formaldehyde (to give formic acid) and with cyanide (giving cyanate and formic acid). Competition between these reactions would account for the observed behaviour, and other competitive reactions of this type are also possible.

Additional disadvantages of the process are the requirement of somewhat elevated temperatures (50°C) and the fact that cyanide is not oxidised beyond cyanate. On balance, therefore, it is not immmediately apparent that the use of formaldehyde or aldehydes in general offers any significant advantages over the alternative cyanide treatment processes involving oxidation with chlorine or hypochlorites, hydrogen peroxide in the presence of catalysts, or peroxymonosulphuric acid.

THE PROCESS CONTROL OF OXIDATION REACTIONS

Redox electrodes

The fact that redox reactions are characterised by electron transfer suggests that monitoring of such reactions should be realisable with electron-responding indicators. Electron conductors such as metals are thus obvious candidates, inasmuch as these are capable of accepting or releasing electrons according to local conditions. Thus if a metal is immersed in a solution in which a reaction involving electron transfer is proceeding, the metal may be expected to take up or release electrons, leaving it more or less negatively charged. Measurement of the zero current potential of the metal (by comparing the potentials with that of a standard reference electrode in open circuit conditions) will thus indicate the level of progress of the redox reaction. A typical cell configuration corresponds closely to that shown in Fig. 2.2 (p. 49), with the indicator metal replacing the glass electrode.

In practice, it is necessary to use a noble metal, i.e. one which is unaffected by side reactions such as oxidation, and so the common redox electrode indicators are platinum and gold. Even these, it should be noted, are not completely inert, and it cannot be assumed that a true indication of redox conditions is always obtained with them (as discussions in the next section will demonstrate). Furthermore, the electrode is only responding correctly when the reaction

$$\text{oxidised species} + \text{electron(s)} \rightleftharpoons \text{reduced species}$$

is thermodynamically reversible. The electrode responds by taking into the Nernst double layer the participant reaction species, and its electrons then are released or taken up according to the activity ratios of the participants. This leads to the development of a surface potential according to the Nernst equation:

$$E = E^{\ominus} + \frac{RT}{nF} \log_e \frac{a_{ox}}{a_{red}}$$

where n is the number of electrons involved in the reaction, and the 'a' terms refer to activities of the oxidised and reduced forms. If the reaction does not permit ready interchange of electrons between the species, the Nernst equation will not apply, and so the observed potential will not be truly representative of the redox condition.

It must be appreciated that many reactions of interest involve not only the specific oxidised and reduced forms of reactant, but also other entities, e.g. hydrogen ions. The potential at the redox electrode surface depends on the activity ratios of all the reaction participants, and so for measurements to be significant and specific only to a_{ox}/a_{red}, the activities of the other species must be constant. This has important implications for process control, e.g. it may be necessary simultaneously to control other parameters of which pH is a common one.

It should also be recognised that in the absence of one of two components of a reduction–oxidation couple, a redox indicator electrode will not adopt meaningful or reproducible potentials. Thus, for example, in reference to the Fe^{3+}/Fe^{2+} couple, $a_{Fe^{3+}}$ cannot be monitored if $a_{Fe^{2+}}$ is zero, and vice versa. This of course follows from the mathematical implications of the Nernst equation, and means that redox electrodes cannot be used as monitors of individual components of a redox couple. They are therefore of limited value as direct pollution indicators.

Oxidation–reduction potential titration curves
General principles

The progress of an oxidation–reduction reaction can be followed by the execution of a titration between oxidant and reductant, using a redox indicator electrode and stable reference system. To avoid polarisation effects, the potential difference between these electrodes is measured under virtual open circuit conditions to give the change in redox potential throughout the course of the titration. For this, a suitable instrument is an electrometer, such as a commercial pH meter suitably calibrated in millivolts.

A typical theoretical titration curve is shown in Fig. 3.10, for which we may take as an example the oxidation of ferrous iron to ferric iron by hypochlorite. The two half-reactions involved are

$$Fe^{3+} + e \rightleftharpoons Fe^{2+} \quad E^{\ominus} = +0.77 \text{ V}$$

$$\tfrac{1}{2}Cl_2 + e \rightleftharpoons Cl^- \quad E^{\ominus} = +1.36 \text{ V}$$

Summation gives

$$Fe^{2+} + \tfrac{1}{2}Cl_2 \rightarrow Fe^{3+} + Cl^-$$

and the Nernst equations for the redox potentials are

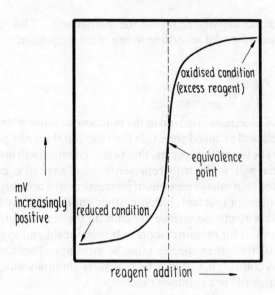

Fig. 3.10 — Ideal redox titration curve for oxidation of reduced form of species by oxidant reactant.

$$E' = E^{\ominus\prime} + \frac{RT}{F} \log_e \frac{a_{Fe^{3+}}}{a_{Fe^{2+}}}$$

and

$$E'' = E^{\ominus\prime\prime} + \frac{RT}{F} \log_e \frac{(a_{Cl_2})^{\frac{1}{2}}}{a_{Cl^-}}$$

As the curve shows, the addition of oxidising agent (chlorine) to ferrous iron makes the redox electrode more positive, due to an increase in the ratio $a_{Fe^{3+}} : a_{Fe^{2+}}$ up to the equivalence point of reaction. Beyond the equivalence point, the ratio of significance is $(a_{Cl_2})^{\frac{1}{2}} : a_{Cl^-}$; this also increases as chlorine is added. Detailed analysis of the curve characteristics is not relevant here, being available in standard texts: it suffices to say that the sigmoid shape of the redox titration curve is the base from which it is possible to observe the state of a redox reaction, i.e. whether the reaction has been completed or not, and also the equivalence point. The equivalence point, E_{eq}, is theoretically determinable for this single-electron reaction (multiple-electron reactions are more complicated) from

$$E_{eq} = \tfrac{1}{2}(E_{Fe^{3+}/Fe^{2+}} + E_{Cl_2/Cl^-})$$

In practice the situation is not so simple. Firstly, in this particular example, and indeed in most of those redox reactions employed in waste water treatment, the reaction is not truly thermodynamically reversible, with the result that E_{eq} does not occur exactly as predicted. Secondly, because of this element of thermodynamic irreversibility, the measured redox potential difference does not necessarily reflect the true redox potential condition; and this may mean that the apparent equivalence point, conventionally defined as the point where $\partial E/\partial V$ is at a maximum, is not a true one. Thirdly, the redox potential measurements, again due to thermodynamic irreversibility, may not be very reproducible. The summarising conclusion is that for most reactions used in waste water treatment, the redox curve does not accurately reflect the progress of the reaction. From a practical point of view this is not of major importance, however, because oxidative treatment of a pollutant demands that there be effectively no polutant left (or minimum, dependent on reaction effectiveness). It is not realistic in plant practice to attempt to add a precise amount of reagent to achieve a theoretical reaction equivalence, and so inevitably a small amount of overdosing must occur. For this, the redox curve as measured is normally an adequate indicator of acceptable treatment.

Standardisation of redox cells
In the same way that pH as measured cannot be unambiguously translated into hydrogen ion activities or concentrations (see Chapter 2), so redox millivolt measurements cannot be translated into redox activity or concentration ratios without making extra-thermodynamic assumptions. Indeed, the situation is frequently more complicated by virtue of the involvement of other species in the redox reaction being measured. For most redox measurements applied to waste water treatment control this essential uncertainty is not usually of major importance, because, as has been shown before, the control operation normally looks for relatively large changes in redox ratio, and hence in millivolt values. It must be appreciated, however, that if redox measurements are made with a view to identifying change in activity or concentration of a reactant well away from the titration endpoint—in the region beyond the 'knee' of the curve—all the uncertainties of interpretation discussed in Chapter 2 apply. This means that redox potential measurements are not really satisfactory as means of *determining* redox ratios.

The great variety of the possible range of reduction–oxidation reactions that can be monitored with redox sensors not only makes interpretation difficult, but also any associated attempt to prepare standards against which the measured values can be compared. Nevertheless from a practical viewpoint it is useful on occasions to have available standard redox solution compositions that give rise to defined millivolt outputs with redox electrode cells, rather as with pH. With pH measurements the various drifts and losses of sensitivity that can occur necessitate the frequent use of standard pH solutions as a check on the validity of the measurements; with redox electrodes there are fewer sources of drift, but loss of sensitivity due to precipitate or oil coating can be just as serious, and so there is merit in having standards available.

The attention that has been paid in the study of pH standards has not been echoed for redox standards. Light (1972) has given a value for a 0.1 M ferrous ammonium sulphate, 0.1 M ferric ammonium sulphate, 1 M sulphuric acid solution of $+439\,mV$ for a platinum electrode vs. 1 M KCl silver–silver chloride reference, but in practice two standard solutions are desirable as a minimum. One suitable system, defined in Table 3.8, is based on the quinhydrone redox system, consisting of an equimolar mixture of quinone and hydroquinone (formerly used as a pH indicator):

$$\text{(quinone)} + 2H^+ + 2e \rightleftharpoons \text{(hydroquinone)}$$

quinone   hydroquinone
(1,4-cyclohexadienedione)   (1-4)benzenediol)

Addition of a small quantity of quinhydrone to a standard pH buffer solution (pH$\not>$7) will provide a solution the pH of which defines the millivolts output of a redox cell placed in the solution.

Table 3.8 — E.m.f. values of redox cells using standard pH buffer solutions with quinhydrone

The following data relate to the cell using a saturated KCl calomel reference electrode:

$$\text{Hg; Hg}_2\text{Cl}_2 \mid \text{satd. KCl} \parallel \text{pH buffer, quinhydrone} \mid \text{Pt}$$

The values are given in millivolts. In all cases the platinum electrode is positive with respect to the reference electrode with the two standard pH buffers quoted (see Table 2.1, p. 43, for preparation details).

(1) 0.05 mol kg^{-1} potassium hydrogen phthalate
(2) 0.025 mol kg^{-1} disodium hydrogen phosphate, Na$_2$HPO$_4$, 0.025 mol kg^{-1} potassium dihydrogen phosphate.

	Temperature, °C				
Solution	5	10	15	20	25
(1)	236.1	231.8	227.4	222.8	218.5
(2)	73.9	68.1	61.9	55.7	49.7

If a reference electrode different from satd. KCl calomel is used, a correction for the cell e.m.f. value can be made using the data of Table 3.9. Calculate the millivolt difference between the potentials of the electrode used ($= E_{ref}$) and the satd. KCl calomel ($= E_{cal}$) at the appropriate temperature as ($E_{cal} - E_{ref}$), and add this to the cell values given in the table above, taking account of the sign of ($E_{cal} - E_{ref}$). For example, at 15°C for solution (1), a 3.5 M KCl Ag/AgCl reference electrode should give a cell output of $227.4 + (251.1 - 211.7) = 266.8$ mV.

The procedure for standardisation follows the same pattern as for pH electrode cells. The redox cell system is stood in one of the solutions, and the millivolt output noted. After removal and rinsing with water or a small quantity of the second solution, the cell is transferred to the second solution, and the output again noted. The difference between the two readings should correspond to the defined difference for the solutions at the temperature used. Whether or not the actual millivolt output corresponds exactly to the values defined for the solution used will depend on several factors, including particularly the type and condition of the reference electrode used. For most waste water applications, precise correspondence is not important; but if the cell system is behaving properly it should give the correct *difference* between two standardising solutions for the temperature of measurement. As a check between redox measurements made with different types of reference electrode, all values can be brought to a common reference potential, such as that of the standard hydrogen electrode. Table 3.9 gives some reference electrode potentials, defined with respect to the standard hydrogen electrode, to aid this normalisation procedure.

Table 3.9 — Standard reference electrode potential values

The following data, given in millivolts, refer to the e.m.f. values of the cell Pt; $H_2 \mid a_{H^+} = 1 \parallel$ reference electrode.

	Temperature, °C				
Reference electrode	5	10	15	20	25
3.5 M KCl calomel	257.4	255.6	253.8	252.0	250.1
3.8 M KCl calomel	255.5	253.6	251.8	249.9	248.8
Satd. KCl calomel	257.5	254.3	251.1	247.9	244.4
3.5 M KCl Ag/AgCl	218.7	215.2	211.7	208.2	204.6
3.8 M KCl Ag/AgCl	211.8	213.5	209.2	205.5	210.3
Satd. KCl Ag/AgCl	218.7	213.8	208.9	204.0	198.9

In all of the above cases, the reference electrode is positive with respect to the hydrogen electrode. To convert a measured redox cell millivolt value obtained with one of the above reference electrode to the value vs. the standard hydrogen electrode (Pt, $H_2/a_{H^+} = 1$) add the appropriate millivolt value from the table. Thus, for example, a measured redox value at 10°C where the redox indicator electrode is + 400.0 mV with respect to the satd. KCl calomel reference electrode would be + 654.3 mV vs. the standard hydrogen electrode.

As has already been indicated, redox cells do not necessarily give millivolt outputs that are highly reproducible with many of the reduction–oxidation systems they are used to follow. It is of importance therefore that where they are to be used for control of a titration form of treatment the millivolt change characteristics of the titration curves should be established empirically for the system to be controlled. There are so many factors that can influence the actual millivolt values observed, and indeed also the shape of the titration curve, with the magnitudes of millivolt change and millivolt values at the titration endpoint, that to adopt quasi-theoretical control redox millivolt values is unrealistic (although general levels can be adopted as a starting point where no other information is available). The practical aspects of redox control are exemplified in the following sections.

The cyanide–hypochlorite/chlorine system
A widely used oxidation reaction is that involving the destruction of cyanide by chlorine/hypochlorite, discussed earlier in this chapter. This shows all the characteristics mentioned: the oxidation of cyanide is irreversible, the redox potential

measurements are not highly reproducible, and the reaction endpoint is not unambiguously determinable from the titration curve. Having said that, one must also add that in practice redox systems are usable, given an understanding of certain limitations. Fig. 3.11 shows curves for the oxidation and hydrolysis of cyanide ions by

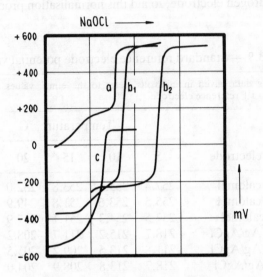

Fig. 3.11 — Potentiometric titration curves for the oxidation of cyanide by hypochlorite.
a. Platinum indicator electrode. b_1. Gold indicator electrode with stirring. b_2. Gold indicator electrode without stirring. c. Silver indicator electrode. Reference electrode saturated KCl calomel (curves are displaced along abscissa for clarity). (After Mattock and Uncles 1962.)

sodium hypochlorite at pH 11, plotted from observed millivolt values using both platinum and gold as indicator electrodes, measured with respect to a saturated potassium chloride/calomel reference electrode (Mattock and Uncles 1962). The characteristics of the platinum curve are indicative of an irreversible reaction; but the equilibrium value for an excess of hypochlorite of approximately $+0.77$ V (with respect to the hydrogen electrode) is consistent with the standard potential for the OCl^-/Cl^- couple of $+0.88$ V, taking into account the pH conditions and concentrations involved, suggesting that the electrode may be following this reaction in excess hypochlorite conditions. The same is true of the curve for the gold indicator, but here the pre-endpoint characteristics are quite different. It is clear that with excess cyanide, the gold is responding according to

$$Au(CN)_2^- + e \rightleftharpoons Au + 2CN^-$$

This reaction has a standard potential of -0.67 V with respect to the hydrogen electrode, which is consistent with the curve in Fig. 3.11. Attack on gold by cyanide is quite significant, such that not only is the observed potential affected by agitation of the solution, but also the gold quite rapidly dissolves.

Also shown in this figure is the response of a silver electrode, which although not a redox indicator in these circumstances has an interest as a sensor. In excess cyanide conditions, the electrode is responding according to

$$Ag(CN)_2{}^- + e \rightleftharpoons Ag + 2CN^-$$

while beyond the endpoint the response is probably initially to

$$AgCl + e \rightleftharpoons Ag + Cl^-$$

Quite soon after the appearance of excess hypochlorite, the potential becomes constant, indicating that the silver chloride layer is insensitive to ClO^-. This means that although silver can be used as a cyanide indicator (it is in fact better for this than gold, because it is not so sensitive to solution agitation), it is of limited value as a chlorine indicator, being only useful for low chlorine concentrations.

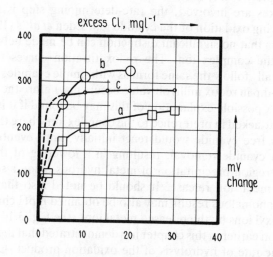

Fig. 3.12 — Relative response characteristics of potentiometric indicator electrodes in cyanide–hypochlorite systems containing excess available chlorine, Cl. a. Platinum indicator electrode. b. Gold indicator electrode. c. Silver indicator electrode. (After Mattock and Uncles 1962.)

Fig. 3.12 shows the relative response characteristics of the three metals in cyanide–hypochlorite systems containing an excess of available chlorine (Mattock and Uncles 1962). It is apparent that if monitoring is needed for conditions beyond the stage of full cyanide destruction (and this is in general desirable in treatment control), importance attaches to response in this region. Silver is clearly of limited value, and perhaps is only applicable to concentrations up to about 5 mg l^{-1} available chlorine. It must be stressed that in any case the responses are not sufficiently reproducible to be used as reliable concentration indicators (quite apart from the theoretical problems of interpreting potential measurements in terms of concentrations), and their value is only semi-quantitative at best. The millivolt change from the apparent reaction endpoint is however quite significant, and is extremely useful to indicate that cyanide has been treated. This leads on to another set of possible complications; but before these are discussed it is worth reviewing the response–time characteristics of the indicators.

Redox indicator electrodes do not always respond rapidly; and indeed, the less well defined the reaction system, or the greater the thermodynamic irreversibility, the slower is the response. With platinum and gold, there is an immediate response, but full equilibration takes some 10 minutes following hypochlorite addition, and approximately 3 minutes following introduction of cyanide to lower the potential. In other terms, these periods imply time constants of about 2 minutes and $\frac{1}{2}$ minute respectively. Silver is much faster, however.

All of the above studies refer to simple cyanide solutions. In the majority of waste waters, the cyanide is present in complex form with, for example, copper, cadmium, zinc or iron and sometimes nickel. In these circumstances, the chemistry of cyanide treatment by hypochlorite may change, and the redox titration curves may change also. In the simplest cases, for example those where copper, cadmium and zinc cyanide complexes are involved, the rate-determining step is probably still the hydrolysis following oxidation of the cyanide ion: Eden *et al.*'s (1950) experimental evidence suggests that no significant distinction can be made between direct attack and attack on the complex ion. The redox titration curves for metal cyanide complexes generally follow the same form as with simple cyanide, but it is frequently found that a pre-dip in redox millivolt value occurs just before the titration endpoint (see Fig. 3.11). A possible explanation of this may be found if it is assumed that the free CN^- ion is attacked in preference to the complex ion where there is a mixture of the two. Excess free cyanide would react initially, until eventually the complex dissociates from cyanide removal, resulting in a lowering of the millivolt value; around this dip region precipitation of metal hydrous oxide also occurs, which is consistent with metal ion release. It should be noted also that if the pH is not controlled then anomalous results may also be obtained if pH changes occur on the uptake of hydroxyl ions by the released metal ions (see Fig. 3.14).

The discussion earlier in this chapter has demonstrated that the overall reaction is slow because the rate of hydrolysis of the oxidation product, CNCl, is slow—not because the rate of oxidation of CN^- to CNCl is (it is not). The redox electrode does not, it may be assumed, monitor the hydrolysis reaction; but it is diverting to ruminate on the fact that the time constant for the electrode response for the hypochlorite addition is of the same order of magnitude as the time constants for the cyanide destruction process in comparable conditions. That aside, the fact that (a) the cyanide oxidation is not pH-dependent (within the pH region 10–13) and (b) the indicator electrode is, suggests that another influence is present. It is possible that the pH response is due to the formation of (molecular level) oxide films on the platinum indicator electrode. (In this respect, a gold electrode should be better, because it is less susceptible to oxidation; although its other disadvantages in the presence of excess cyanide means that it is rarely used for monitoring and control of the process.) A further interesting point of practical significance is that as the pH increases, so the response times tend to become longer. This may be due to partially to adsorption of 'poisons' on the electrode surface, because the electrode can be rejuvenated by cathodic cleaning. This is both a reductive and a desorptive procedure, carried out by applying a low voltage to the electrode, e.g. by connection to the negative terminal of a 4.5 V battery, with completion of the circuit by a simple wire or other connection, both electrodes being immersed in dilute sulphuric acid (pH 3–5). (As a corollary to this, it can be noted that the practice sometimes recommended of cleaning platinum electrodes with nitric acid will have the reverse effect, and will tend to induce oxidation of the surface.) Cathodic cleaning has to be used with intelligence: the process activates the surface and probably leaves some sorbed hydrogen, so a period of about 2–3 hours must be allowed for the electrode to re-equilibrate in the medium being measured. In general, cleaning procedures to be adopted for metal indicator electrodes follow those for glass electrodes given in Chapter 2 (p. 54)—although, of course, for silver the use of hydrochloric acid is to be avoided.

It was implied earlier that the apparent titration endpoint or inflection may not necessarily correspond to complete cyanide destruction. With more stable metal cyanide complexes the position is certainly more complicated. Fig. 3.13 shows

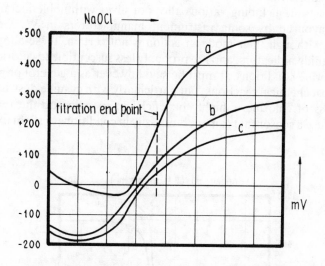

Fig. 3.13 — The effect of ferrous iron on the redox titration curve for the CN^-/OCl^- system using a platinum indicator electrode. a. 50 mg l^{-1} CN$^-$. b. 50 mg l^{-1} CN$^-$ plus 3 mg l^{-1} Fe^{2+}. c. 50 mg l^{-1} CN$^-$ plus 10 mg l^{-1} Fe^{2+}. Reference electrode saturated KCl calomel. (After Mattock 1967.)

titration curves with platinum electrodes for the treatment of cyanide in the presence of ferrous iron. It can be seen that the characteristic S-shapes are progressively flattened as the metal content increases (nickel has a similar effect). This could be ascribed to a diminution of the degree of oxidation, due to the resistances of nickelocyanide and ferrocyanide ions to chlorinative breakdown. With the former, the reaction is slow, and may be incomplete, while with the latter, oxidation proceeds only to ferricyanide. However, the curves are not fully consistent with these causes, because some cyanide destruction undoubtedly occurs where only trace amounts of iron substantially affect the titration curve. It is apparent that there is substantial interference with the redox response, although whether this is due to preferential adsorption of the complex species on to the electrode's active sites, as may be possible, is not clear. The consequence, however, is clear enough: it means that the platinum electrode is not a reliable indicator for the cyanide–hypochlorite reaction when nickel or iron is present.

There can be other interferences, such as those arising from the co-presence of different redox couples. With some electroplating solutions, for example, certain additives that may be present, e.g. aromatic aldehydes, may be pre-oxidised, and the titration curve can then show a peak or inflection before the cyanide treatment endpoint is reached. This can have important consequences in automatically controlled dosing systems, and can vitiate the use of platinum electrodes altogther. In these circumstances, the disadvantages of silver as an indicator are outweighed, and it represents a sensible alternative.

Other metal indicator electrodes have been employed to follow the cyanide–hypochlorite reaction, including various alloys of silver with noble metals, and these have been favoured by some instrument manufacturers in W. Germany and Switzerland. Although their mode of action is not certain, these alloys appear to behave as cyanide indicators, and are probably less susceptible to chloride poisoning in excess chlorine conditions. It must be said however as a general principle that in control of most chemical reactions, but particularly with cyanide destruction (where a slight excess of chlorine is highly beneficial in speeding up the reaction), it is important to use a sensor that can operate easily in conditions of slight excess dosing.

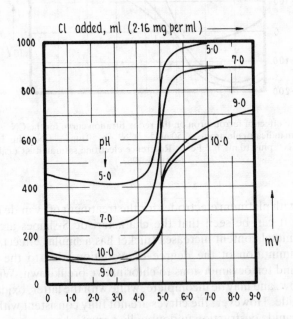

Fig. 3.14 — The effect of pH on the redox titration curve for the CN^-/OCl^- system using a platinum indicator electrode. Starting cyanide concentration = 87 mg l^{-1}. Reference electrode 3.8 M KCl silver–silver chloride. (Cook and Mattock 1986.)

To conclude this section, it is interesting to examine the behaviour of redox electrodes in the cyanide–hypochlorite systems at lower pH values than those conventionally employed, viz., 10–12. Within this region, one should expect that with a simple set of mechanisms there should be no significant pH effect, but that this is not so is shown in Fig. 3.14. It is known that at high pH values platinum, for example, exhibits some pH-responsive properties, which could confuse the picture, but quite apart from this there is a clear influence of pH on the redox curves down to pH values as low as 5. It is particularly interesting also that the titration endpoints are similar, even allowing 30 minutes reaction time, implying no variation in chlorine demand. This is inconsistent with the chemistry of chlorine uptake, as reviewed earlier in this chapter: a higher chlorine : cyanide ratio certainly occurs in neutral and acid conditions, but this is not reflected in the titration curves. More investigation of this is merited.

Chlorination of ammonia and metal ammonia complexes

The chemistry of these reactions has been reviewed earlier in the chapter. A recent study (Eilbeck and Mattock 1982, Eilbeck 1984) has demonstrated that at pH 8 redox control of the chlorination of both free and metal-complexed ammonia is feasible. The potential of a gold or platinum indicator electrode changes dramatically (400 mV increase) at the breakpoint when incremental amounts of chlorine (as sodium hypochlorite) are added to solutions of ammonia or of the ammines of Cu^{2+}, Ni^{2+}, Cr^{3+}, Zn^{2+}, Ag^+ or Cd^{2+}. This is illustrated in Fig. 3.15 together with the results of a parallel experiment in which the conventional breakpoint curve was determined under identical conditions of pH and temperature. It is significant that the electrometric endpoint coincides with the breakpoint of the lower curve in Fig. 3.15,

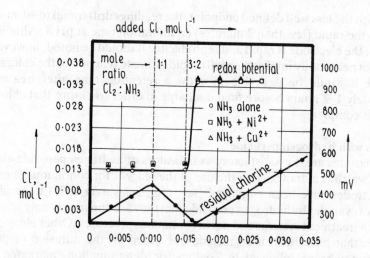

Fig. 3.15 — Redox potentials exhibited by gold indicator electrode during chlorination of ammonia and metal ammines, and breakpoint chlorination curve for ammonia under identical conditions. 0.01 M NH₃ starting concentration, 0.005 M metal ion concentrations, pH = 8.0, temperature = 20°C. Reference electrode saturated KCl calomel.

and the reader is referred to the original papers for a discussion of this. The measured redox potential is very dependent on pH, which also varies through the course of a titration, so for effective redox control simultaneous pH control is essential. Fig. 3.16 shows the much poorer response of the electrode at pH 10 where,

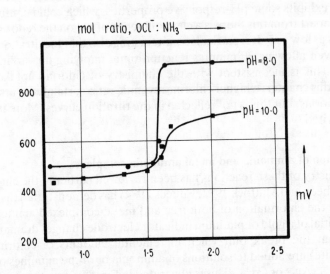

Fig. 3.16 — The effect of pH on the redox potential curves for a gold indicator electrode following the chlorination of ammonia. 0.01 M NH_3 starting concentration, temperature = 20°C. Reference electrode saturated KCl calomel.

in addition to the less well defined endpoint, the readings drift for up to 40 minutes in contrast to the rapid (less than 2 minutes) equilibration time at pH 8. Although the response of the electrode is rapid and reproducible it should be noted, however, that this does not necessarily imply an equally rapid ultimate reaction of the chloramines. In practice it would be desirable to allow a retention time after treatment of approximately 1–2 hours before final discharge in order to ensure that chloramine oxidation is complete.

Oxidations with hydrogen peroxide
Redox electrodes are not satisfactory as indicators of hydrogen peroxide electrode reactions, which is a drawback in the use of the reagent for continuous operations. Silver electrodes have been employed (see, e.g., Schwarzer 1975) to monitor the oxidation of cyanides by hydrogen peroxide, but they are apparently only suitable for use in batch treatment systems and not in continuous processes. Other electrochemical devices that have been used include one based on the diffusion of peroxide through a membrane followed by coulometric determination, marketed by the Envirotech Corporation in the USA, and another which measures the oxygen evolved from reaction with hypochlorite by an oxygen gas electrode (from Solvay, France). In the UK, a continuous colorimetric analyser based on the reaction of peroxide with potassium titanyl oxalate has been constructed by Laporte Industries.

Occasionally it may be possible to use secondary reaction effects to follow peroxide concentrations. For example, if NO_2^- is oxidised to NO_3^- in acid solution, a fall in pH is observed as the buffering effect of the relatively weak acid HNO_2 is removed as it is converted to the fully dissociated strong acid HNO_3. This permits the use of conventional pH electrodes to monitor the reaction (Rodenkirchen 1983). One example where redox monitoring is possible is in the removal of residual silver from photographic fix bath effluent (Sims 1983), where there is a large millivolt change at completion of silver/silver oxide deposition.

Oxidations with peroxymonosulphuric acid

Here again the preferred indicator in continental European practice is the silver electrode when the application is to cyanide oxidation. The silver acts as a cyanide indicator and would be subject to similar limitations here as when chlorine is the oxidant if chlorides are co-present with cyanide. The same system (silver electrode) can also be used to follow the oxidation of thiocyanates by peroxymonosulphate at pH values of 5.5 or higher.

Specific reagent or pollutant monitoring

It is not within the scope of this book to review the use of specific pollutant monitors such as are used to check water quality. Here it is relevant only to consider such devices as they may be appropriate to the control of treatment processes, and with this limitation there are remarkably few that have either realistic potential value or are used in practice. The problem with most specific (or selective) ion electrodes, e.g. those for metals and other anions of interest in the water pollution sphere, is that they are subject to a variety of interactions which it is necessary to suppress before the measurements can be trusted (see, e.g., Bailey 1976). This requirement demands relatively expensive associated instrumentation, the use of which is rarely justified. Possibly the most interesting are the sensing electrodes that employ a membrane to separate the sensor from the solution.

The oxygen electrode

Oxygen is reducible to hydroxide ions at a platinum or gold cathode by the application of a potential of between -700 and -1000 mV with respect to a silver/silver chloride reference anode. This potential ensures diffusion-controlled conditions but is below that which water is reduced. If the cathode potential is held constant, then the current delivered by such a cell with an unpolarised reference electrode will be linearly related to the rate at which oxygen or other reducible species diffuses to the cathode. In a given medium and at a fixed temperature,

$$i \propto pO_2$$

where i is the diffusion current, and pO_2 is the partial pressure of dissolved oxygen. The theoretical form of the relationship, upon which polarographic determinations are generally based, is discussed in Chapter 5 in the review of electrode processes. From a direct practical point of view the primary necessities are to ensure a stable diffusion layer thickness (by thorough constant agitation at the cathode surface), temperature control (or temperature compensation), avoidance of interfering side reductions, and the minimisation of electrode poisoning.

A major contribution to the construction of an oxygen electrode cell meeting these requirements made by Clark (1959), who introduced the use of a polyethylene film covering a platinum cathode (see Fig. 3.17a), the anode being a

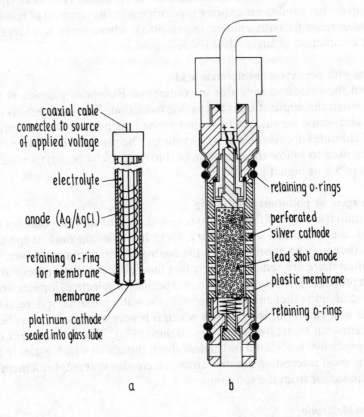

coaxial cable
connected to source
of applied voltage

electrolyte

anode (Ag/AgCl)

retaining o-ring
for membrane

membrane

platinum cathode
sealed into glass tube

retaining o-rings

perforated
silver cathode

lead shot anode

plastic membrane

retaining o-rings

a   b

Fig. 3.17 — The membrane oxygen electrode. a. Clark type. b. Mackereth design.

stable reference electrode that is not significantly polarised by the currents passing through the cell. The function of the membrane (PTFE, silicone and a variety of other materials have been used) is that it prevents the passage of ions to the cell, limiting transfer to substances, such as gases, that are membrane-permeable. It also assists in maintaining a defined diffusion layer at the cathode surface (although controlled stirring at the membrane is still needed), and significantly reduces poisoning of the cathode from species present in the sample being tested. The cell is thus established with an inert cathode and a stable reference anode (frequently a standard Ag/AgCl electrode) in a buffered electrolyte with a gas-permeable membrane separating the working cell from the test solution.

One of the first and certainly successful oxygen gas electrodes suitable for plant purposes was described by Mackereth (1964), commercial versions of which are still available. An illustration is given in Fig. 3.17. As in the cell described by Mancy and Westgarth (1962), the cathode of the Mackereth electrode is of silver and the anode is of lead. The potential developed galvanically between these two metals in the potassium carbonate/bicarbonate aqueous electrolyte is approximately 0.7 V, sufficient to effect diffusion-controlled oxygen reduction without the need for an externally applied polarising voltage as is necessary with the Clark electrode. The current output of a Mackereth electrode will clearly depend on the physical dimensions of the cathode and on the membrane thickness and permeability, but typical commercial cells give approximately 200 mA at 100% oxygen saturation, with a linear response from 0% to 200% and an accuracy of about ±5%.

A wide range of oxygen electrodes is now commercially available, mostly using platinum cathodes, although gold has also been utilised (Poole and Morrow 1977). Gold does, however, offer some advanges in being less subject to adsorptive poisoning, which can still occur even with membrane protection. Temperature-compensating bridge circuits are often built into the cells, and stirrers are sometimes also included to ensure a rapid flow of the test solution over the membrane surface to maintain reproducible diffusion-controlled conditions. Standardisation is carried out using solutions having essentially the same electrolyte composition as the sample to be tested. A zero oxygen content is first established by addition of a small amount of an oxygen-consuming reagent, such as sodium sulphite, to an appropriate water sample. A similar composition sample is saturated with air at a known pressure and the diffusion current measured. The oxygen concentration in the saturated sample is determined analytically, e.g. by titration, and a straight-line graph drawn between two diffusion current values against the two oxygen concentrations. (Tables for oxygen solubilities in pure water and sea water composition saline solutions are given in International Oceanographic Tables (1973), but these are not necessarily representative of the composition applying in most waste waters.) It should be appreciated that where the measurements are made at a barometic pressure different from the standardisation pressure, a correction factor needs to be applied. For most work where extreme pressure variations are not involved, a correction factor, F, calculated from

$$F = \frac{P}{P_s}$$

can be applied, where P is the air pressure in the sample determination, and P_s is the pressure used in standardisation:

$$[O_2] = FS$$

where S is the solubility determined at pressure P_s.

Regular maintenance is essential with oxygen electrodes to ensure that measurements are reliable. This includes particularly a programme of membrane cleaning, especially important in conditions of heavily fouled waters or where depositions can occur. Membranes should also be replaced at regular intervals. Standardisations should be carried out at least weekly, and after membrane cleaning or replacement.

Oxygen electrodes have found particular application in controlling aeration of activated sludge plants as well as in monitoring water quality, but they are also suited to any chemical aerobic systems.

For a fuller discussion of oxygen electrodes see Hitchman (1978).

The ammonia electrode

This, although also a membrane-covered electrode, works on a quite different principle from the oxygen electrode, and derives from the development of a carbon dioxide gas electrode by Stow *et al.* (1957) and Severinghaus and Bradley (1958). In this case (Bailey and Riley 1975), the membrane is applied over a pH glass electrode sandwiching a thin film of an ammonia–ammonium chloride buffer solution (see Fig. 3.18). As ammonia diffuses across the membrane, so the pH response of the glass

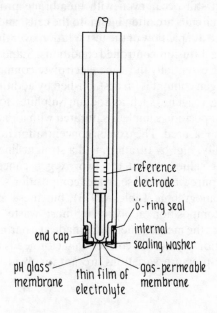

reference electrode

o-ring seal

internal sealing washer

end cap

pH glass membrane

thin film of electrolyte

gas-permeable membrane

Fig. 3.18 — The membrane ammonia or carbon dioxide electrode.

electrode reflects the ammonia partial pressure in the test solution as

$$E = E^{\ominus} - 2.303 \frac{RT}{F} \log[NH_3]$$

The ammonia electrode has been used for monitoring water qualities, but can also be applied to waste waters. Other developments based on the same principles measure sulphur dioxide and nitrogen oxides (see Bailey 1976).

THE PRACTICAL APPLICATION OF CHEMICAL OXIDATION TECHNIQUES

Reactor systems

The effective realisation of a chemical reaction in a plant situation is highly dependent on the design of the reactor system, as has been shown in Chapter 1.

Moreover, the choice between continuous and batch treatment has to be made on the basis of several considerations: the volumes to be treated, the degree of stringency of treatment imposed by the discharge limits, the time scale for completion of reaction(s), and the feasibility of control (automatic or manual).

With many oxidation processes, as this chapter has demonstrated, the time necessary to complete reactions can be substantial, ranging from 5–10 minutes to hours. It is rarely practical to entertain continuous treatment when the time for effective completion of the reaction exceeds 4–5 hours, because (particularly for large flow rates) this results in cumbersome systems; in these circumstances it is usually better to have two batch treatment systems, operated alternately.

A more significant problem arises when there is no effective means for monitoring the reaction process by instrumental methods, such that periodic analytical checks need to be made. This situation definitely points to the use of batch procedures, if only to ensure that treatment is completed.

Automatic control
The comments on the practicalities of applying automatic control in waste water treatment processes made in Chapter 1 are particularly well demonstrated by many oxidation processes. Consider, for example hydrolysis at pH 11 of cyanogen chloride formed by the oxidation of cyanide with hypochlorite for which the half-reaction time (see earlier in this chapter) is approximately

$$t_{\frac{1}{2}} = \frac{0.693}{k[10^{-3}]} = 693/k$$

At $10°C$, $k \simeq 200 \, l \, mol^{-1} \, min^{-1}$, implying a half-reaction time of 3.5 minutes, or, for 99% reaction, $t_{99} = 23$ minutes. In the presence of 10 mg l^{-1} excess available chlorine, the value of k increases to $1500 \, l \, mol^{-1} \, min^{-1}$, with reductions in $t_{\frac{1}{2}}$ and t_{99} to 0.5 and 3.1 minutes respectively. If we consider a continuous process then the residence time for 99% reaction can be calculated from CFSTR theory (Chapter 1, p. 30, equation (1.3)) as

$$t_{99} = \frac{1}{k[10^{-3}]} \cdot \left(\frac{100}{1} - 1\right) = \frac{99}{1500 \, [10^{-3}]} = 66 \text{ minutes}$$

This is a much longer time, which is to be expected, since the treated material is continually being mixed with untreated influent. If the process is considered to involve plug-flow rather than continuous mixed flow then the calculated residence time

$$t_r \text{ (plug-flow)} = \frac{1}{k[10^{-3}]} \log_e 100 = \frac{4.6}{1500[10^{-3}]} = 3.1 \text{ minutes}$$

which, as expected, is the same as calculated above from the kinetic treatment. In practice, of course, the actual operation of the process will fall between the two extremes of CFSTR and plug-flow conditions, and nominal retention times of 10–30 minutes are commonly used. The slowness of response of some platinum indicator electrodes also requires adequate retention times. There is a case for two-stage processing with automatic control of chemical dosing in the first reactor, followed by a second reaction completion stage, although this is rarely used in practice.

The same arguments apply in principle to redox control of chlorination of ammonia and amines, where although the redox response by gold and platinum electrodes is fast, the time scale for reaction completion is much longer, and is essentially a function of pH.

Oxygen electrodes, when used for automatic control of air oxygen dosing, are usually to be found in slow oxidation rate systems, and somewhat different operating conditions prevail. In the oxidation of sewage, for example, oxygen electrodes are sometimes used to ensure efficient use of oxidant, and to maintain an optimum level of dissolved oxygen within an activated sludge medium. Here again, as should generally be the case in effluent treatment, the system is established to ensure that there is always some excess of reagent, the difference here being that the reagent is not toxic.

Electrode poisoning

Redox and membrane electrodes are susceptible to physical poisoning by precipitate, grease or oil deposition in the same way that pH electrodes are, and the same care and cleaning procedures need to be employed. Suitable cleaning techniques have been indicated in Chapter 2 (p. 55) and on p. 161.

Reagent dosing

It is not appropriate here to go into detail on engineering aspects of dosing oxidation reagents, and only general observations of relevance need to be made.

The fact that many of the stronger oxidants in common use are highly corrosive means that care has to be exercised in the selection of equipment to ensure that appropriate materials of construction are incorporated. Generally, it is necessary to use plastics materials such as pvc, abs, pvdc, ptfe and the like: stainless steel in most grades is usually unsatisfactory, the only suitable metals normally being the titanium alloys, e.g. Hastelloy. This caveat applies to all components of a valve, e.g. electromagnetic or pneumatic (and its associated seals) or pump that is to be used to dispense reagent, and likewise applies to pipework and reaction and storage vessels. Mixer materials have to be chosen with care—frequently pvc-coated or rubber-coated steel shafts and blades are employed.

An important factor in the choice of reagents is cost. This chapter has reviewed the most important oxidation reagents in plant use, and some possible oxidants have not been included because they are not readily available or are expensive. Even within the range of reagents considered, only air, oxygen and chlorine/hypochlorite are extensively employed: the others are used only where they offer specific advantages by being more powerful oxidants, able to effect reactions where the others cannot. Hydrogen peroxide is finding favour in some otherwise difficult applications, but ozone is quite expensive in terms of the capital costs of equipment needed to generate it, while chlorine dioxide, although an extremely effective oxidant in many circumstances where other reagents will not work, presents severe safety problems.

This highlights an aspect of great significance in oxidation treatment—operating safety. Practically all of the very effective oxidants are by their very nature dangerous and demand stringent control for their safe transportation, storage and dosing. In recent years, for example, the use of chlorine gas (with the aid of commercial chlorinator equipment to form an aqueous solution for dosing) has diminished dramatically in favour of sodium (and sometimes calcium) hypochlorite, and a large part of the reason is safety. The problems of transportation on urban highways are always a factor, and *in situ* generation has much to commend it, provided the equipment needed is not prohibitively costly and demanding of special skills. Chlorine dioxide is generated *in situ* (suitable equipment is available commercially), as is ozone, and sodium hypochlorite can also be electrolytically generated in some small-demand applications (see Chapter 5, p. 218). Manufacturers provide detailed safety information on storage and handling of the chemicals they supply, and it cannot be emphasised too strongly how important it is that their recommendations be followed. This applies equally to the modes of dosing, reactor design and treatment plant operating procedures, all of which have to take into account the characteristics of the chemical being used, of the reaction products, and of course the materials being treated.

The final note to this chapter can be that in waste water oxidation treatment it is not only the chemistry that has to be considered when a reaction process is chosen: frequently what may appear on paper to be the most suitable choice can be discarded purely on the grounds of cost or operating risks. Overall effectiveness must be the goal, not individual theoretical arguments.

REFERENCES

Acher, A. J., and Rosenthal, J. (1977) *Wat. Res.* **11** 557.

Aieta, E. M., Berg, J. D., Roberts, P. V., and Cooper, R. C. (1980) *J. Wat. Pollut. Control Fed.* **52** 810.

Aloy, M., Folachier, A., and Vulliermet, B. (1976) *Tannery and Pollution,* Centre Technique due Cuir.

Anderson, R., and Greaves, G. F. (1983) *Wat. Poll. Control* **82** 18.

Andrews, C. C. (1980) *U.S. Govt. Rept. Announce Index* **80** 3561.

APHA/AWWA/WPCF (1985) *Standard Methods for the Examination of Water and Wastewater,* 16th edn, American Public Health Association, American Water Works Association, Water Pollution Control Federation.

API (1969) *API Manual on Disposal of Refinery Wastes,* American Petroleum Institute.

Arisman, R. K., Musick, R. C., Crase, T. C., and Zeff, J. D. (1980a) *A.I.Ch.E. Symp. Ser. No. 197* 169.

Arisman, R. K., Musick, R. C., Zeff, J. D., and Crase, T. C. (1980b) *Proc. 35th Ind. Waste Conf. Purdue Univ.* 802.

Atkins, P. F., Jr., Scherger, D. A., Barnes, R. A., and Evans III, F. L. (1973) *J. Wat. Pollut. Control Fed.* **45** 2372.

Autelo, J. M., Arca, F., Barbadillo, F., Casado, J., and Varela, A. (1981) *Environ. Sci. Technol.* **15** 912.

Baden, H. C., and Webster, N. Y. (1972) *U.S. Patent* No. 3, 772, 194.

Bailey, D. A., and Humphreys, F. E. (1967) *J. Soc. Leather Trades Chem.* **51** 154.

Bailey, P. L. (1976) *Analysis with Ion-Selective Electrodes,* Heyden.

Bailey, P. L., and Bishop, E. (1973) *J. Chem. Soc. Dalton Trans.* **9** 912; 917.

Bailey, P. L., and Riley, M. (1975) *Analyst* **100** 145.

Barnes, D., O'Hara, M., Samuel, E., and Waters, D. (1981) *Environ. Technol. Letters* **2** 85.

Basolo, F., and Pearson, R. G. (1967) *Mechanisms of Inorganic Reactions,* 2nd edn, Wiley.

Bauch, H., Burchard, H., and Arsovic, H. M. (1970) *Ges. Ing.* **91** 258.

Benson, D. (1968) *Mechanisms of Inorganic Reactions in Solution,* McGraw-Hill.

Benson, S. W., and Axworthy, A. E. (1959) *ACS Advances in Chemistry Ser. No. 21* 398.

Beszedits, S. (1980) *Am. Digest. Rept.* **69** 37.

Beychok, M. R. (1967) *Aqueous Wastes from Petroleum and Petrochemical Plants,* Wiley.

Bober, T. W., and Dagon, T. J. (1975) *J. Wat. Pollut. Control Fed.* **47** 2114.

Bober, T. W., Dagon, T. J., and Slovonsky, I. (1973) *U.S. Patent* No. 3, 767, 572.

Borkowski, B. (1967) *Wat. Res.* **1** 367.

Buescher, C. A., Dougherty, J. H., and Skrinda, R. T. (1964) *J. Wat. Pollut. Control Fed.* **36** 1005.

Cannon, R. D. (1980) *Electron Transfer Reactions,* Butterworths.

Chen, J. W. (1973) *A.I.Ch.E. Symp. Ser. No. 129* 61.

Chen, K. Y., and Morris, J. C. (1972) *Environ. Sci. Technol.* **6** 529.

Clark, L. C. (1959) *U.S. Patent* No. 2, 913, 386.

Cole, C. A., Ochs, D., and Funnell, F. C. (1974) *J. Wat. Pollut. Control Fed.* **46** 2579.

Connick, R. E., and Chia, Y.-T. (1959) *J. Am. Chem. Soc.* **81** 1280.

Cook, J. E., and Mattock, G. (1986) unpublished work.

Cuthbertson, A., and Robinson, E. (1980) *Anal. Proc. (London)* **17** 423.

Department of Environment (1979) *Methods for the Examination of Waters and Associated Materials. Dissolved Oxygen in Natural and Waste Waters,* Her Majesty's Stationery Office.

Dobson, J. G. (1947) *Sew. Wks. J.* **19** 1007.

Doré, M., Langlais, B., and Legube, R. (1978) *Wat. Res.* **12** 413.

Ebara-Infilco Co. (1981) *Jap. Patent* Nos. 81 44091; 81 44098.

Eden, G. E., and Wheatland, A. B. (1950) *J. Soc. Chem. Ind.* **69** 166.

Eden, G. E., Hampson, B. L., and Wheatland, A. B. (1950) *J. Soc. Chem. Ind.* **69** 244.

Eden, G. E., Freke, A. M., and Melbourne, K. V. (1951) *Chem. Ind. (London)* 1104.

Edwards, J. O. (1964) *Inorganic Reaction Mechanisms,* Benjamin.

Eigen, M., and Kustin, K. (1962) *J. Am. Chem. Soc.* **84** 1355.

Eigen, M., and Wilkins, R. G. (1965) *Mechanisms of Inorganic Reactions. ACS Advances in Chemistry Ser. No. 49* 55.

Eilbeck, W. J. (1983) unpublished work.

Eilbeck, W. J. (1984) *Wat. Res.* **18** 21.

Eilbeck, W. J., and Mattock, G. (1982) *Chem. Ind. (London)* 807.

Eisenhauer, H. R. (1965) *J. Wat. Pollut. Control Fed.* **37** 1567.

Eisenhauer, H. R. (1968) *J. Wat. Pollut. Control Fed.* **40** 1887.

Eisenhauer, H. R. (1971) *J. Wat. Pollut. Control Fed.* **43** 200.

Elmghari-Tabib, M., Laplanche, A., Venien, F., and Martin, G. (1982) *Wat. Res.* **16** 233.

Erndt, E., and Kurbiel, J. (1980) *Environ. Prot. Eng.* **6** 19.

Evans, F. L. (ed.) (1972) *Ozone in Water and Wastewater Treatment*, Ann Arbor Science.

Eye, J. D., and Clement, D. P. (1972) *J. Am. Leather Chem. Assoc.* **67** 256.

Fair, G. H., Geyer, J. C., and Okun, D. (1968) *Water and Waste Enginering*, Wiley.

Fenton, H. J. H. (1894) *J. Chem. Soc.* **65** 899.

Fenton, H. J. H. and Jones, H. O. (1900) *J. Chem. Soc.* **77** 69.

Fertik, H. A. (1978) *Inst. Soc. Am. Trans.* **17** 53.

Fertik, H. A., and Sharpe, A. (1980) *Inst. Soc. Am. Trans.* **19** 3.

Freund, T. (1960) *J. Inorg. Nucl. Chem.* **15** 371.

Frost, A. A. (1951) *J. Am. Chem. Soc.* **73** 2680.

Games, L. M., and Staubach, J. A. (1980) *Environ. Sci. Technol.* **14** 571.

Garrison, R. L., Prengle, H. W., Jr., and Mauk, C. B. (1975a) *U.S. Patent* No. 3, 920, 547.

Garrison, R. L., Prengle, H. W., Jr., and Mauk, C. B. (1975b) *Metal Progr.* **108** 61.

Gjertsen, L., and Wahl, A. C. (1959) *J. Am. Chem. Soc.* **81** 1572.

Gomaa, H. M., and Faust, S. D. (1971a) *Proc. 161st A. C. S. Meeting, Pesticide Chemistry Symp.*

Gomaa, H. M., and Faust, S. D. (1971b) *J. Agric. Food Chem.* **19** 302.

Gould, J. P., and Weber, W. J., Jr. (1978) *J. Wat. Pollut. Control Fed.* **48** 47.

Grainger, R. K., and Mattock, G. (1973) unpublished work.

Gurol, M. D., and Singer, P. C. (1982) *Environ. Sci. Technol.* **16** 377.

Gurol, M. D., and Singer, P. C. (1983) *Wat. Res.* **17** 1173.

Haber, F., and Weiss, J. (1934) *Proc. Roy. Soc. (London)* **A147** 332.

Hamilton, C. E., Teal, J. L., and Kelly, J. A. (1969) *U.S. Patent* No. 3, 442, 802.

Hann, V. A. (1983) *Ozone: Sci. Eng.* **5** 69.

Heist, J. A. (1974) *A.I.Ch.E. Symp. Ser. No. 136* 456.

Hewes, C. G., and Davison, R. R. (1973) *A.I.Ch.E. Symp. Ser. No. 129* 71.

Hillis, M. R. (1977) *Proc. 3rd Int. Symp. Ozone Technology*, Int. Ozone Inst.

Hitchman, M. L. (1978) *Measurement of Dissolved Oxygen*, Wiley–Interscience.

Hofmann, M. (1977) *Environ. Sci. Technol.* **11** 61.

Hoigné, J., and Bader, H. (1976) *Wat. Res.* **10** 377.

Hoigné, J., and Bader, H. (1983) *Wat. Res.* **17** 173.

Hoigné, J., Bader, H., Haag, W. R., and Stachelin, J. (1985) *Wat. Res.* **19** 993.

International Oceanographic Tables (1973) Vol. 2, National Institute of Oceanography of Great Britain/UNESCO.

Junkermann, H., and Schwab, H. (1978) *U.S. Patent* No. 4, 104, 162.

Katzer, J. R., Ficke, H. H., and Sadana, A. (1976) *J. Wat. Pollut. Cont. Fed.* **48** 921.

Kearney, P. C., Zeng, Q., and Ruth, J. M. (1984) *A.I.Ch.E. Symp. Ser. No. 259* 195.

Keating, E. J., Brown, R. A., and Greenberg, E. S. (1977) *Proc. 32nd Waste Conf. Purdue Univ.* 464.

Kennedy, M. V., Stojanovic, B. J., and Shuman, F. L., Jr. (1972) *J. Agric. Food Chem.* **20** 341.

Khandelwal, K. K., Barduhn, A. J., and Grove, C. S. (1959) *ACS Advances in Chemistry Vol. 21. Kinetics of Ozonation of Cyanides.*

Kieszkowski, M., and Krajewski, S. (1968) *Trait. Surface* **9** No. 82 3.

Kimura, M., Suzuki, T., and Ogata, Y. (1980) *Bull. Soc. Japan* **53** 3198.

Kinoshita, S., and Sumada, T. (1972) *Advances in Water Pollution Research. Proc. 6th IAWPR Symp. 607.* Pergamon.

Kochi, J. K. (1973) *Free Radicals,* Vol. II, Wiley.

Konbeck, E. (1977) *U.S. Patent* No. 4, 012, 321.

Krebs, H. A. (1929) *Biochem. Z.* **204** 343.

Ladbury, J. W., and Cullis, C. F. (1958) *Chem. Rev.* **58** 403.

Lawes, B. C., Fournier, L. B., and Mathre, O. B. (1973) *Plating* **60** 902.

Legue, B., Langlais, B., and Doré, M. (1981) *Wat. Sci. Technol.* **13** 553.

Li, K. Y., and Kuo, C. H. (1980) *A.I.Ch.E. Symp. Ser. No. 197* 142.

Light, T. S. (1972) *Anal. Chem.* **44** 1038.

Longley, K. E., Moore, B. E., and Sorber, C. A. (1980) *J. Wat. Pollut. Control Fed.* **52** 2098.

Mackay, K. M., and Mackay, R. A. (1981) *Introduction to Modern Inorganic Chemistry,* International Textbook Co.

Mackereth, F. J. H. (1964) *J. Sci. Inst.* **41** 38.

Mancy, K. H., and Westgarth, W. C. (1962) *J. Wat. Pollut. Control Fed.* **34** 1037.

Martell, A. E., and Sillén, L. G. (1964) *Stability Constants. Chem. Soc. Spec. Pub. No. 17.*

Martin, J. L., and Rubin, A. J. (1977) *Proc. 32nd Ind. Waste Conf. Purdue Univ.* 814.

Masschelein, W. J. (1979) *Chlorine Dioxide,* Ann Arbor Science.

Mattock, G. (1967) *Proc. 19th Ont. Ind. Waste Conf.* 39, Ontario Water Resources Commission.

Mattock, G., and Uncles, R. W. (1962) *The 1962 Effluent and Water Treatment Manual,* Thunderbird, 182.

Mauk, C. E., and Prengle, H. W., Jr. (1976) *Pollut. Eng.* **8** 42.

McAuley, A., and Hill, J. (1969) *Quart. Rev.* **23** 18.

Meunier, L., and Kapp, M. (1931) *Cuir tech.* **20** 365.

Moses, D. V., and Smith, E. A. (1954) *U.S. Patent* No. 2, 690, 425.

Murphy, J. S., and Orr, J. R. (1975) *Ozone Chemistry and Technology,* Franklin Institute.

Nebel, C., and Stuber, L. M. (1976) *Proc. 2nd Int. Symp. Ozone Technology* 336, Int. Ozone Inst.

Nebel, C., Gottschling, R. D., Holmes, J. L., and Unangst, P. C. (1976) *Proc. 31st Ind. Waste Conf. Purdue Univ.* 940.

Netzer, A. (1976) *Progr. Wat. Technol.* **8** 25.

Ng, K. S., Mueller, J. C., and Walden, C. C. (1978) *J. Wat. Pollut. Control Fed.* **50** 1742.

Norrish, R. G. W., and Wayne, R. P. (1965) *Proc. Roy. Soc.* **A288** 361.

Novak, F., and Sukes, G. (1981) *Ozone: Sci. Eng.* **3** 61.

O'Neill, E. T., Castrantas, H. M., and Keating, E. J. (1978) *Proc. 33rd Ind. Waste Conf. Purdue Univ.* 471.

Paschke, R. A., Hwang, Y. S., and Johnson, D. W. (1977) *J. Wat. Pollut. Control Fed.* **49** 2445.

Patterson, J. W. (1975) *Wastewater Treatment Technology,* Ann Arbor Science.

Pauling, L. (1960) *The Nature of the Chemical Bond,* 3rd edn, Cornell Univ. Press.

Peleg, M. (1976) *Wat. Res.* **10** 361.

Peyton, G. R., Huang, F. Y., Burleson, J. L., and Glaze, W. H. (1982) *Environ. Sci. Technol.* **16** 448; 454.

Poole, R., and Morrow, J. (1977) *J. Wat. Pollut. Control Fed.* **49** 422.

Posselt, H. S. (1972) *Physicochemical Processes for Water Quality Control,* Weber, W. J., Jr., 392, Wiley–Interscience.

Posselt, H. S., Anderson, F. J., and Weber, W. J., Jr. (1968) *Environ. Sci. Technol.* **2** 1087.

Prengle, H. W., Jr., and Mauk, C. E. (1978) *A.I.Ch.E. Symp. Ser. No. 178* 228.

Prengle, H. W., Jr., Mauk, C. E., Legan, R. W., and Hewes, C. G., III (1975) *Hycrocarbon Processing* **54** 82.

Pressley, T. A., Bishop, D. F., and Roan, S. G. (1972) *Environ. Sci. Technol.* **6** 622.

Price, C. T., Larson, T. E., Beck, K. M., Harrington, F. C., Smith, L. C., and Stephanoff, L. (1947) *J. Am. Chem. Soc.* **69** 1640.

Prober, R., Melnyk, P. B., and Mansfield, L. A. (1977) *Proc. 32nd Ind. Waste Conf. Purdue Univ.* 17.

Rav-Acha, Ch., and Blitz, R. (1985) *Wat. Res.* **19** 1273.

Reutskii, V. A., Ovechkin, V. S., Sazhin, B. S., and Zhuravleva, T. Yu (1981) *Khim. Tekhnol. Vody* **3** 169.

Rice, R. G., and Browning, M. E. (1981) *Ozone Treatment of Industrial wastewater,* Noyes Data Corp.

Rice, R. G., and Netzer, A. (eds.) (1982) *Ozone Technology and its Practical Applications,* Ann Arbor Science.

Robinson, E. (1976) *Brit. Patent* No. 1, 436, 700.

Robinson, E. (1978) *New Processes of Waste Water Treatment and Recovery,* ed. Mattock, G., Ellis Horwood.

Rodenkirchen, M. (1983) *Galvanotechnik (Saulgau)* **74** 924.

Röhrer, E. (1977) *Chem. Ind. (London)* 816.

Rolla, E., and Chakrabarti, C. L. (1982) *Environ. Sci. Technol.* **16** 852.

Rossotti, F. J. C., and Rossotti, H. (1961) *The Determination of Stability Constants,* McGraw-Hill.

Roth, J. A., and Sullivan, D. E. (1983) *Ozone: Sci. Eng.* **5** 37.

Roth, J., Moench, W. L., Jr., and Debalek, A. (1982) *J. Wat. Pollut. Control Fed.* **54** 135.

Schmitt, R. J. (1982) *Proc. 36th Ind. Waste Conf. Purdue Univ.* 375.

Schwarzer, H. (1975) *Galvanotechnik (Saulgau)* **66** 22.

Selm, R. P. (1959) *Ozone Chemistry and Technology,* American Chemical Society.

Severinghaus, W., and Bradley, A. F. (1958) *J. Appl. Physiol.* **13** 515.

Shambaugh, R. L., and Melnyk, P. B. (1978) *J. Wat. Pollut. Control Fed.* **50** 113.

Sheppard, J. C., and Wahl, A. C. (1957) *J. Am. Chem. Soc.* **79** 1020.

Sidwick, J. M., and Barnard, R. (1981) *Chem. Ind. (London)* 277.

Sims, A. F. E. (1981) *Effl. Wat. Treat. J.* **21** 109.

Sims, A. F. E. (1983) *Chem. Ind. (London)* 555.

Singer, P. C., and Gurol, M. D. (1983) *Wat. Res.* **17** 1163.

Snider, E. H., and Porter, J. J. (1974) *J. Wat. Pollut. Control Fed.* **45** 886.

Stackelln, J., and Hoigné, J. (1982) *Environ. Sci. Technol.* **16** 377.

Stein, G., and Weiss, J. (1951) *J. Chem. Soc.* 3265.

Stern, E. W. (1971) *Transition Metals in Homogeneous Catalysis,* ed. Schrauzer, G. N., Marcel Dekker.

Stewart, R. (1964) *Oxidation Mechanisms,* Benjamin.

Stewart, R., and Van der Linden, R. (1960) *Can. J. Chem.* **58** 2237.

Stirling, C. J. M. (1965) *Radicals in Organic Chemistry,* Oldbourne Press.

Stow, R. F., Baer, R. F., and Randall, B. F. (1957) *Arch. Phys. Med. Rehabil.* **38** 646.

Stumm, W., and Lee, F. G. (1961) *Ind. Eng. Chem.* **53** 143.

Sullivan, D. E., and Roth, J. A. (1980) *A.I.Ch.E. Symp. Ser. No. 197* 142.

Taube, H. (1965) *J. Gen. Physiol.* **49**, Pt 2, 29.

Taube, H. (1970) *Electron Transfer Reactions of Complex Ions in Solution,* Academic Press.

Teletzke, G. G. (1964) *Chem. Eng. Progr.* **60** (1) 33.

Teramoto, M., Imamura, S., Yatagai, N., Nishikawa, Y., and Teranishi, H. (1981) *J. Chem. Eng. Japan* **14** 383.

Tokyo Electric Co. (1972) *Jap. Patent* No. 9, 075, 481.

Twigg, G. H. (1962) *Chem. Ind. (London)* 4.

Ventron Technology (1977) *Factsheet No. 2,* Ventron Technology (Cheshire, UK).

Wajon, J. E., Rosenblatt, D. H., and Burrows, E. P. (1982) *Environ. Sci. Technol.* **16** 396.

Weber, W. J., Jr., (1972) *Physicochemical Processes for Water Quality Control,* Wiley–Interscience.

Weiss, J. (1935) *Naturwiss.* **23** 64.

Zabban, W., and Helwick, R. (1980) *Plat. Surf. Finish.* **67** 56.

Zeevalkink, J. A., Visser, D. C., Arnoldy, P., and Boelhauser, C. (1980) *Wat. Res.* **14** 1375.

4

Chemical reduction

REDUCTION REACTIONS IN POLLUTION ABATEMENT

It was pointed out in Chapter 3 that one of the objectives of chemical oxidation is to accelerate the natural process occurring in an aerobic aquatic environment that lead eventually to the oxidative degradation of pollutant material. Reduction processes would in this sense be undesirable, in that they would create an undesirable C.O.D. loading. However, reduction reactions are employed in waste water treatment, where the aim is usually to remove the pollutant by phase transfer, e.g. to solid form, as in the reductive recovery of metals: or through precipitation of an insoluble material that can be disposed to a landfill site; or to gaseous form, as in the reduction of nitrate to nitrogen. It is of course necessary to ensure that an excess of reductant is not discharged, so in practice careful control has to be applied in treatment operations.

In chemical terms, distinction between oxidation and reduction reactions is without meaning, and the theoretical discussions in Chapter 3 apply equally in this chapter: they will therefore not be reiterated here. Nevertheless, despite their chemical inseparability from oxidation processes, reduction reactions are different in practice from the point of view of waste water treatment, and so justify separate consideration.

MONITORING AND CONTROL OF REDUCTION REACTIONS

The general principles outlined in Chapter 3 concerning the application of sensors to oxidation reactions apply in like manner to reduction processes. Thus the principal tool is the potentiometric redox electrode cell, and its application is based on the characteristics of the redox curve for the reaction being followed. So far as direct monitoring of reactants is concerned, there is a dearth of suitable plant electrodes. The sulphide-ion selective electrode has some application in anaerobic biological processes; and a sulphur dioxide membrane electrode, based on diffusion of the gas through a membrane to a sulphite–bisulphite electrolyte to provide pH changes that can be measured with a glass electrode (cf the ammonia and carbon dioxide membrane electrodes), can be used for monitoring reactions using SO_2 as a reductant, and is commercially available; but except for the oxygen electrode (discussed in Chapter 3) there is little else. Secondary reaction changes that may occur, such as those of pH, can be used as secondary indicators; and of course colorimetric reactions can be employed to monitor a reaction. These latter are, however, rarely used on-line in waste water treatment.

Redox curve characteristics as they relate to reactions of interest are considered below within the various sub-sections with particular reductants.

COMMON REDUCTANTS AND THEIR APPLICATIONS

Sulphur dioxide and its salts
General properties
Sulphur dioxide is an easily liquefied gas (boiling point approximately $-10°C$) with a characteristic pungent odour; it is both unpleasant and physiologically harmful as a choking gas, and consequently demands care in use. It is also corrosive to metals when moist. Commercial supply is as a liquid under pressure, in cylinders or drums. For application it is commonly dosed by means of a sulphonator (similar to the chlorinators used for chlorine), where an aqueous solution of the gas is prepared automatically for injection into the solution being treated.

The gas is moderately soluble in water, giving acidic solutions which are weakly reducing:

$$SO_4^{2-} + 4H^+ + (x-2)H_2O + 2e \rightarrow SO_2.xH_2O \qquad E^{\ominus} = +0.17 \text{ V} \qquad (4.1)$$

The acid solutions of SO_2 have been referred to as solutions of sulphurous acid (H_2SO_3). However, there is no evidence for the existence of this species in any significant concentration, and the equilibria in aqueous solutions are best described in terms of the equations

$$SO_2 + xH_2O \rightleftharpoons SO_2.xH_2O$$

$$SO_2 \cdot xH_2O \rightleftharpoons HSO_3^- + H_3O^+ + (x-2)H_2O$$

The first dissociation constant of 'H_2SO_3' is therefore properly defined as:

$$K_1 = \frac{a_{HSO_3^-} \cdot a_{H^+}}{a_{total\ diss.\ SO_2} - a_{HSO_3^-} - a_{SO_3^{2-}}} = 1.3 \times 10^{-2}$$

Aqueous solutions of sulphur dioxide react with alkalis to give two series of salts: the hydrogen sulphites (HSO_3^-) and the sulphites (SO_3^{2-}). The hydrogen sulphite ion undergoes dimerisation to give $S_2O_5^{2-}$, metabisulphite:

$$2HSO_3^- \rightleftharpoons S_2O_5^{2-} + H_2O$$

Connick (1982) has recently re-measured the equilibrium constant for this dimerisation and obtained a value of $8.8 \times 10^{-2}\,l\,mol^{-1}$; there is clearly a preponderance of the HSO_3^- form in solution.

Sodium salts for all these species are commercially available and all are used as reducing agents in particular waste water treatment processes as alternatives to sulphur dioxide: their advantages are a greater ease of handling and cheaper dosing equipment for storage and pumping of the solutions than applies with sulphur dioxide. The solutions are, however, similarly corrosive, and smell of sulphur dioxide, and so demand similar care in use, with well-ventilated environments. Sodium metabisulphite and sodium sulphite are available commercially as solids, but $NaHSO_3$ does not exist as a solid, and is sold as a concentrated solution.

Aqueous solutions of sulphites are stronger reducing agents than aqueous solutions of sulphur dioxide. The pH effect is quite marked, the reducing power increasing with increasing pH:

$$SO_4^{2-} + H_2O + 2e \rightleftharpoons SO_3^{2-} + 2OH^- \qquad E^\ominus = +0.93\ V \qquad (4.2)$$

(cf. equation (4.1)). In view of this it might at first be supposed that all reductions using these reagents should be carried out at high pH. The true situation is not so straightforward however, and it must be remembered that the ease of reduction of the substrate species may also be pH-dependent. The pH dependence of the oxidation and reduction half-reactions may well operate in reverse directions, e.g., in acid solution:

$$Cr_2O_7^{2-} + 14H^+ + 6e \rightleftharpoons 2Cr^{3+} + 7H_2O \qquad E^\ominus = +1.33\ V$$

whereas in alkaline solution

$$CrO_4^{2-} + 4H_2O + 3e \rightleftharpoons Cr(OH)_3 + 5OH^- \qquad E^\ominus = -0.13\ V$$

Hence, although SO_2 (as SO_3^{2-}) in alkaline solution is much better reducing agent than it is in acid solution (as $SO_2{\cdot}xH_2O$), Cr^{VI} is reduced with much greater difficulty in alkaline solution. This is also obvious if one calculates the overall free energy change for the reduction of Cr(VI) by SO_2 in acid and in alkaline solution from the respective standard electrode potentials using the equation

$$\Delta G^\ominus = -nE^\ominus F$$

as described in Chapter 3.

Bisulphite ions, a common reagent form for sulphur(IV) reductions, can engage in reduction–oxidation reactions via either one-electron or two-electron free radical steps, as for example with ferric iron, Fe^{3+}:

$$\cdot SO_3 + H^+ \xrightarrow{\quad H_2O \quad} SO_4^{2-} + 3H^+$$

$$HSO_3^- \quad \xrightarrow{-2e}$$

$$\xrightarrow{-e} \quad \cdot HSO_3 \quad \xrightarrow{\quad} S_2O_6^{2-} + 2H^+$$

$$\cdot HSO_3 \quad \xrightarrow[H_2O]{Fe^{III}} \quad Fe^{II} + SO_4^{2-} + 3H^+$$

In fact one-electron paths are usually followed, and the stoicheiometry is variable, with the nett reaction being a combination of

$$HSO_3^- + 2Fe^{3+} + H_2O \rightarrow 2Fe^{2+} + 3H^+ + SO_4^{2-}$$

and

$$2HSO_3^- + 2Fe^{3+} \rightarrow 2Fe^{2+} + 2H^+ + S_2O_6^{2-}$$

The relative importance of these two reactions in deciding the overall stoicheiometry depends on the conditions under which the reaction occurs. In the presence of a large excess of Fe(III) the first reaction predominates, with a mechanism that may be

$$Fe^{3+} + HSO_3^- \rightleftharpoons Fe(HSO_3)^{2+}$$

$$Fe(HSO_3)^{2+} \rightarrow Fe^{2+} + \cdot HSO_3$$

$$HSO_3 + Fe^{2+} \rightarrow Fe^{3+} + HSO_3^-$$

$$HSO_3^- + Fe^{3+} + H_2O \rightarrow SO_4^{2-} + Fe^{2+} + 3H^+$$

Where there is a deficiency of Fe(III), competition for the $\cdot HSO_3$ radical occurs via the reaction

$$2 \cdot HSO_3 \rightarrow 2H^+ + S_2O_6^{2-}$$

giving dithionate as a reaction product. This can account for practical observations of variable bisulphite consumption in waste water treatment, since the reaction course will be determined to some extent by the mode of reagent dosing and by mixing efficiencies. However, it must also be borne in mind that sulphite oxidation occurs also simply from the presence of oxygen, the more so the lower the pH. This demands that for efficient reaction the reactor vessels should be closed. Taylor and Qasim (1984) studied the effects of oxygen interference in chromate reduction by sulphite and showed it can be significant at pH values below 4.

Applications
Reduction of hexavalent chromium
This is one of the major uses for sulphur(IV) compounds in waste water treatment, where reduction to Cr(III) is necessary to permit subsequent precipitation of the hydrous oxide by pH adjustment. Cr(VI) is a common pollutant in metal-finishing waste waters, where it is not precipitated simply by pH adjustment.

Two ionic forms exist in solution, in labile equilibrium with each other: CrO_4^{2-} predominates in alkaline solution. $HCrO_4^-$ and $Cr_2O_7^{2-}$ co-exist between pH 6 and pH 2, and below pH 1, H_2CrO_4 is the main species:

$$H_2CrO_4 \rightleftharpoons HCrO_4^- + H^+ \qquad\qquad K = 4.1$$

$$HCrO_4^- \rightleftharpoons CrO_4^{2-} + H^+ \qquad\qquad K = 10^{-5.9}$$

$$Cr_2O_7^{2-} + H_2O \rightleftharpoons 2CrO_4^- + 2H^+ \qquad K = 10^{-2.2}$$

As indicated earlier, acid solutions of Cr(VI) are more readily reduced than alkaline ones, and with SO_2 or its salts the reaction is normally carried out in the pH range 2–3. Fig. 4.1, from Chamberlin and Day (1956), shows the relative effects of pH on the rate of reduction of Cr(VI) by SO_2 in the presence of excess SO_2: above pH 2–3 the reaction slows significantly.

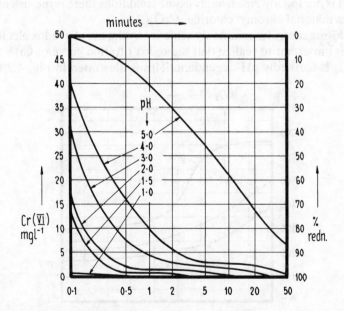

Fig. 4.1 — The effect of pH on rate of reduction of chromium(VI) by sulphur dioxide, in the presence of excess SO_2 (after Figure 5 in Chamberlin and Day 1956).

Acid is consumed in the reaction, the overall stoicheiometry being represented according to the reagent employed by the equations

$$Cr_2O_7^{2-} + 3SO_2 + 2H^+ \rightarrow 2Cr^{3+} + 3SO_4^{2-} + H_2O$$

$$Cr_2O_7^{2-} + 3NaHSO_3 + 8H^+ \rightarrow 2Cr^{3+} + 3NaHSO_4 + 4H_2O$$

$$Cr_2O_7^{2-} + 3Na_2SO_3 + 8H^+ \rightarrow 2Cr^{3+} + 3Na_2SO_4 + 4H_2O$$

$$2Cr_2O_7^{2-} + 3Na_2S_2O_5 + 16H^+ \rightarrow 4Cr^{3+} + 6NaHSO_4 + 5H_2O$$

(although it should be appreciated that in the pH range 2–3 the predominant reducing species is $SO_2 \cdot H_2O$, there being little undissociated HSO_3^- present). The reduction mechanism may involve free radicals (Haight *et al.*, 1965, Benson 1968), as already reviewed above for Fe(III). Here there is also likely to be a variation in the stoicheiometry, with competing nett reactions that may be represented by

$$2HCrO_4^- + 4HSO_3^- + 6H^+ \rightarrow 2Cr^{3+} + 2SO_4^{2-} + S_2O_6^{2-} + 6H_2O \qquad (4.3)$$

$$2HCrO_4^- + 3HSO_3^- + 5H^+ \rightarrow 2Cr^{3+} + 3SO_4^{2-} + 5H_2O \qquad (4.4)$$

Reaction (4.4) may predominate, particularly in the presence of excess Cr(VI), but with excess HSO_3^- the former route could be followed, resulting in less efficient use of the reagent. Good control of reagent dosing is therefore called for. Of course it is also essential to apply pH control simultaneously to maintain reaction speed (because without acid addition the pH will drift upwards from acid consumption in the reaction). Sulphuric acid is commonly employed, although hydrochloric can be used if the pH is not too low: in strongly acidic conditions there is the risk of forming the dangerous material chromyl chloride, CrO_2Cl_2.

Reagent dosing control is commonly achieved by the use of a redox electrode cell. However, it is important to realize that the redox titration curve of Cr(VI) against HSO_3^- or SO_2 is markedly pH-dependent. This is illustrated in Fig. 4.2 (Mattock

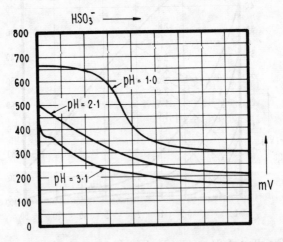

Fig. 4.2 — The effect of pH on the redox titration curve for the Cr(VI)/HSO_3^- system using a platinum indicator electrode. Reference electrode saturated KCl calomel. (After Mattock and Uncles 1962.)

and Uncles 1962), and demonstrates, independently of reaction considerations already stated, the need for simultaneous pH and redox control in carrying out the reaction. Electrode response is also affected by pH (Mattock and Uncles 1962), slowing as the pH increases (see Fig. 4.3). It is noteworthy that the titration curve flattens above pH 5 to virtual uselessness, which together with the slower electrode response is consistent with the reaction rate falling to negligible magnitude in neutral and alkaline solutions. Comparison of the reaction rates indicated in Fig. 4.1 with the electrode response characteristics shown in Fig. 4.3 suggests that both are of the same order, which may indicate that the redox electrode follows the reaction rapidly. The implications for automatic treatment of a continuously flowing stream in a CFSTR vessel (discussed in Chapter 1) are that the reactor retention time should not be less than approximately 10 minutes at pH values of up to 2.0, and more at high values.

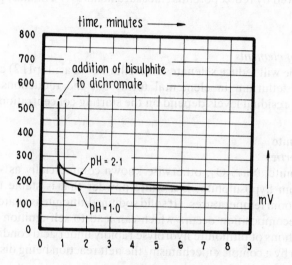

Fig. 4.3 — The effect of pH on speed of response of platinum redox indicator electrode in the $Cr(VI)–HSO_3^-$ system, using excess bisulphite reagent. Reference electrode saturated KCl calomel. (After Mattock and Uncles 1962.)

In strongly acid conditions (pH < 1.5) platinum redox electrodes frequently show symptoms of being poisoned, which may be ascribable to adsorption of Cr(VI) species blocking the active exchange sites. As with alkaline poisoning (discussed in Chapter 3, for high pH cyanide chlorination) the remedy may be cathodic cleaning (the technique for which is also discussed in Chapter 3, p. 160), but sometimes even this is only a short-term remedy, particularly where the chromium concentrations are high. Gold can be used as an alternative indicator, being apparently less readily poisoned. Control is most suitably applied at the lower knee of the curve, with a slight excess of reductant to ensure full Cr(VI) reduction, bearing in mind the fact that with significant excess bisulphite conditions there could be a disproportionately higher reagent consumption (from reaction (4.3) occurring).

Dechlorination

Dissolved chlorine in excess of any acceptable concentration level may be removed by reduction with sodium bisulphite or sulphur dioxide. The reaction is

$$OCl^- + HSO_3^- \rightarrow Cl^- + HSO_4^-$$

and can be followed by a redox electrode system.

Reduction of permanganate
Permanganate is reduced by bisulphite to manganese dioxide. the reaction (over the pH range 3–12) can be written

$$2MnO_4^{2-} + 3HSO_3^- + 2H^+ \rightarrow 2MnO_2 + 3HSO_4^- + H_2O$$

It can be followed by redox potential measurements (W. Thomas, personal communication, 1984).

Precipitation of elements
Sulphur dioxide will reduce selenates and tellurates in acid (pH 2) conditions to give selenium and tellurium in elemental form, but the reductions are not always complete, and residual levels depend on the starting concentrations.

Sodium dithionite
General properties
Sodium dithionite, $Na_2S_2O_4$, otherwise known commercially as sodium hydrosulphite or sodium hyposulphite, is a white powder that is stable under anhydrous conditions at room temperatures. (If solid sodium dithionite is heated to 150°C, rapid exothermic decomposition occurs which can lead to self-ignition (Bostian 1981).) Aqueous solutions of dithionite hydrolyse rapidly under acid conditions (Burlamacchi *et al.* 1969) by a complex mechanism, the nett reaction being disproportionation:

$$2S_2O_4^{2-} + H_2O \rightarrow S_2O_3^{2-} + 2HSO_3^- \tag{4.5}$$

In alkaline solution hydrolysis is slower, and some sulphide is also produced:

$$3S_2O_4^{2-} + 6OH^- \rightarrow 5SO_3^{2-} + S^{2-} + 3H_2O \tag{4.6}$$

Under alkaline conditions sodium dithionite solutions are powerful reducing agents, reacting according to

$$S_2O_4^{2-} + 4OH^- \rightarrow 2SO_3^{2-} + 2H_2O + 2e \tag{4.7}$$

the oxidation potential for which reaction is $+1.12$ V. In acid solutions, however, the reducing power of dithionite is much less, and hydrolytic decomposition (equation (4.5) above) occurs competitively, resulting in non-stoicheiometric reductions and complex products. Atmospheric oxygen oxidises dithionite solutions producing, initially, a mixture of hydrogen sulphate and hydrogen sulphite ions according to the equation

$$S_2O_4^{2-} + H_2O + O_2 \rightarrow HSO_3^- + HSO_4^-$$

In the presence of β-anthraquinone sulphonate (Fieser's solution), this property is enhanced, and is the basis of a well-known procedure for removing traces of oxygen from gases. It is obvious, therefore, that reagent solutions of dithionite should be protected from atmospheric oxygen so far as is reasonably practicable.

Applications
Reduction of hexavalent chromium
Sodium dithionite will reduce Cr(VI) to Cr(III) in alkaline solutions (in contrast to SO_2 and its salts), and this fact is extensively used in the electroplating industry to remove excess chromic acid from chromium-plated surfaces by immersing them in a solution containing $Na_2S_2O_4$ and Na_2CO_3 (pH approximately 11.5). Reduction occurs according to the equation

$$2CrO_4^{2-} + 3S_2O_4^{2-} + 2OH^- + 2H_2O \rightarrow 2Cr(OH)_3 + 6SO_3^{2-}$$

The advantage of this reaction in waste water treatment is that with it there is no need to lower the pH of neutral or alkaline solutions containing chromate to achieve reduction prior to precipitation of the hydrous oxide. A treatment operation is thus eliminated, and reagent costs are reduced (although the extra cost of sodium dithionite offsets the extent of savings on acid, bisulphite and alkali). However, reduction is not necessarily complete: as much as 20 mg l^{-1} Cr(VI) can remain after treatment of 2000 mg l^{-1} Cr(VI) at pH 9 even with a 3-fold excess of dithionite over the theoretical requirement (J. E. Entwistle, personal communication, 1984). Another disadvantage is the fact that the reaction cannot be followed in alkaline conditions by 'measurement of the redox potential: the titration curve is nearly horizontal, showing little or no inflection in the particularly useful pH region 7–10. As a consequence it is necessary to add a large excess of reagent to ensure an adequate degree of reaction completion, limiting sensible application to batch treatment operations or to the continuous treatment of dilute or known stable concentrations of Cr(VI).

Precipitation of metals and insoluble metal salts
The potential for dithionite reduction in alkaline solution (equation (4.7)) suggests that the reagent should be capable of reducing several types of metal ion to the metal—notably silver, nickel and copper (see Appendix 1 for details of metal–metal ion half-reaction standard potentials)—and in fact this occurs. In practice, the metal when formed may be in a colloidal condition, and so requires further treatment for separation. A polyacrylamide polyelectrolyte may be used, if necessary also with a coagulating agent such as ferric or aluminium sulphate (Göckel 1972), but of course with the latter reagents some of the potential recovery benefits are lost. Alternatively, the solution can be heated to 50°C (Dart Industries 1971).

The reduction procedure can be useful when the metal ions concerned are bound with complexing agents, rendering normal pH adjustment for precipitation of the hydrous oxide inadequate for full treatment. In the presence of complexes, the reaction may be slow in alkaline conditions, and a lower pH condition may be needed if in such an environment the complex dissociates. However, at lower pH values, the reducing power of dithionite is less, so a compromise has to be adopted. Eilbeck and Potter (1980) found that with copper co-present with ammonia and citrate it is nevertheless possible to operate at a pH of 5, giving residual soluble copper concentrations of only a few milligrams per litre. Control of the reaction was possible with redox electrodes, the millivolt shift at the equivalence point being of the order of 300–400 mV. The precipitate formed consisted of a mixture of metallic copper, cuprous oxide and copper sulphide. The formation of sulphide does represent some disadvantage in relation to sludge disposal if further processing, e.g. for metal recovery, is not practised. Nickel can also be removed from its complexed condition, e.g. Ni–ethylenediamine, by dithionite addition. Again, some sulphide is formed.

As may be expected, gold and mercury salts are readily reduced to the metals with dithionite. Selenium and tellurium can also be precipitated (in finely divided form) from selenate and tellurate in neutral or acid conditions, although not necessarily completely.

Ferrous iron
General properties
Ferrous iron is well-known reducing agent in analytical chemistry, and its oxidation to the ferric condition has been studied exhaustively (see Chapter 3 for discussions). The standard electrode potential for

$$Fe^{3+} + e \rightleftharpoons Fe^{2+}$$

is $+0.77$ V, and Fe^{2+} thus provides intermediate reducing power in terms of the more widely used reagents. Ferrous ion is oxidised by air more readily in alkaline than acid conditions (see Chapter 3), although its selective use as a reductant for pollutants can be more efficient in acid conditions.

The properties of ferrous salts suitable for reduction reactions are described in Chapter 6 (p. 287): it is important to maintain acid solution conditions to prevent hydrolysis and retard oxidation by air during reagent storage.

Applications
Reduction of hexavalent chromium
Ferrous iron readily reduces chromates in both acid and alkaline conditions. The general equations for reduction are:

acid conditions:

$$Cr_2O_7^{2-} + 6Fe^{2+} + 14H^+ \rightarrow 2Cr^{3+} + 6Fe^{3+} + 7H_2O$$

alkaline conditions:

$$CrO_4^{2-} + 3Fe^{2+} + 4OH^- + 4H_2O \rightarrow Cr(OH)_3 \downarrow + 3Fe(OH)_3 \downarrow$$

As already discussed earlier in this chapter, the reduction consumes acid, but when it is carried out with ferrous iron in alkaline medium there is a nett alkali demand, due to the precipitation of the chromic and ferric hydrous oxides. This is of some significance in the practical application of the reaction in waste water treatment, because if reduction is carried out at, say, pH 3 (the condition usually applied with SO_2 or HSO_3^- reduction), and then the pH is raised to 9–10 to precipitate the hydrous oxides in the usual way, there is an initial acid demand followed by an alkali demand (including the requirements for both pH elevation and the precipitation). However, if alkaline conditions are maintained, throughout, only an alkali requirement exists, and one that is in principle less than in the other case. It may therefore be more economic to carry out the reduction in alkaline conditions, which is also sensible because the hydrous oxides are then simultaneously precipitated in one instead of two treatment operations. Obviously, where the waste water is already alkaline, there is no need to reduce the pH, and separation of Cr(VI) streams from others containing precipitable metal ions is unnecessary.

There is, however, a penalty in this approach, in that the reduction is relatively slow in alkaline conditions. This fact is reflected in the response characteristics exhibited by redox electrodes: whereas equilibrium millivolt readings can be obtained in $\frac{1}{2}$–2 minutes in the pH region 1–3 during execution of a redox titration curve, at pH 9 the time increases to approximately 30 minutes. This means that the retention time in a CFSTR vessel operating at pH 9 needs to be at least an hour, as opposed to 5–10 minutes when the control pH is 3.

Another factor of importance in operating automatic redox control on reagent dosing is the magnitude of the millivolt change at the titration endpoint. Fig. 4.4, depicting equilibrium redox titration curves for the Cr(VI)–Fe(II) system at different pH values, shows how the endpoint millivolt change reduces as the pH increases.

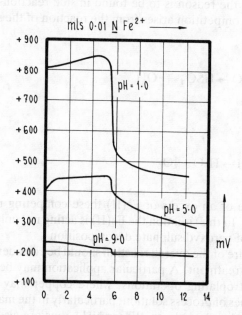

Fig. 4.4 — The effect of pH on the equilibrium redox potential curve for the Cr(VI)–Fe(II) system using a platinum indicator electrode. Starting Cr(VI) concentration = 100 mg l^{-1}. Reference electrode 3.8 M KCl silver–silver chloride. (Cook and Mattock 1986.)

There are advantages in using ferrous iron as a reductant for Cr(VI), including cheapness (with respect to bisulphite or sulphur dioxide), and an improved settlement quality in the hydrous oxide flocs eventually produced, but the extra sludge also generated must be borne in mind.

Reduction of peroxydisulphate
Ferrous iron reduces the peroxydisulphate (persulphate) ion, $S_2O_8^{2-}$, by a mechanism which may be initially

$$Fe^{2+} + S_2O_8^{2-} \rightarrow Fe^{3+} + \cdot SO_4^{-} + SO_4^{2-}$$

followed by a rapid reaction of the sulphate radical ion with another Fe(II) ion:

$$\cdot SO_4^{-} + Fe^{2+} \rightarrow Fe^{3+} + SO_4^{2-}$$

It would therefore appear that the reaction is simply two one-electron transfers, resulting in an overall stoicheiometry involving 2 mol Fe(II):1 mol $S_2O_8^{2-}$. However, stoicheiometric ratios as low as 1:1 have been observed, and, as with the oxidation of HSO_3^{-}, the reason is to be found in side reactions competing for the radical. In this case competition arises from the reaction of the sulphate radical ion with water:

$$\cdot SO_4^{-} + H_2O \rightarrow HSO_4^{-} + \cdot OH$$

followed by

$$\cdot OH + \cdot OH \rightarrow H_2O + \tfrac{1}{2}O_2$$

Thus in the absence of an excess of Fe(II) these competing reactions lead to an evolution of oxygen. In this situation the Fe(II) is acting not only as a reductant, but also as an initiator of peroxydisulphate decomposition.

Clearly this feature of the overall reaction should be considered in its operational use in waste water treatment. A particular application may be in the treatment of effluents from electroplating operations, where copper may be complexed with $S_2O_8^{2-}$ in certain types of process solution, particularly in the manufacture of printed circuit boards and related products. Where NH_4^+ ions are also present, decomposition may be by chlorinative oxidation (see Chapter 3, p. 133); otherwise Fe(II) can be used to reduce the persulphate. From the foregoing mechanism evaluation the optimum control in treatment would be to ensure that reaction mixing is efficient by introducing initially a stoicheiometric deficiency of ferrous iron, and then to follow this by a second-stage addition to complete the reaction if necessary. Redox control would thus be applied to provide initially a redox condition that is on the oxidised portion of the titration curve; as reaction proceeds the redox millivolt value will fall.

Removal of mercury

Mercury salts are reduced to the elemental state by reaction with ferrous sulphate at pH values above 5, which constitutes one of the many approaches available for removing this metal from waste waters. However, as is often the case in reductions of metals, the product is largely colloidal, and additional procedures have to be used for breaking the colloid, e.g. by heat, or addition of coagulants and/or polyelectroytes.

Metals
Reductive properties

It is well known that electropositive metals such as iron, zinc and aluminium will reduce the ions of metals below them in the electrochemical reactivity series. (A discussion of the principles is given in Chapter 3, and the subject is reviewed by Power and Richie (1975).) The table of electrode potentials in Appendix 1 gives $E^{\ominus}$ values for the reaction

$$M^{n+} + ne \rightarrow M$$

and so provides a means of identifying the likelihood of an exchange reaction.

Probably the most familiar example is the deposition of copper metal from solution on to the surface of steel:

$$Cu^{2+} + Fe \rightarrow Fe^{2+} + Cu$$

The process is generally known as 'cementation', and is used to some extent commercially for copper recovery. Application of the principle to waste water streams has some attractions, because not only does it remove undesirable heavy metals but can also present recovery opportunities. Another and potentially worth-while use of metals is as reductant for traces of refractory toxic organic materials, where significant success has been reported in dealing with waste water streams (see below).

Iron
Removal of metals from solution

Cementation of copper on to iron has been studied by several authors. The rate of deposition may be assumed to be controlled by the rate of diffusion of copper ions to the surface of the metal, with an activation energy of approximately 21 kJ mol^{-1} under most conditions (Nadkarni et al. 1967, Nadkarni and Wadsworth 1967, Miller and Beckstead 1973). However, the first-order rate constant for the reaction depends on the concentration of cupric ions, which conflicts with a diffusion-controlled mechanism. This is explicable in terms of the deposited copper retarding the iron dissolution rate, particularly when dense deposits are formed, as demonstrated by Schalch et al. (1976). If the deposits are rough and powdery the deposition rate is faster, as it also is at elevated temperatures. From the normal considerations applying to diffusion-controlled mechanisms (see Chapter 5), optimisation of mass transfer and thus deposition should occur by providing agitation or stirring at the metal surface; and enhanced mass transfer also occurs in packed bed systems (Kubo *et al.* 1979, Bravo de Nahui *et al.* 1986). However, if the conditions give rise to a smooth deposit there can be a detrimental effect, because the rate-controlling step can then be the anodic dissolution of the base metal. Patterson and Jancuk (1977) observed that excess iron dissolution above the anticipated stoicheiometric level may occur, presumably by secondary galvanic action between the deposited copper and the iron substrate. This excess is, however, reduced by agitation (through flaking of copper, from the surface), and by maximising the surface area:volume ratio of the iron. Clearly the conditions for efficient application need to be controlled carefully.

Southgate and Grindley (1945) used a horizontal continuous flow tank containing baled scrap iron wire in two compartments to remove copper from a copper mill effluent. Deposited sludge fell to the bottom of each compartment, from which it was collected periodically. The sludge contained about 30% Cu, 1–2% Fe, water and other impurities such as soap and grease from the manufacturing operations, so substantial refining would have been required to recover the copper as a usable product.

Higher concentrations of copper can be expected to give rougher deposits than dilute ones, and so should ensure a faster initial reaction rate. pH control is also desirable; a value of 4 is probably optimal, since at lower values competitive attack by H^+ ions will dissolve the iron, while at higher ones copper hydrous oxides will form to a significant extent. Southgate and Grindley used retention times of 1–4 hours in their experiments, although beyond 1 hour little improvement was gained. Copper recoveries of the order of 95–99% can be achieved, for example from starting concentrations of 500 mg l^{-1}. Small amounts of chloride improve the cementation rate (Yao 1944), but higher concentrations can retard it.

Arcilesi *et al.* (1983) have described equipment for copper removal consisting of replaceable canisters containing steel wool. They found that the technique is usefully applicable to waste waters containing copper in complexed form with EDTA and Quadrol, as for example from electroless plating operations, after prior adjustment of the pH to 4. An influent copper concentration of 58 mg l^{-1} was reduced to 0.5–0.1 mg l^{-1}, with an effluent Fe concentration of 60 mg l^{-1}, but it is not clear from their description what retention times were used. The potential usefulness of the iron cementation reaction for treating solutions containing complexed metals has also been emphasised by Mayenkar and Lagvankar (1984), who applied it to chelated nickel solutions, although these authors acknowledged that reaction rates are slow.

Silver ions, as may be expected from the position of silver in the electrochemical series, are readily reduced by iron, and this is used in the photographic industry to recover silver. Steel wool shavings are held in a bed through which the silver-containing solution is passed after adjustment of the pH to between 4 and 6.5. The efficiency is not necessarily high, and residual silver levels of 50–100 mg l^{-1} may obtain, offering limited advantage over electrolytic methods (Chapter 5), and often demanding post-treatment, e.g. with hydrogen peroxide (see later, p. 190).

Reduction of hexavalent chromium

Metallic iron can be used to reduce Cr(VI) to Cr(III) in acidic conditions. In a laboratory study, Grainger and Mattock (1974) investigated the effects of pH and residence time on the reduction rates and efficiencies using an upward flow columnar reactor. At pH values of 1–2 the reaction rate is fast, but substantial quantities of hydrogen are evolved from the competitive acid reaction on the iron, while above pH 5 the reaction slows down significantly; a pH of 3–4 would appear to be optimal. Residence times of up to 10 minutes are appropriate. Table 4.1 summarises results

Table 4.1 — Results of plant experiments on Cr(VI) reduction by iron turnings in a columnar reactor

pH	Influent chromate concentration, mg l^{-1}	Residence time, s	Flux through bed, m h^{-1}	%	Effluent chromate concentration mg l^{-1}
1.0	79.4	313	2.36	100	<0.02
1.0	79.4	10.3	72.9	99	0.79
2.0	70.3	5.6	123.0	96.2	2.81
3.2	27	17.6	41.19	99.4	0.16
3.2	27	9.2	80.2	92.4	2.0
5.3	70.4	60.4	12.3	80.9	13.5
5.3	70.4	136.4	5.40	82.5	12.3
5.3	70.4	716	1.04	90.3	7.0
5.3	70.4	11.6	63.7	29.4	49.7
4.1	36.5	6.7	11.1	83.3	6.9
4.1	36.5	64.8	1.13	98.3	0.62
4.1	36.5	9.2	80.2	36.1	23.4
4.1	36.5	57.2	12.9	80.2	7.22

Diameter of column = 50 mm.   Depth of iron turnings in column = 220 mm.

obtained on rinse water samples taken from an electroplating plant, and Fig. 4.5 shows the influence of retention time on efficiency of chromate reduction. An advantage of metallic reduction of Cr(VI) in this manner would be that reagent dosing and associated automatic control equipment is not necessary, making plant operation simpler.

Degradation of refractory organic materials

A particularly interesting application for metals has been described by Wolfe *et al.* (1977) and by Sweeney (1981), who used catalytically activated metal powders to reduce chlorinated hydrocarbon residues. These materials can be highly resistant to conventional oxidation techniques (although attack is possible with strong oxidants-

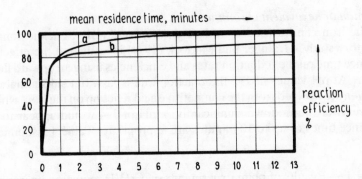

Fig. 4.5 — Efficiency of reduction of Cr(VI) by metallic iron as a function of nominal residue time in columnar reactor. a. Starting Cr(VI) concentration 36.5 mg l^{-1}, pH= 4.1. b. Starting Cr(VI) concentration 70.4 mg l^{-1}, pH= 5.3.

—see Chapter 3). The attractiveness of the process is that it is apparently relatively inexpensive and simple.

Sweeney suspended catalytically activated metal powder, e.g. of iron (although other metals such as zinc and aluminium may be used) in a 2 m deep sand-diluted column bed at weight concentrations of 5–40%, and passed filtered simulated waste solutions through at rates of 0.02–0.12 m^3 min^{-1} m^{-2}, corresponding to contact times of 16.7–100 minutes. Fluidised bed reactors can alternatively be used in upward flow, with two- to three-fold bed expansion, at flow rates of up to 0.9 m^3 min^{-1} m^{-2}, when pre-filtration is not necessary. Hexachlorocyclopentadiene and p-nitrophenol were reduced from 50 μg l^{-1} and 1000 μg l^{-1} to 0.1 μg l^{-1} and 1.5 μg l^{-1} concentrations respectively. A pH of between 5 and 8 is described as being appropriate.

Other applications tested by Sweeny were to trihalomethanes; C_2HCl_2, $C_2H_2Cl_2$ and $C_2H_3Cl_3$, where reductions from 250 μg l^{-1} to <5 μg l^{-1} were observed; PCB wastes; chlorinated phenoxyacetic acid derivatives; di- and trinitrophenols; N-nitrosodimethylamine (to detection limits); and an s-triazine derivative.

Wolfe et al. (1977) reported reductions down to levels as low as 0.01–1 μg l^{-1} from starting concentrations of up to 500 μg l^{-1} of chlorinated pesticides. The method is claimed to be effective on aldrin, chlordane, dieldrin, endrin, heptachlor, kepone, DDT, and lindane.

Aluminium, magnesium and zinc
Aluminium, magnesium and zinc are more reactive than iron, as may be anticipated from their electrochemical potentials, and are particularly effective as powders presenting a large surface area.

Removal of metals from solution
Zinc dust will reduce mercury compounds to the elemental state, and Richard and Brookman (1971) have described a process in which a solid stable complex is formed with an excess of zinc that can be removed by conventional separation methods. Static or fluidised beds of Al, Mg or Zn powder are capable of reducing other metals,

such as Au, Ag, Pd, Pt, Cu, Cd, Pb, Cr and Ni, to degrees dependent on operating conditions. Aluminium foil, strip or thread coated with a few microns of Al_2O_3 has been recommended for recovering noble metals as well as copper and zinc (Wallace 1977, 1978). Mixtures of reductants, e.g. Al powder plus sodium dithionite, can also be effective, and may be used where metal complexes are present, e.g. Ni–ethylene-diamine. One problem can be the formation of colloidal products (as for example in the well-known case of reduction of gold salts), but polyelectrolytes and if necessary coagulants can be used to break these down to provide a filterable material.

Reduction of hexavalent chromium
Zinc dust will reduce Cr(VI) in acid conditions. Hisamatsu and Kitajima (1968) studied the reaction in the pH range 2–3, finding that the rate is faster at lower pH values (cf. Fe reduction, described earlier), but slower at higher Cr(VI) concent-rations. In practical waste water treatment the use of zinc, or even aluminium, is less attractive than alternatives, being relatively expensive and of course generating an additional toxic component in the waste.

Hydrazine
Hydrazine, N_2H_4, has been used for several types of waste water treatment because of its powerful reducing properties, but it does present certain hazards. Hydrazine and its alkyl derivatives are used in the aerospace industry as rocket fuels, where its reactivity provides a powerful source of energy. A recent book by Schmidt (1984) presents a comprehensive review of the properties of hydrazine and its derivatives. Residues of hydrazine can be regarded as polluting (the substance is also potentially carcinogenic), and removal of excess material may be necessary, e.g. by oxidation (see Chapter 3, pp. 117, 145).

General properties
Pure hydrazine, N_2H_4 (melting point 2°C; boiling point 114°C) is a colourless, fuming liquid that burns in air, evolving a great amount of heat. In view of the extreme reactivity of the substance, it is normally supplied to industry and used as an aqueous solution of the hydrate, or as the sulphate or chloride salts. In the latter context it should be noted that hydrazine is a bifunctional base and two series of hydrazinium salts are obtainable:

$$N_2H_4 + H_2O \rightleftharpoons N_2H_5^+ + OH^- \qquad (pK = 6.07)$$
$$N_2H_5^+ + H_2O \rightleftharpoons N_2H_6^{2+} + OH^- \qquad (pK = 15.05)$$

It is apparent from the above equilibrium constants that $N_2H_6^{2+}$ salts will be extensively hydrolysed in aqueous solution, although $N_2H_5^+$ salts (the usual form of supply) are stable. Aqueous solutions of hydrazine and its salts are powerful reducing agents, e.g. as shown by

$$N_2 + 4H_2O + 4e \rightleftharpoons 4OH^- + N_2H_4 \qquad E^\ominus - 1.16 \text{ V}$$

Under some circumstances hydrazoic acid or metal azides may be produced when hydrazine or its salts react with oxidising agents, e.g.

$$N_2H_5^+ + HNO_2 \rightarrow HN_3 + H^+ + 2H_2O$$

Azides of the heavy metals, (e.g. lead, mercury) are explosive and subject to detonation on impact (hence their use in percussion caps and detonators). It is important, therefore, to consider the possibility of the formation of such hazardous materials before employing hydrazine as a treatment reagent.

Applications

One advantage that hydrazine offers as a waste water reactant is that it does not increase the dissolved salt level on decomposition, the reaction products being nitrogen and water. This compares favourably with other reductants, and can be of assistance when salinity levels in discharging waters have to be controlled tightly.

Recovery of metals

Hydrazine has been used in the treatment of metal-finishing effluents, e.g. those containing copper, by reduction to the cuprous state in alkaline medium. It has been applied in the Lancy 'integrated' treatment process (see Chapter 1, p. 36) involving closed loop recirculation of a first rinse stage after acid copper plating, whereby excess hydrazine is not discharged directly to waste, but only as a much diluted residue by dragout from the first rinse to a following one. The sludge is dense (more so than that formed with sodium dithionite, for example) and is yellow in the presence of a small amount of excess hydrazine, turning to orange with a large excess. Typical soluble copper residuals are of the order of 10 mg l^{-1}, but with an excess of the reagent this may rise to 20–30 mg l^{-1}.

Hydrazine is effective in precipitating metal compounds from their complexes. Cuprammine sulphate solutions, for example, after adjustment to pH 6 and treatment with hydrazine, will deposit a blue-violet precipitate after 4–30 minutes, leaving a copper residual concentration of the order of 1 mg l^{-1} (Ebara-Infilco Co. 1973). The process may be accelerated by heating to about 60°C, and is claimed to be effective for treating rinse waters from electroless copper plating (Kampermann 1973). Mercury is also efficiently reduced to the metallic state, but in both processes colloidal products are formed.

Reduction of hexavalent chromium

Hydrazine will reduce Cr(VI) compounds in alkaline conditions:

$$4CrO_4^{2-} + 3N_2H_5^+ + 5H^+ \rightarrow 4Cr(OH)_3 \downarrow + 3N_2 + 4H_2O$$

Acid is consumed in the reaction, which will therefore cause the pH to rise in dilute solutions.

Main application of the reaction has been in the Lancy closed loop rinsing system, and particularly in the treatment of rinse waters following chromate passivation operations in metal finishing. Use of bisulphite reductant in acid conditions would reduce the passivate film on the metal surface, but hydrazine in alkaline conditions does not cause this.

Sodium borohydride
General properties

Sodium borohydride, $NaBH_4$, is an extremely powerful reducing agent, as is indicated by the standard electrode potential of -1.24 V for the reaction

$$B(OH)_4^- + 4H_2O + 8e^- \rightleftharpoons BH_4^- + 8OH^-$$

It will reduce many metal ions, either as simple aquo-ions or as complexes, to lower oxidation states or even to the elemental metal. The solid, a white crystalline powder, is moderately soluble in water (25 g per 100 g H_2O at 0°C), but the solutions, although stable at high pH, undergo rapid hydrolysis in neutral or acid conditions, liberating hydrogen according to

$$BH_4^- + 4H_2O \rightarrow B(OH)_3 + OH^- + 4H_2$$

Since OH^- is released during the reaction it is both to be expected and in fact observed that the rate of decomposition falls as the reaction proceeds. Stability data for aqueous solutions of sodium borohydride are summarised in Figs. 4.6 and 4.7. It

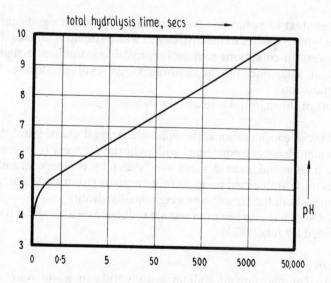

Fig. 4.6 — Hydrolysis of sodium borohydride as a function of pH, based on data given by Wade (1981).

is also noteworthy that the hydrolytic decomposition of the alkaline solutions, although normally slow, is catalysed by traces of transition metal ions and is accelerated at higher temperatures. Commercially, sodium borohydride is available

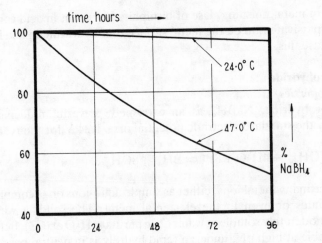

Fig. 4.7 — Hydrolysis of sodium borohydride as a function of temperature (reproduced from
Wade 1971).

either as a dry powder or as a stabilised solution in aqueous sodium hydroxide. In this
context it has been reported (Wade 1981) that an aqueous solution containing 12%
$NaBH_4$ in 41% NaOH decomposes by only approximately 0.002% per day at 54°C.

In terms of its potential as a waste water treatment reagent sodium borohydride
has several drawbacks:

(1) In the context of pollution abatement the addition of significant quantities of
    soluble borate species to the discharged waste stream is undesirable.
(2) Under certain conditions sodium borohydride introduces a significant fire or
    explosion risk due to the hydrogen gas evolved during its hydrolytic
    decomposition.
(3) The cost of the material is relatively high.

In view of these considerations, the most widespread use of borohydride in waste
water treatment has been connected with material recovery (e.g. precious metals),
precipitation of metals where these are bound by complexes, and in removing
particularly undesirable contaminants (e.g. mercury) from waste water streams prior
to discharge, where the benefits outweigh the disadvantages.

A useful summary review of inorganic reductions with sodium borohydride has
been provided by Jula (1974).

Applications
The principal application of sodium borohydride in waste water treatment, as
already mentioned, is in recovery of valuable metals. The reaction for a monovalent
metal such as silver can be written

$$8Ag^+ + BH_4^- + 2H_2O \rightarrow 8Ag \downarrow + BO_2^- + 8H^+$$

with corresponding equations for polyvalent metals. Metals whose compounds are reducible include cobalt, copper, gold, iridium, lead, mercury, nickel, palladium, platinum, rhodium and silver (Cook and Lander 1981). Although in many cases the metal is precipitated as the element, in the case of cobalt and nickel compounds the corresponding borides, Co_2B and Ni_2B, may also be formed. The reaction is somewhat pH-dependent, usually being initially faster at pH 6.5–8, but more complete at pH 10–11: a minimum reaction time of 15–30 minutes is desirable. The reaction can be followed by redox measurements or, with batch treatments (commonly used), can be considered complete when the pH remains stable for 5–10 minutes.

Interest has been shown in the reagent as a means of removing mercury from chloralkali cell waste waters. Mercury compounds are reduced to the metallic state in alkaline solution, and some of this settles out. A significant proportion is formed in a colloidal condition, however, and has to be removed for efficient operation. Coulter and Bell (1974) have suggested the use of an inert gas at an elevated temperature as a carrier to remove the particles, while De Angelis *et al.* (1978) propose filtration with powdered anthracite coal.

Another application has been for the reductive treatment of organic lead compounds appearing in waste waters from the manufacture of tetraethyl lead compounds (Lores and Moore 1973).

Organic reductants
Mention may be made of various organic reductants that may be employed, particularly for removal of metals from solution. Examples include aldehydes (formaldehyde, acetaldehyde) and sugars (dextrose, sucrose). Apart from cost, a major disadvantage of such reagents is that they introduce alternative pollution into wastes by creating a C.O.D., which must then be satisfied.

Hydrogen peroxide
Hydrogen peroxide, normally regarded as an oxidant (see Chapter 3 for a full discussion of the properties of the compound), does also show reducing properties, particularly in alkaline conditions:

$$O_2 + 2H^+ + 2e \rightarrow H_2O_2 \qquad E^\ominus = +0.695 \text{ V}$$
$$O_2 + H_2O + 2e \rightarrow HO_2^- + OH^- \qquad E^\ominus = -0.076 \text{ V}$$

Applications
Dechlorination
Although bisulphite is a convenient reagent for dechlorination (see earlier in this chapter), hydrogen peroxide does offer some advantages, in that it introduces no additional salts and generates oxygen as a reaction product, which is of course beneficial in waste waters; any excess rapidly decomposes to give oxygen and water. It is not, however, an effective dechlorination reductant in all circumstances.

In aqueous solution chlorine disproportionates according to

$$Cl_2 + H_2O \rightarrow Cl^- + HOCl + H^+$$

Hypochlorous acid, HOCl, is a weak acid with a pK of 7.25, providing on dissociation

$$HOCl \rightleftharpoons OCl^- + H^+$$

Hydrogen peroxide reacts rapidly with OCl^-, but only slowly with HOCl. This means that where HOCl predominates hydrogen peroxide will be unsuitable for dechlorination. Thus the pK value indicates that at pH 6.25 the acid will be only 10% dissociated, although it will be 90% so at pH 8.25, so alkaline conditions are called for. Furthermore, in highly acid conditions there is a tendency for the hydrogen peroxide to oxidise chloride back to chlorine:

$$2Cl^- + H_2O_2 + 2H^+ \rightarrow Cl_2 + 2H_2O$$

Other problems may arise if the solution also contains chloramines, since these are attacked only very slowly by hydrogen peroxide. Significant loss of reagent by decomposition occurs before adequate dechlorination is achieved.

Reduction of hexavalent chromium
Hydrogen peroxide will reduce Cr(VI) compounds provided the pH is maintained in a slightly alkaline condition (7–8.5):

$$2CrO_4^{2-} + 3H_2O_2 + 2H_2O \rightarrow 2Cr(OH)_3 \downarrow + 3O_2 + 4OH^-$$

The reaction appears to have been little studied, although a patent claim (Furukawa Electric Co. 1973) refers to it as a means of reducing Cr(VI) to Cr(III) with simultaneous precipitation of the hydrous oxide.

Reduction of metal complexes
It has been reported by Sims (1981) that silver complexed with thiosulphate, as in photographic fixing solutions, can be removed by treatment with hydrogen peroxide. Initial electrolysis (see Chapter 5) will remove most of the silver, but the residues can be precipitated as a silver–silver oxide sludge with hydrogen peroxide. This is a reduction reaction, although Sims also states that the C.O.D. is simultaneously reduced by about 50% from 50 000–70 000 mg l^{-1} starting levels. The reaction can be followed by redox millivolt measurement, a fall of approximately 350 mV occurring at the endpoint. The pH also drops to a value of 1–2. Residual silver concentrations quoted by Sims are of the order of 4 mg l^{-1} from starting levels of 350 mg l^{-1} and 900 mg l^{-1}.

Sulphamic acid (amidosulphonic acid)
Sulphamic acid, $NH_2.SO_3H$, is an amide of sulphuric acid, and can be represented by the structure

$$H_3N^+ .SO_3^-$$

It is crystalline (melting point 206°C), and readily soluble in water. Hydrolysis occurs in acid solutions, by the mechanism

$$NH_3^+.SO_3^- + H^+ \rightleftharpoons NH_3^+.SO_3H \xrightarrow{H_2O} NH_4^+ + H_2SO_4$$

In alkaline medium hydrolysis is slow, the rate decreasing as the pH rises:

$$[NH_2SO_3]^- + OH^- \rightarrow NH_3 + SO_4^{2-}$$

It is susceptible to oxidation by reagents such as chlorine and chlorate, with eventual formation of nitrogen and sulphuric acid, but its apparently sole area of application in waste water treatment is in the oxidation of nitrite. This proceeds by

$$NH_2.SO_3H + HNO_2 \rightarrow N_2 + H_2SO_4 + H_2O$$

in acid conditions (pH 3.5 is suitable in practice), and has the advantage that nitrogen rather than nitrate is formed as an end product. The reaction is not fast, being governed by a slow step involving reaction of the undissociated nitrous acid with sulphamate (see Hughes 1967 and Biddle and Miles 1968).

An extensive review of sulphamic acid chemistry has been given by Benson and Spillane (1980).

Reductive ion exchange resins

Reductive resins of a polymeric amine–borane type have been produced commercially by the Rohm & Haas Co. (USA), coded as Amborane 345 and 355 (Rohm & Haas 1981). These have been proposed as reductants for the recovery of precious metals from solution, for which the exchange reaction can be written

$$n\underbrace{[P]BH_3}_{\text{resin}} + 6M^{n+} \rightarrow \underbrace{[P]H^+ + 6M^+}_{\text{resin}} + 5nH^+ + nH_3BO_3$$

After effective exhaustion of the resin with metal, any excess unreacted borane functionality in the resin is removed by treatment with an acid (pH = 2) solution of acetaldehyde for 10 hours, filtration, resin washing and incineration of the treated resin at 500–800°C to recover the metal. If anionic base metal complexes or high concentrations of precious metal complexes are present, a further treatment with 4% sodium hydroxide, followed by washing, is recommended before incineration.

The cost of these materials precludes their use for other then precious metal recovery, and the production of borates in the waste water represents something of a disadvantage. Even with precious metals it would appear that there are several other cheaper alternatives for recovery—even using standard ion exchange resins.

Activated carbon
Activated carbon can be effective in removing noble metals, where it may act partially as a reductant and also provide a large surface area on which the reduced metal (which may be in colloidal form) can deposit. It is suitable for recovering palladium and platinum from rinse waters, and has been applied for removing traces of mercury down to levels of the orders of 0.2–20 $\mu g\ l^{-1}$. Sorption of mercury is optimal in the pH range 4–5 (Huang and Blankenship 1984). An advantage can be that the surface-active properties help to remove colloidal suspensions.

Hydroxylamine
Hydroxylamine, NH_2OH, provides some reducing action, and has been suggested as a reagent for removal of mercury (Patron *et al.* 1978), whereby it serves both as a reductant and as a precipitant. As with hydrazine, however, the reagent would generally be undesirable, and has not been applied significantly.

REFERENCES

Arcilesi, D. A., Spearot, R. H., and Peck, J. V. (1983) *Brit. Patent Appl.* No. 8, 316, 410.
Benson, D. (1968) *Mechanisms of Inorganic Reactions in Solution,* McGraw-Hill.
Benson, G. A., and Spillane, W. J. (1980) *Chem. Rev.* **80** 151.
Biddle, P., and Miles, J. H. (1968) *J. Inorg. Nucl. Chem.* **30** 1291.
Bostian, L. C. (1981) *Speciality Inorganic Chemicals, Spec. Pub. No. 40,* ed. Thompson, R. C., Royal Society of Chemistry, 60.
Bravo de Nahui, F., Hooper, R. M., and Wragg, A. A. (1986) *Chem. Ind. (London)* 571.
Burlamacchi, L., Guarini, G., and Tiezzi, E. (1969) *Trans. Farad. Soc.* **65** 496.
Chamberlin, N. S., and Day, R. V. (1956) *Proc. 11th Ind. Waste Conf. Purdue Univ.,* 129.
Connick, R. E. (1982) *Inorg. Chem.* **21** 103.
Cook, J. E., and Mattock, G (1986) unpublished work.
Cook, M. M., and Lander, J. A. (1981) *Pollut. Eng.* **13,** 36.
Coulter, M. O., and Bell, D. (1974) *U.S. Patent* No. 3, 847, 598.
Dart Industries (1971) *U.S. Patent* No. 2, 242, 473.
De Angelis, P., Morris, A. R., and MacMillan, A. L. (1978) *U.S. Patent* No. 4, 098, 697.
Ebara-Infilco Co. (1973) *Jap. Patent* No. J4 9, 119, 819.
Eilbeck, W. J., and Potter, D. M. (1980) unpublished work.
Furukawa Electric Co. (1973) *Jap. Patent* No. J4 9, 089, 346.
Göckel, H. (1972) *German Patent* No. 2, 122, 415.
Grainger, R., and Mattock, G. (1974) unpublished work.
Haight, G. P., Perchonock, E., Emmenegger, F., and Gordon, G. (1965) *J. Am. Chem. Soc.* **87** 3835.
Hisamatsu, Y., and Kitajima, Y. (1968) *J. Metal Finish. Soc. Japan* **19** 466.
Huang, C. P., and Blankenship, D. W. (1984) *Wat. Res.* **18** 37.
Hughes, M. N. (1967) *J. Chem. Soc. A* 902.
Jula, T. F. (1974) *Inorganic Reductions with Sodium Borohydride,* Ventron Corp. (USA).

Kampermann, D. R. (1973) *U.S. Patent* No. 3, 770, 630.

Kubo, K., Mishina, A., Aratani, T., and Yano, T. (1979) *J. Chem. Eng. Japan* **12** 495.

Lores, C., and Moore, R. B. (1973) *U.S. Patent* No. 3, 770, 423.

Mattock, G., and Uncles, R. W. (1962) *The 1962 Effluent and Water Treatment Manual* 182, Thunderbird.

Mayenkar, K. V., and Lagvankar, A. L. (1984) *Proc. 38th Ind. Waste Conf. Purdue Univ.* 457.

Miller, J. D., and Beckstead, L. W. (1973) *Metall. Trans.* **4** 1967.

Nadkarni, R. M., and Wadsworth, M. E. (1967) *Trans. Metall. Soc. A.I.M.E.* **239** 1066.

Nadkarni, R. M., Jelden, C. E., Bowles, K. C., Flanders, H.E., and Wadsworth, M. E. (1967) *Trans. Metall. Soc. A.I.M.E.* **239** 581.

Patron, G., Napoli, D., Ratti, G., and Tubiello, G. (1978) *U.S. Patent* No. 4, 087, 359.

Patterson, J. W., and Jancuk, W. A. (1977) *Proc. 32nd Ind. Waste Conf. Purdue Univ.* 853.

Power, G. P., and Richie, J. M. (1975) *Modern Aspects of Electrochemistry. No. 11,* ed. Conway, B. E., and Bockris, J. O'M., Prentice-Hall.

Richard, M. D., and Brookman, G. (1971) *Proc. 26th Ind. Waste Conf. Purdue Univ.* 713.

Rohm & Haas (1981) *Amborane 345 and Amborane 355 Reductive Resins.*

Schalch, E., Nicol, M. J., Balestra, P. E. L., and Stapleton, W. M. (1976) *Rept. No. 1799 Nat. Inst. Metallurgy (S. Africa).*

Schmidt, E. W. (1984) *Hydrazine and its Derivatives,* Wiley.

Sims, A. F. E. (1981) *Effl. Wat. Treat. J.* **21** 109.

Southgate, B. A., and Grindley, J. (1945) *Ind. Chemist* **21** 144.

Sweeney, K. H. (1981) *A.I.Ch.E. Symp. Ser.* **77** 67; 72.

Taylor, C. R., Jr., and Qasim, S. R. (1984) *Wat. Pollut. Control* **83,** 420.

Wade, R. C. (1981) *Speciality Inorganic Chemicals Spec. Pub. No. 40,* 25, ed. Thompson, R. A., Royal Society of Chemistry.

Wallace, R. A. (1977) *U.S. Patent* No. 4, 008, 077.

Wallace, R. A. (1978) *U.S. Patent* No. 4, 082, 546.

Wolfe, N. L., Zepp, R. G., Gordon, J. A., Baughman, G. L., and Cline, D. M. (1977) *Environ. Sci. Technol.* **11** 88.

Yao, Y.-L. (1944) *Trans. Electrochem. Soc.* **86** 371.

5

Electrochemical processes

THE NATURE OF THE ELECTROCHEMICAL PROCESS

Basic concepts

An electrochemical or electrolytic cell is simply a device that uses electrical energy to effect a chemical change; the converse process of transforming chemical energy into electrical energy takes place in a galvanic cell or battery. In their simplest forms electrolytic cells consist of two electrodes (anode and cathode) immersed in an electrically conducting solution (the electrolyte), and connected together, external to the solution, via an electrical circuit which includes a current source and control devices. The chemical processes occurring in such cells are reduction–oxidation ones, and take place at the electrode/electrolyte interfaces. The electrode at which reduction occurs is referred to as the cathode and, conversely, the anode is the electrode at which oxidation processes occur. It is clear that there must be a stoicheiometric relationship between the processes occurring at the anode and those at the cathode, since an identical number of electrons will be released to the anode by species undergoing oxidation as will be released by the cathode to species being reduced.

The fundamental ideas on oxidation and reduction reactions are dealt with in some detail in Chapter 3 and will not be reiterated here. Also, it is not possible to present a comprehensive review of the complete electrochemical picture in a text of this type, where the principles only are presented as they are of particular significance in the context of waste water applications. For fuller theoretical discussions the reader is referred to texts such as those of Bockris and Reddy (1970), Newman (1973), Crow (1979), and Pletcher (1982).

In an electrolytic cell such as that depicted diagrammatically in Fig. 5.1 there are three distinct processes by which electricity is conducted:

(1) Metallic conduction occurs in the external circuit, where the current is carried solely by electrons.
(2) Electrolytic conduction takes place within the bulk of the solution, where the charge carriers are anions and cations. All ionic species will contribute to this

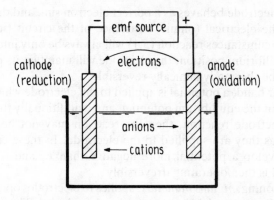

Fig. 5.1 — Diagrammatic representation of an electrolytic cell.

process, whether or not they are involved directly in the oxidation or reduction reactions; their relative contributions are dictated by their concentrations and transport numbers

(3) At the electrode/electrolyte interfaces oxidation and reduction processes occur by which the ionic conduction of the solution is coupled to the electronic conduction within the electrode and external circuit. It is here that electrical energy is converted into chemical work. Electrical losses occurring elsewhere in the cell appear as ohmic heating.

Reversible and irreversible electrode reactions—kinetic polarisation

As an illustrative example, consider the electrolytic cell of Fig. 5.1 to have a silver metal cathode in contact with an electrolyte containing Ag^+ ions. In the absence of an applied potential, i.e. under open-circuit conditions, the electrode will spontaneously adopt a potential determined by the activity of the Ag^+ ions in the solution with which it is contact. This potential, the equilibrium potential E_e, is calculable from the Nernst equation (5.2) for the electrode reaction (5.1):

$$Ag^+_{(aqu)} + e \rightleftharpoons Ag_{(s)} \tag{5.1}$$

$$E_e = E^{\ominus}_{Ag^+/Ag} + \frac{RT}{F} \log_e a_{Ag^+} \tag{5.2}$$

If the electrode is included in a circuit such that a potential very slightly more negative than E_e is applied, then electrons will flow from the external circuit into the electrode. The equilibrium (5.1) will be disturbed and a nett reaction will take place in an attempt to re-establish the equilibrium position. Reaction (5.1) will thus proceed from left to right, electrons will be taken up by Ag^+ ions, and silver metal will be plated out on the electrode surface. If this reaction is able to take place sufficiently rapidly to remove electrons as fast as they are supplied from the external circuit, then the electrode behaves as a perfect electron sink and the current will be limited only by the electrical (ohmic) resistance of the circuit (including the cell itself). In these circumstances reaction (5.1) will always be only infinitesimally away from the true equilibrium position, the electrode will maintain its potential E_e, and the cell is said to be thermodynamically reversible.

If, on the other hand, a potential is applied to the electrode which is significantly more negative than the equilibrium potential, and if additionally there is a slow step in the overall electrode reaction, then the reaction may not be able to remove electrons as fast as they are supplied to the electrode. In these circumstances the electrode will develop a potential more negative than E_e and is then said to be polarised; the cell is then operating irreversibly.

The same reasoning, *mutatis mutandis*, applies to electrodes operating as anodes, i.e. when the applied potential is more positive than the equilibrium potential, and oxidation reactions occur. It should be noted that polarisation always makes cathodes more negative and anodes more positive than the respective equilibrium potentials would suggest. The effect of polarisation on practical electrolytic cells therefore makes it necessary to apply a larger voltage than that calculated from standard electrode potentials before electrolysis can occur. The difference between the actual working potential of an electrode and the equilibrium (Nernst) potential is called the 'overpotential', and is usually given by the symbol η.

It is important to realise that equilibrium (5.1) is dynamic, not static, and that at the position of equilibrium, although there is no nett reaction and hence no nett current flow, the forward (cathodic) and reverse (anodic) reactions are both proceeding at finite velocities, and hence there are associated finite electron currents. These currents are referred to as the cathode current and the anode current, and are given the symbols i_c and i_a respectively. Since no nett current flows at the equilibrium potential then of course $i_a = -i_c = i_0$ at this point. This current i_0, is called the exchange current; it provides a measure of the rates at which the cathodic and anodic reactions are occurring at any particular electrode. There are anodic and cathodic currents associated both with cathodes and with anodes: in a working cell $i_c > i_a$ at the cathode whereas at the anode $i_c < i_a$.

The exchange current is characteristic both of the particular electrode reaction and of the nature of the electrode surface, and is an important parameter in the design of electrochemical cells. In general it is the exchange current density rather than the exchange current which is used as a design parameter, since the nett current and hence the rate of formation of product (or removal of reactant) is obviously also a function of electrode surface area.

If one considers a general electrode reaction, of which reaction (5.1) is an example:

$$ox + ne \underset{k_1}{\overset{k_{-1}}{\rightleftharpoons}} red$$

then the rates of the forward (reduction) reaction, r_1, and the reverse (oxidation) reaction r_{-1}, may be written

$$r_1 = k_1[\text{ox}]_e = I_c/nF$$
$$r_{-1} = k_{-1}[\text{red}]_e = I_a/nF$$

where $[\text{ox}]_e$ and $[\text{red}]_e$ are the concentrations of the oxidised and reduced forms of the reactant near the electrode surface, and I_c and I_a are the forward (cathodic) and reverse (anodic) current densities. If the rate of the forward reaction is greater than the rate of the reverse reaction then reduction occurs more rapidly than oxidation, the potential of the electrode is displaced from the equilibrium value (by η, the overpotential), and the electrode behaves as a cathode. The effect of η is to make the reduction process more rapid and the reverse oxidation reaction less rapid, as is reflected in the changed shape of the reaction co-ordinate/energy profile shown in Fig. 5.2. The transfer coefficient, α, is the fraction of the overpotential assisting

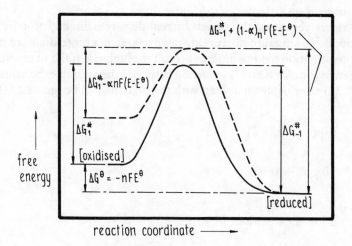

Fig. 5.2 — Free energy changes during electrode reactions.

reduction, and hence $1 - \alpha$ is the fraction assisting oxidation. In this case it is clear that the activation energy for the forward (reduction) process $\Delta G_1^{\#}$ is decreased by $\alpha F n \eta$, whilst the activation energy for the oxidation reaction $\Delta G_{-1}^{\#}$ is increased by $(1 - \alpha)F n \eta$, where η is equal to $(E - E_e)$.

The Butler–Volmer equation is fundamental in relating the operationally import-ant exchange current density to the electrode overpotential, and for a one-electron reaction takes the form

$$I = I_0 \left[\exp(-\alpha F \eta/RT) - \exp((1 - \alpha)\, \eta F/RT)\right] \tag{5.3}$$

where I = nett current density; I_0 = exchange current density; and α is the transfer coefficient or symmetry factor, and is related to the shape of the activation energy barrier for the slow step of the electrode reaction as previously described. The numerical value of α lies between 0 and 1 and frequently has a value of approximately 0.5. Equation (5.3) can be written in simplified forms for certain limiting conditions often encountered in practical applications. If the overpotential is small (< 10 mV, say) then

$$I = I_0 \left[1 - \frac{\alpha F\eta}{RT} - \left(1 + \frac{(1-\alpha)F\eta}{RT} \right) \right] \tag{5.4}$$

Since $\exp(x) \approx 1 + x$ if x is small, hence

$$I = -I_0 F\eta / RT \tag{5.5}$$

Therefore, for low overpotentials the nett current density is directly proportional to the overpotential. However, in many real systems large overpotentials are encountered and in this situation a quite different result is obtained. If the overpotential is large and negative (i.e. for a highly polarised cathode) the second exponential term in equation (5.3) is negligible compared with the first and can be ignored. Then

$$I = I_0 \exp(-\alpha F\eta / RT) \tag{5.6}$$

or

$$\log_e I = \log_e I_0 - \alpha F\eta / RT \tag{5.7}$$

Hence,

$$\eta = \frac{RT}{\alpha F} \log_e I_0 - \frac{RT}{\alpha F} \log_e I \tag{5.8}$$

Since for a given electrode reaction I_0 and α are constants, equation (5.8) simplifies to

$$\eta = a + b \log I \tag{5.9}$$

where a and b are constants. Equation (5.9) is identical in form to the empirical equation proposed by Julius Tafel in 1905, and called, in deference to him, the Tafel equation. Plots of η vs. $\log I$ are known as Tafel plots (see Fig. 5.3).

A similar result is arrived at for large positive values of η, i.e. for a highly polarised anode, but in this case it is the first exponential term in equation (5.3) which is negligible compared with the second, so that

$$\eta = \frac{RT}{(1-\alpha)F} \log_e I_0 + \frac{RT}{(1-\alpha)F} \log_e I \tag{5.10}$$

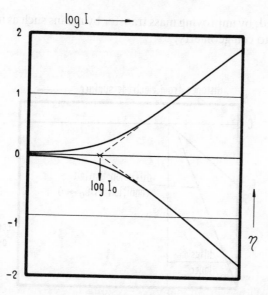

Fig. 5.3 — Tafel plots of overpotential, η, against log I (nett current density) for electrode reactions, including anodic and cathodic polarisation curves. The broken extrapolation lines cross the zero overpotential line at the exchange current density for the reversible potential.

and again

$$\eta = a' + b' \log I$$

Since the assumptions made in deriving the Tafel equation from equation (5.3) are not valid for small values of η, it is not to be expected that the experimental Tafel lines of Fig. 5.3. would continue linearly to cut the log I axis: as $\eta \to 0$, $I \to 0$ (and hence $\log I \to -\infty$), from equation (5.5). However, the asymptotes to the experimental Tafel plots do intersect the log I axis at $\log I_0$ (see Fig. 5.3). This incidentally provides an experimental method for determining I_0, the exchange current density, and the slopes of the lines enable an estimate of the symmetry factor to be made.

Concentration polarisation

It is perhaps self-evident that any reaction at an electrode surface in which solute species are discharged can only be sustained if reactant is transported from the bulk solution phase to the electrode surface so as to replace that consumed in the electrochemical reaction. At low overpotentials, the various processes by which this mass transport occurs (e.g. diffusion, ionic migration, convective mixing) can keep pace with the demands of the reaction, and mass transport is not a rate-limiting step. However, with increasing overpotential and reaction rate, a point will eventually be reached where, no matter how efficient the electron transfer process, the nett reaction rate will be limited by the mass transport requirement, and the electrode is then said to be concentration-polarised. This effect can be reduced, but never entirely eliminated, by improving mass transfer by means such as increased agitation and by attention to cell geometry.

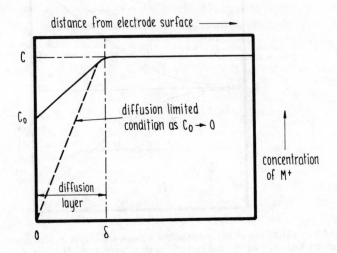

Fig. 5.4 — Diagrammatic representation of ionic concentration gradients in the vicinity of a cathode.

Figure 5.4 is a representation of the concentration gradients in the vicinity of an electrode where reduction of M^+ is occurring. For the situation where the cell is producing a constant current the concentration of oxidised species at the electrode surface will fall as it is removed by the reduction process, and will be made good by ionic migration (due to the electric field) and by diffusion due to the concentration gradient generated across the diffusion layer. A steady state will be reached when the concentration of the oxidised species at the electrode surface, C_0, will become constant at a value less than the bulk concentration of the species. The degree to which C_0 is less than the bulk concentration, C, will depend on the efficiencies of the migration and diffusion processes that dictate the rate at which reactant can be delivered to the electrode surface. The rate of arrival of reactant by diffusion alone is given by Fick's law, which, if one assumes a linear concentration gradient over the diffusion layer, may be written

$$r_{\text{diff}} = D(C - C_0)/\delta \tag{5.11}$$

where D is the diffusion coefficient and δ is the thickness of the diffusion layer. The current at any instant is given by the equation

$$i = nFDA(C - C_0)/\delta \tag{5.12}$$

The rate of ionic migration of the reactant M^+ is given by

$$r_{mig} = t_+ I/nF \qquad (5.13)$$

where I is the current density and t_+ is the transport number of the M^+ ion. (Note the transport number of an ion is defined as the fraction of the total current carried by that ion.) From equations (5.11) and (5.13) it follows that the rate of delivery of M^+ to the electrode across the diffusion layer will be

$$r_{total} = \frac{I}{nF} = t_+ I/nF + D(C - C_0)/\delta \qquad (5.14)$$

As the voltage is increased so the current density increases, so long as the concentration gradient across the diffusion layer can increase sufficiently to supply M^+ at the required rate. However, a situation will eventually arise where the concentration gradient across the diffusion layer will become constant as C_0 tends to zero and the rate of the electrode reaction can no longer increase. At this point the electrode reaction is diffusion-limited, and

$$r_{total} = t_+ I_{lim}/nF + DC/\delta \qquad (5.15)$$

It should be noted that the diffusion current is proportional to the bulk concentration of the electroactive species, and hence as electrolysis proceeds so the reaction rate will decrease.

In the context of waste water treatment this mass transfer limitation is particularly significant where the electroactive species are in relatively low concentration; for example, in the later stages of electrolytic processes for the exhaustive removal of metal ions, where diffusion-limited conditions lead to marked decreases in cell efficiencies.

Cell resistance

In addition to overpotentials generated by kinetic and concentration polarisations, a third source of overpotential or overvoltage can be traced to the finite resistance of the electrolytic solution within the cell. Because of this cell resistance an ohmic (or 'iR') voltage drop between the electrodes will be observed. The higher the cell resistance, the greater will be the voltage necessary to pass a given current and the greater the energy losses due to resistive heating.

Current–potential relationships

In Fig. 5.5 the electrolysis current is plotted against the potential applied to the electrode of interest: such a diagram is known as a polarogram. In conventional polarography (an electrochemical analytical technique) the electrode of interest is usually, although not always, a dropping mercury cathode, and the anode may be a

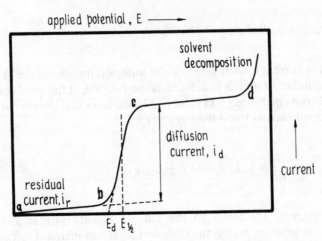

Fig. 5.5 — Characteristic plot of electrolysis current against applied electrode potential for a reversible electrode reaction. $E_\frac{1}{2}$ = half-wave potential; E_d = decomposition potential (not thermodynamically significant).

pool of mercury of relatively large surface area. The principles governing the current–potential relationships are, however, the same in all electrolytic processes and are particularly pertinent to the behaviour of electrolytic cells designed to process waste water streams, as for example those cells used to recover metals by cathodic reduction.

Before electrolysis can occur the applied potential must overcome the equilibrium potentials of the anode and cathode, any associated overpotentials, and the ohmic drop between them. The decomposition voltage (labelled E_d in Fig. 5.5) where electrolysis occurs has therefore very limited direct theoretical significance, but is of course extremely important from a practical viewpoint. In the region b–c of Fig. 5.5 electrolysis is occurring freely and the increasing mass transport requirement of the electrode process is satisfied. At point c, however, the current becomes diffusion-limited and further increases in applied potential have little effect on the electrolysis current until point d is reached, when a second electrode reaction involving other species occurs. This second reaction may for example involve a more electropositive metal ion or perhaps the electrolytic decomposition of the solvent, water.

The form of Fig. 5.5 is schematic and only approximates to the situation obtaining in practical cells. Nevertheless certain important points emerge which relate directly to the operation of such cells:

(1) The diffusion-limited current, i_d, and hence the efficiency of electrolysis is proportional to the bulk concentration of the electroactive species and will hence decrease as electrolysis proceeds.
(2) Attempts to maintain a constant current by increasing the applied voltage must ultimately fail as the electrode behaviour moves on to and along the diffusion limited plateau c–d.

(3) Failure to control the potential, particularly with complex solutions, will lead to loss of control of the desired electrochemical reaction and may ultimately lead to the formation of undesired products and the decomposition of the solvent. This introduces energy wastage and the hazards associated with the unplanned generation of hydrogen and oxygen.

The precise shape of the current–potential curve, and particularly the magnitude of $E_{\frac{1}{2}}$, the 'half-wave potential', in relation to the standard Nernstian potentials of the half-cells, is very much a function of the thermodynamic reversibility or otherwise of the electrode reactions. For a thermodynamically reversible electrode process $E_{\frac{1}{2}}$ is approximately equal to $E^{\ominus}$, the Nernstian potential, whereas for irreversible processes associated with large overpotentials $E_{\frac{1}{2}}$ will be considerably displaced from the equilibrium (Nernstian) potential (see Fig. 5.6). For cathodic processes $E_{\frac{1}{2}}$ will be

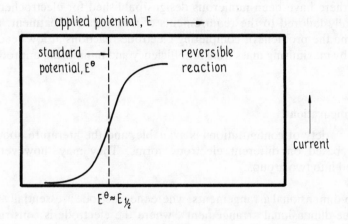

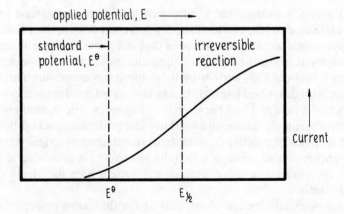

Fig. 5.6 — Characteristics of thermodynamically reversible and irreversible current–potential curves.

more negative than $E^\ominus$ and for anodic processes $E_{\frac{1}{2}}$ will be more positive than $E^\ominus$. Large overpotentials associated with the electrodes will hence lead to a large increase in the potential required to produce electrolysis.

ELECTROCHEMICAL REACTOR DESIGN

The primary requirements

The design of an electrochemical cell for waste water treatment must take into account a number of constraints that are evident from the discussion in the preceding section. While these constraints apply in general to any electrochemical cell, the particular objective of waste water treatment that one is seeking, namely to complete an electrode reaction rather than merely to cause it to occur efficiently, means that extra demands are made. For this reason the majority of electrowinning and preparative cell designs are inappropriate to waste water treatment. For the same reason there have been numerous designs published for electrochemical reactors specifically tailored to the requirements of waste water treatment, all seeking to overcome the problem of completing electrode reactions at low reactant concentrations by maximising mass transfer efficiency and minimising unproductive energy losses.

Cell configuration

A wide variety of configurations is available, and the literature abounds with cell designs based on different electrode forms. They may, however, be broadly classified into two groups:

(1) two-dimensional arrangements, where the electrode is essentially in sheet form;
(2) three-dimensional arrangements, where the electrode is constructed so as to provide opportunity for electrochemical reaction from three axial sources.

The first group is exemplified by the most common commercial cell type, using parallel plate electrodes, in which anodes and cathodes are usually stacked alternately to give a total working surface area that will provide a suitable current density. Mass transfer is maximised by providing electrolyte flow between the electrodes either via a recirculation system or by induced agitation; alternatively, as in the 'Chemelec' cell described later, diffusion-controlled conditions are sought by using inert beads in a fluidised bed between the electrodes. The second group is exemplified by arrangements where the working electrode is a packed-bed of electrode material granules or powder (e.g. carbon) maintained in a rigid configuration by a suitable enclosure and through which the electrolyte is circulated to optimise mass transfer. An alternative three-dimensional system uses the electrode particles in fluidised form.

It is an essential element of any cell design that some means of separating the anolyte from the catholyte be achieved. It is common, although by no means universal, practice to employ a diaphragm of some kind to prevent mixing of the two sets of reaction products; in simpler forms physical separation of the anode and cathode zones may be adequate but at the cost of some increase in cell resistance. If the electrolyte is highly conducting, with a high concentration of salts, a membrane separator may not be essential, although almost invariably its inclusion is helpful. The selection of a particular membrane needs care because they too can introduce unwanted ohmic resistance in the circuit with an associated ohmic drop. Classically, porous diaphragms have been employed in electrowinning cells, but the newer conducting membranes, e.g. du Pont (USA) 'Nafion', have shown distinct advantages in providing physical flexibility and hence design adaptability, as well as the significant feature of inhibiting unwanted ionic transport.

When the separating membrane is of an ion-exchange type, the cell obtained is then essentially an electrodialysis one. Electrodialysis cells have been employed for the desalination of brackish waters and sea water, and many large-scale plants of this type have been installed for this purpose. For waste water treatment, however, the peculiar factors associated with the general need for operation at low current densities means that the emphasis is changed more towards electrode configuration than is usual with desalination designs.

Some examples of particular cell electrode designs that have been applied to waste water treatment can be briefly described to provide a picture of the approaches that have been used.

Parallel plate system
One of the more successful commercial examples of electrochemical reactor design applied to waste water treatment has been the 'Chemelec' cell, developed in the UK (Lopez-Cacicedo 1976) and marketed under licence by BEWT (UK). This uses the principle of alternate cathodes and anodes but incorporates, as a means of minimising ion depletion at the working electrode surface, a fluidised bed of inert glass beads of controlled diameter (0.5–2.0 mm). The working electrode is of expanded metal (used to promote turbulence), with electrolyte being pumped through a slotted disc to maintain the beads in a fluidised state. Fig. 5.7 illustrates in diagrammatic form the features of the cell and the mode of operation. This cell has been particularly applied to the removal and recovery of metal ions from dilute solutions (discussed in a later section). In this case the metal is deposited on to electrode materials such that the coated electrodes can be used directly in other applications.

The 'Chemelec' cell design represents an advance over the more customary approach of circulating electrolyte at a rapid rate through the cell chamber to maintain good dispersion conditions. Other forms of agitation are of course possible, as for example by using ultrasonic probes or air/gas sparging. Such simpler cells have been extensively used for less demanding recovery operations, such as those involving the noble metals gold and silver, and are well established commercially.

The spiral wound configuration
In order to provide maximum electrode surface area in a small volume, flexible electrodes have been proposed where the working electrode, a separator and the counter electrode are sandwiched together as sheets and rolled on to an axis to give a tubular form that can be housed conveniently in a modular unit (see Fig. 5.8). This 'Swiss-roll' arrangement, which is not unlike its counterpart used in reverse osmosis equipment, has been described by Robertson, Ibl and their co-workers (Robertson *et al.* 1975, Robertson and Ibl 1977, Robertson *et al.* 1978), and is the basis of the

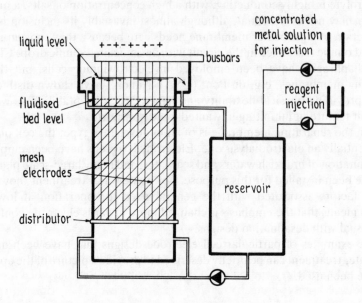

Fig. 5.7 — Features of the 'Chemelec' parallel plate electrochemical cell used for metal recovery.

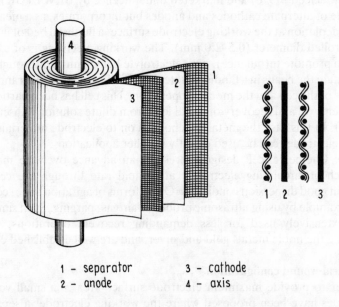

1 - separator   3 - cathode
2 - anode   4 - axis

Fig. 5.8 — Construction of the spiral wound electrochemical cell (reproduced from Robertson
et al. 1978).

'extended surface electrode' system described by Williams (1975).The electrodes may be either a continuous sheet or a net, while the separating material may be of porous cloth or similar material, or an ion-exchange membrane.

Advantages clearly apparent from this form of construction are not only a large surface area (200 cm³ per cm³ of electrode volume) but also a low ohmic drop between the electrodes, uniform electrical flux distribution, high mass transfer rates and good promotion of turbulence by the separator. On the other hand there must be a proneness to blockages and difficulties in removing electrodeposited materials (such as metals).

Rotating electrodes

In rotating electrodes agitation is provided by moving the electrode rather than the solution, and in many fundamental studies the working electrode has been rotated. Rotating disc electrodes have been extensively investigated but seldom applied in plant-scale equipment, although commercial cells incorporating rotating cylinder electrodes have been developed. The theoretical basis for these has been reviewed by Gabe and Walsh (1983), who also tabulate some of the applications that have been realised by proprietary equipment. Only a few applications have been reported for waste water, notably for silver and copper recovery. The principal commercial development was by Holland in the UK of the 'Eco-cell' (reviewed by Holland 1978) for cathodic reductions. The Eco-cell rotating cylinder electrode is shown in Fig. 5.9,

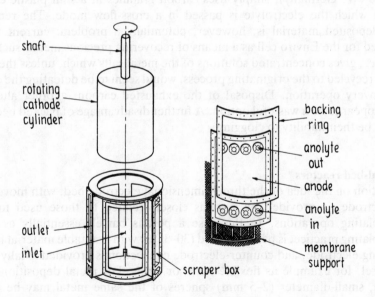

Fig. 5.9 — The 'Eco-cell' rotating cylinder electrode (reproduced from Holland 1978).

illustrating the concentric arrangement of cathode cylinder, ion-exchange membrane separator and anode: the device also incorporates a scraper to dislodge electrodeposited material from the cathode. The cell operates at high efficiency, and its performance can be optimised by potentiostat control of the applied cathode potential. As a commercial device it has received negligible acceptance, however.

A variation on the idea of electrode rotation is to use a bipolar pump, as described by Jansson and Ashworth (1977). Electrolyte is passed through a pump consisting of a magnetically driven rotor and a stator to both of which current is applied to drive the electrochemical reaction. Commercial pumps of this kind have been marketed (Totton Electrical Pumps, UK), but no significant application to waste water treatment appears to have been made.

Packed bed electrodes

As a way of reducing current densities while providing an adequately large surface area to ensure that sufficient current can be passed to give high reaction rates, there is the option of using a packed bed of conducting materials through which the electrolyte passes. This idea was examined by Bennion and Newman (1972) and Chu et al. (1974), and more recently Agarwal et al. (1984) applied the principle with 'reticulated vitreous' carbon electrodes, these being glassy and self-supporting, with a honeycomb structure to allow passage of fluid.

At least two systems are commercially available. One, introduced by Das Gupta and Fleet (1976), uses carbon fibres 5–15 μm in diameter, providing an extremely large surface area to mass ratio. Another, the 'Enviro' cell marketed by Deutsche Carbone (W. Germany), simply uses carbon granules in a slim plastic container through which the electrolyte is passed in a cross-flow mode. The removal of electrodeposited material is, however, potentially a problem: current reversal, suggested for the Enviro cell as a means of recovering precious metals such as gold and silver, gives concentrated solutions of the metal salts which, unless they can be directly recycled to the originating process, would seem to be defeating the object of the recovery operation. Disposal of the exhausted carbon, another alternative, would appear to be a wasteful option. A further disadvantage of packed beds would seem to be their liability to clogging.

Tumbled-bed reactors

A variation on the idea of the three-dimensional electrode bed, with movement of the electrode, is provided by systems closely related to those used for barrel electroplating operations. In this case a porous barrel, essentially as used in electroplating practice, is partially filled (30–35%) with a suitable material for use as a working electrode, and counter-electrode connection is provided axially through the barrel, for example as flexible wires or strips. For metal deposition on to a cathode, small-diameter (2–5 mm) spheres of the same metal may be used or, alternatively, steel substrate balls pre-plated with the metal can be employed. Obviously re-use of the recovered material has to be considered and, where economics permit, growth on the same metal substrate is to be preferred.

Such equipment for use in waste water application, primarily for removal of metals, has been described by Kammel and Lieber (1978) and by Tison (1981). The electrochemical efficiency is probably not high, and the electrical flux characteristics are not likely to be good, but the deposits are smooth (presumably because of the tumbling action in the barrel) and the system is not prone to clogging. It has the considerable merits of simplicity, low capital cost and ease of maintenance, and would seem to present an option worthy of greater commercial examination.

Fluidised-bed systems
Considering the foregoing descriptions it is a fairly logical development that fluidisation of a particular electrode bed should be applied to electrochemical cell design, and several publications have described such systems (Backhurst *et al.* 1969, Tarjanyi *et al.* 1971, Barker and Plunkett 1976, Van der Heiden *et al.* 1978) with primary reference to cathodic deposition of metals. Fig. 5.10 shows the basic form

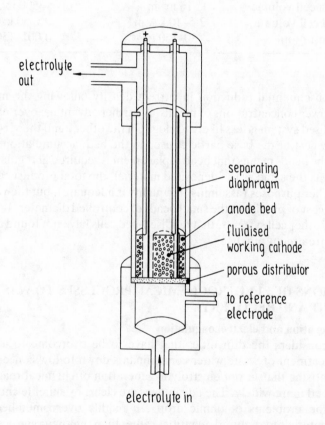

Fig. 5.10 — Form of fluidised bed electrode system.

that the system may take, with a cylindrical anode, a membrane to separate the anolyte from the catholyte, and a cathode consisting of small beads of conductive material maintained in a fluidised condition by an upflow of the electrolyte to be treated. The cathode current is conveyed through the fluidised bed, a proportion of which is always in physical contact to ensure electrical conduction. The cell described by Van der Heiden *et al.* uses a series of anode rods concentrically enclosed by a low porosity, low electrical resistance diaphragm, these tubes themselves being enclosed by a cylinder accommodating the fluidised bed of particles.

Van der Heiden *et al.* (1978) have pointed out some of the advantages of fluidised-bed systems over planar electrodes as summarised in Table 5.1. Notably

Table 5.1 — Comparison of fluidised-bed and conventional planar electrodes (after Van der Heiden *et al.* 1978).

Characteristic	Planar	Fluidised bed
Cathode area/cell volume	$16 \text{ m}^2 \text{ m}^{-3}$	$3500 \text{ m}^2 \text{ m}^{-3}$
Current load/cell volume	$2.5\text{--}10 \text{ kA m}^{-3}$	$15\text{--}50 \text{ kA m}^{-3}$
Metal concentration	$30\text{--}150 \text{ g l}^{-1}$	$0.001\text{--}150 \text{ g l}^{-1}$

there is a consequential reduction in current density, allowing the more efficient removal of lower concentrations of metals. A further advantage over packed beds is that the fluidised system is less likely to clog. On the other hand there is the problem of the supply cost of the base particles used in the bed: accumulation of deposited material on the beads means that bead replacement is required at regular intervals. If the substrate is the same as the deposited material, the total product can be re-used or sold for other purposes (assuming the quality is adequate), but the value of this is offset by the cost of providing the small beads of controlled diameter. It is significant, as with many other cell systems, that fluidised-bed cells have not found acceptance in waste water treatment practice.

APPLICATIONS OF ELECTROCHEMICAL PROCESSES TO WASTE WATER TREATMENT AND RECOVERY

Reagent generation and electrocoagulation

When one considers the difficulties that have to be overcome to achieve direct electrodic treatment of waste water contaminants down to low concentrations it is perhaps surprising that *in situ* electrolytic generation of chemical reagents has not been employed more widely. The advantages are clear: by suitable choice of source electrolyte the problems of ohmic drop are readily overcome; because one is generating materials to required quantities rather than removing species down to low concentrations the cells can be of conventional types, without the need for low current densities, special electrode configurations, and so on; and by generating substances which may otherwise be dangerous or expensive to transport there can be an operational simplification. Yet in practical terms only the generation of hypochlorite appears to have become a serious commercial reality, particularly where sea water is available as a primary material source. Such cells have found application for disinfection purposes, as for example on board ships, and for slime and algae control in industrial cooling water circuits. Even so, use of these cells for, e.g., cyanide destruction (see later in this chapter for a fuller discussion), does not seem to have developed (at least in the Western world), although at first sight there would seem to be an obvious way of applying them. Thus the control of hypochlorite generation and use in a chlorination reactor could be by means of the redox techniques discussed in Chapter 3. One disadvantage would certainly be the introduction of strongly alkaline solution (carrying the hypochlorite) into the effluent, but this is a matter of scale, and may not be too important in urban situations where disposal is undertaken to sewer and isolated high background salinities do not generally present a problem. Some limited use of electrochemical chlorine generators has been made in sewage treatment, but this is generally less likely to be appropriate as an application than it is for industrial waste water treatments. Capital equipment costs may be an inhibiting factor, but generation cells should by their simpler nature be less costly than waste water electrochemical reactor systems. Discussions of hypochlorite generation cells are given by Mantell (1960) and Adamson *et al.* (1963), which give an idea of the general design characteristics.

Electrolysis of sea water was used by Føyn (1959) to disinfect and purify sewage in Oslo fjord. Chlorine generated at the anode provided some disinfection, while alkali, also formed, precipitated magnesium in the sea water as hydroxide; this flocculated sewage solids and provided a means for precipitating phosphate as magnesium ammonium phosphate, thus to reduce algal growth. The flocculated mass was floated by the hydrogen formed at the cathode, and was skimmed off (see also Marson 1967 for a description of the process). One plant installation was made (in Guernsey), but the process does not appear to have been applied since.

Some interest has been shown in the reduction of Cr(VI) by secondary electro-chemical generation. Onstott *et al.* (1973) studied the removal of chromate from cooling tower blowdown by reducing it to Cr(III), with simultaneous precipitation as hydrous oxide, by electrochemically generating ferrous hydroxide in the effluent. A flow type of cell was employed, with the sacrificial anode being of low carbon steel and the cathode of stainless steel. A problem arose due to the formation of colloidal products when the amounts of Fe^{2+} generated were between 25 and 200 mg l^{-1}, which was also the concentration band desired for reduction. It was therefore found expedient to operate at the somewhat less economical level of 200 mg l^{-1} Fe generation. Residual Cr levels were of the order of 0.05 mg l^{-1}, and only at low Cr levels was there significant competition by dissolved oxygen for the generated ferrous hydroxide.

Another application of electrochemical generation for the treatment of chromate has been described by Nojiri *et al.* (1980), again using iron electrodes to reduce Cr(VI) to Cr(III). Starting from acid conditions, the pH rose naturally from 2 to 10–12, with formation of a precipitate that settled after addition of either a non-ionic or anionic polyelectrolyte. According to these authors, the fact that the settled sludge blackened is indicative of ferrite formation, and they consider that a spinel type of chromium ferrite is formed. They claim that the method should be applicable to other metals capable of forming ferrites, such as Zn, Cd, Ni, Mn and Pb, with the advantage of extremely low solubilities, thus leaving small dissolved residuals.

A particular form of electrochemical reagent generation, whereby substances are generated primarily to act as coagulants, and known as electrocoagulation, has been used in some industries as an alternative to adding reagent for the purpose. The objective is to generate, for example, Fe^{2+} ions from a sacrificial electrode, the generated ions then coagulating or demulsifying the contaminants present. The technique has been studied by several writers: Ramirez *et al.* (1977), for example, applied it to tannery waste waters; but Kramer *et al.* (1978) observed fouling of iron electrodes when used for coagulation of oil emulsions, necessitating periodic charge reversal and electrode scraping.

One apparent advantage of electrocoagulation would be in the treatment of acid rinse waste waters where large volumes are involved and treatment with chemicals such as lime or limestone can involve large quantitites of these reagents. Jenke and Diebold (1984) made a laboratory investigation of such an application for non-ferrous metal waste waters including aluminium and iron, using a stainless steel anode and an inert cathode. Ferrous ions generated at the anode (pH 4.9) admixed with aluminium and manganese contaminants, and resulted in precipitation and co-adsorption in the high pH conditions prevailing in the cathode zone. However, although virtually 100% Al removal was observed, only 25% Mn removal was achieved.

Sacrificial dissolution of metal anodes has also been advanced as a means of removing phosphate from sewage. A patent claim (Ontario Research Foundation 1969) proposes the use of a.c. or pulsating d.c. electrolysis with either Fe or Al electrodes, with the suggestion that salt can be introduced if necessary to increase the electrical conductivity; and Dobolyi (1978) claims that electrolytic dissolution of Al is more efficient in removing phosphate than using the same amount of aluminium by chemical precipitation, provided the molar ratio of Al:P is greater than approximately 1.6:1.

Other suggestions for forms of secondary generation have been made; for example, hydrogen peroxide generation by reduction of oxygen at an activated carbon cathode has been described (General Electric Co. 1971); and for the oxidation of phenol it has been claimed (Wabner and Grambow 1983) that generation of $\cdot OH$ radicals at a lead dioxide anode consumes less energy than ozonation.

Little commercial use of these various proposals appears to have been made, however, although the China National Aero Technology Instrument Factory produces electrochemical units for chromate reduction (using Fe cathodes) and cyanide destruction (using salt for hypochlorite generation). It does appear that there are opportunities for exploitation of electrochemical reagent generation as a waste water treatment method.

Metals' removal and associated processes

Possibly the most obvious application for electrochemical cells in waste water treatment, and certainly the one that has been most enthusiastically pursued, is the cathodic reduction of metal ions from solution to deposit the metal. It is also unquestionably not simple. The attractions are clear, in that it seems a logical and energy-efficient procedure to recover metals in useful form rather than to precipitate insoluble compounds having far less re-use value; but the fact that in many instances metals are present as mixtures means that there are substantial difficulties in obtaining an adequately pure metallic product, quite apart from the problems associated with bringing ionic concentrations down to the 10 mg l^{-1} or lower levels. Nevertheless, much ingenuity has been applied.

Apart from mining and electrowinning operations for the recovery of residual metals, the most significant area where metal recovery could be practised is in waste waters associated with surface treatment operations. In many of these processes, articles to be treated are immersed in a solution that may contain dissolved metals (often at concentrations of the order of 10%), then withdrawn after processing, and rinsed with water. These rinses run to waste and appear as effluent to be treated with concentrations of the order of 100–1000 mg l^{-1}, and treatment is usually sought to give residual levels of the order of 0.1–10 mg l^{-1}. A great deal of effort has been expended on the problem of lowering concentrations down to these levels, and most of the cell designs described earlier have been developed primarily for metal removal. A number of companies have examined the possibilities offered by carbon bed cathodes, where large surface areas can be provided to ensure the necessary low current densities: Müller (1983), for example, describes application of the 'Enviro' cell (Deutsche Carbone, W. Germany). Agarwal *et al.* (1984) used a reticulated vitreous carbon cathode system for removal of Cu, Cr, Pb, U and Zn, claiming similar low current densities and high mass transfer coefficients. Carbon fibre electrodes have also been promoted (Mohanta and Das Gupta 1982). The alternatives of fluidised bed (used by Heger *et al.* 1982 to reduce Cr(VI) to Cr(III) and rotating cylindrical cathodes have also been tried, the latter principally for copper.

However, one of the commercially successful designs has been the parallel plate 'Chemelec' cell, which does not provide a particularly low current density: its advantages of relative simplicity have been exploited by the mode of its application in metal finishing operations.

As already mentioned, metal finishing and related operations involve immersing the workpiece in a series of process solutions, with intermediate rinsing with water to clean the surfaces of excess process solution before transfer to the next one. One or more rinse stations may be used, and the discharges from these constitute the major source of continuous effluent from surface treatment plants. Where for example electroplating is carried out, the rinses will contain dissolved metals; the mass quantities in these rinses depend on the amounts dragged over from the preceding process solution, but the concentration depends on the quantities of rinse water used. Since for electrochemical recovery low concentrations are a problem, organisation of the rinse discharges is an important prerequisite for effective application.

Consider a metal finishing process and its following rinse(s) as depicted in Fig. 1.5 (p. 35). Suppose that the dragout of metal-containing solution is at a rate D, and that the flow rate through a single rinse station is W. On the assumption of steady state conditions, the dilution of the original process solution in the following rinse will be $(W + D)/D$. In a second rinse this dilution is much greater, being $[(W + D)/D]^2$. Clearly it is far more advantageous to recover from the first rinse than from succeeding ones, where not only will the metal concentration be quite low (in practice usually only a few milligrams per litre as opposed to several hundreds or thousands of milligrams per litre in the first rinse) but the amount to be recovered is correspondingly also reduced. There is moreover a powerful need to keep the rinse types separate, because mixtures of metals or even of one metal type with different anions can create substantial recovery difficulties (apart from the further dilution involved).

Application of the Chemelec cell has taken advantage of the stronger concentration in the first rinse, by recirculating this rinse through the cell and back to the rinse station in a closed loop. Operation of the cell to give a residual metal ion concentration of 100–250 mg l^{-1} can extract some 90% of the metal; the following rinse may then have a concentration of 1–2.5 mg l^{-1} if 100-fold dilution is applied to the dragged-over solution, and this may often be run to waste or mixed with other rinses for further dilution. This technique has been described by Surfleet and Crowle (1972), and is similar to that used in the Lancy 'integrated' chemical treatment approach for metal finishing wastes, where the first rinse is subjected to chemical dosing in closed loop operation, with significant contaminant dilution in succeeding rinses (see Pinner 1967 for a review of the Lancy system).

Use of the Chemelec cell in the particularly favourable case of copper deposition has been described by Lopez-Cacicedo (1975, 1981), where current densities of the order of 50–115 A m^{-2} were used, at copper concentrations (in copper sulphate) in the region of 175–800 mg l^{-1}, with current efficiencies of 70–80%. Optimum performance, using a titanium mesh cathode and 6% Sb lead anode, was obtained with a current density of 113 A m^{-2} and a copper concentration of 475 mg l^{-1} giving 88% current efficiency. pH control is essential and is maintained by automatic addition of sodium hydroxide. With acid zinc solutions the current efficiency is lower, being 30–50% with current densities of 65–150 A m^{-2} and starting zinc concentrations of 520–550 mg l^{-1}, although performance is much better with alkaline zincate solutions. The deposition of nickel from Watt-type nickel-plating solutions also requires pH control by sulphuric acid addition, and Bettley *et al.* (1981) found a quasi-limiting current density around 150 A m^{-2}; efficiencies naturally fall as the metal concentration reduces, down to the order of 30% at 200 mg l^{-1} Ni. Cadmium has also been studied (Tyson 1984), and tin is another candidate.

Electrochemical treatment of copper, silver and gold rinses, again based on working with a first static rinse only, and also using a simple plate cell, have been described by Robertson *et al.* (1983); although current efficiencies were not monitored, cost comparisons with ion-exchange treatment were claimed to be favourable. Chelated copper, as from electroless copper-plating wastes, can also be treated with the advantage that pH control within the range 3–11 is unnecessary (Bishop and Breton 1984).

Most application of electrochemical metal recovery in industrial practice has been for the relatively valuable metals silver and gold, where there is the additional advantage that favourable deposition conditions permit the use of simpler, higher current-density cells and removal to lower concentrations than is possible with less noble metals. Mercury can also be removed well, as has been shown by Matlosz and Newman (1982) with contaminated brine solutions, where Hg reductions from 34 μg l^{-1} down to <5 μg l^{-1} were achieved. On the other hand, the economic advantages of electrolytic recovery of metals such as copper and nickel have been stressed by Lopez-Cacicedo (1975) and indeed there are successful applications, although they have not been as extensive so far as simple economic arithmetic should inspire. One may ask why this is. Certainly in some situations the capital outlay deters potential users; in others there have been doubts as to the purity of the recovered product being adequate to ensure an attractive re-sale price or problem-free re-use of the recovered metal in the plating line, where high quality is essential. It must be appreciated that for metal recovery, as in all recovery operations, economic arithmetic is not the only factor an industrialist takes into account: if the recovered material cannot be safely re-used in his process, or if it is in excess of his requirements, then he must resort to selling it, which brings a new dimension to his business that he may be reluctant to pursue. Prospective purchasers are often unwilling to conclude contracts for the purchase of recovered materials when supplies to them are likely to be variable in quality and quantity, and the establishment of suitable trading relationships can involve a company operating recovery processes in activities both technical and commercial that are outside its normal business expertise. For this reason many companies prefer in practice to have their valuable wastes removed for recovery by others.

This pattern of attitudes can shift the emphasis away from in-house recovery to specialised recovery at treatment centres, and some supply houses selling process solutions will accept the spent solutions back for recovery of the valuable materials, e.g. copper from spent etchants. Recovery is also feasible from metal hydroxides (provided there is only one metal or a simple mixture), by electrolytic treatment on a batch basis after dissolution in an appropriate acid, and a patent by Cannell (1980) describes equipment specifically for that purpose.

There is no lack of interest in process solution recovery associated with removal of accumulated metal, as in the treatment of spent etchant solutions used particularly in the printed circuit manufacturing industry. Cupric chloride, for example, accumulates copper as cuprous chloride and eventually loses its etching qualities. The amount of metal lost by chemical treatment and disposal can be quite significant, and it is common practice to effect partial recovery by chemical oxidation with hydrogen peroxide. Cathodic removal of the copper represents, however, an attractive alternative. Hillis (1983, 1984) describes a cell suitable for such regeneration, using a cation exchange membrane to separate catholyte from anolyte, and graphite cathode and anode. The arrangement for continuous regeneration is shown in Fig. 5.11:

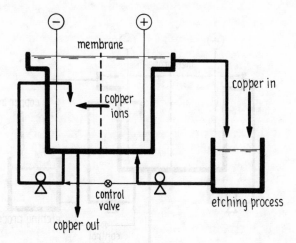

Fig. 5.11 — Diagrammatic representation of Capenhurst electrochemical cupric chloride etchant recovery cell system (also suitable for ferric chloride) (reproduced from Hillis 1984).

etchant is pumped from the etching machine into the anode compartment, where oxidation occurs, e.g.

$$Cu^+ - e \rightarrow Cu^{2+}$$

and the regenerated solution returns to the machine. Migration of cuprous ions across the membrane takes them to the cathode, where they are deposited by reduction as a dendritic powder. The cell is also applicable to the batch regeneration of ferric chloride etchant (Hillis 1979): the spent material, containing ferrous and cupric ions as impurities, is placed in the cathode compartment, with ferrous chloride in the anode compartment. Electrolysis initially forms ferrous iron (from ferric) at the cathode, and then subsequently copper deposition occurs. When most of the copper has been removed the etchant is transferred to the anode compartment and a new batch of spent etchant is placed in the cathode compartment. Further electrolysis oxidises Fe^{2+} at the anode to Fe^{3+}, thus recovering substantially pure ferric chloride. A similar batch treatment principle was also earlier described by Tirrell (1971).

Ammonia/ammonium chloride solutions are also commonly employed as copper etchants. During etching diammino cuprous ions are formed via

$$Cu + Cu(NH_3)_4^{2+} \rightarrow 2Cu(NH_3)_2^+$$

which withdraws ammonia from availability for etching, and so reduces the etching rate. Re-oxidation back to the cupric condition can be achieved by aeration, which occurs naturally in the usual type of spray etching machine. Electrochemical copper removal can, however, additionally be practised, as described by Hillis (1983, 1984) and shown diagrammatically in Fig. 5.12. Once again, a divided cell with cation

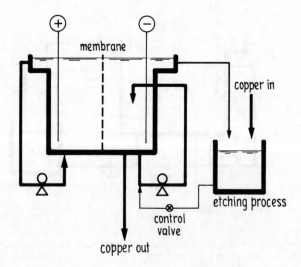

Fig. 5. 12 — Diagrammatic representation of Capenhurst electrochemical ammoniacal etchant recovery system (reproduced from Hillis 1984).

exchange membrane separator is employed, with cooled circulating sulphuric acid as anolyte and cooled etchant solution circulated through the cathode chamber: copper deposits at the graphite cathode with re-formation of oxidised etchant. Because the cation exchange membrane does not completely prevent transfer of chloride ions into the anode compartment, some chlorine is evolved at the anode, which should preferably be absorbed in a scrubber system. This leakage also demands the use of appropriately resistant anode materials, such as platinised titanium.

Another area where electrolytic process solution recovery could be attractive is in the oxidation of Cr^{3+} formed as a degradation product in chromate etchants and passivation solutions. Patents and papers have described equipment for such purposes, e.g. one using an electrodialysis cell (Great Lakes Carbon Corporation 1968), and another study by Apfelbach (1979).

Oxidation of cyanides

Another electric waste water treatment application that has been extensively studied is the anodic oxidation of cyanides. Kuhn (1971) has reviewed work up to 1970; here a summary of salient aspects will be given to highlight the advantages and limitations.

There is no general agreement in the literature on the oxidation mechanism. Schmidt and Meinert (1957) and Sawyer and Day (1963) consider that the primary mechanism involves formation and subsequent decomposition of cyanogen:

$$CN^- \rightarrow CN\cdot + e \qquad (5.16)$$
$$2CN\cdot \rightarrow (CN)_2$$
$$(CN)_2 + 2OH^- \rightarrow CN^- + CNO^- + H_2O$$

The formation of other products found, e.g. ammonia, is ascribed to reactions of cyanogen with the solution. On the other hand it is possible that the anode material may have some influence on the reaction course, with graphite being different from platinum. There are several claims, e.g. by Sperry and Caldwell (1949), Meyer *et al.* (1953) and Drogon and Pasek (1965), that another reaction is involved:

$$CN^- + 2OH^- \rightarrow CNO^- + H_2O + 2e \qquad (5.17)$$

while Easton (1966) also gives

$$2CNO^- + 4OH^- \rightarrow 2CO_2 + N_2 + 2H_2O + 6e$$

Hillis (1975) has suggested that the experimental observation that between 1.7 and 2.0 Faradays are required for oxidation of one mole of cyanide is evidence for mechanisms (5.16) and (5.17) occurring simultaneously, but this clearly cannot be the case, since it is apparent that 2 Faradays are theoretically required for both reactions (the third step of (5.16) regenerates CN^-). However, a one Faraday/mole mechanism is possible (Dart *et al.* 1963):

$$2CN^- \rightarrow (CN)_2 + 2e$$

$$(CN)_2 + 4H_2O \rightarrow 2NH_4^+ + (COO)_2^{2-}$$

The presence of small amounts of oxalate in the reaction products supports this probability. Cyanate formed in the reactions subsequently hydrolyses, possibly via

$$CNO^- + 2H_2O \rightarrow NH_4^+ + CO_3^{2-}$$

and

$$CNO^- + NH_4^+ \rightarrow (NH_2)_2CO$$

Many cells have been described (see Kuhn 1971), frequently employing graphite anodes and steel cathodes, usually operated at warm temperatures (40–90°C), and desirably with good solution agitation. Current efficiencies fall dramatically with cyanide concentrations below about 1000 mg l^{-1}, which has meant that application has mostly been to treatment of spent cyanide concentrates, e.g. hardening salts (containing 10–15% CN^-), and spent plating solutions. General experience with most cells is that the operating efficiency is in the region of 0.006–0.015 kW h g^{-1} CN^- for solutions stronger than 1000 mg l^{-1} CN^-: Fig. 5.13 shows the effect of

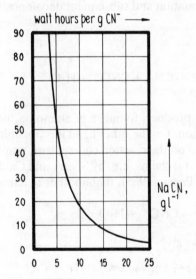

Fig. 5.13 — Effect of cyanide concentration on energy consumption (current efficiency) in the anodic destruction process (reproduced from Entwistle 1976).

cyanide concentration on current efficiency as found by Entwistle (1976) with his cell design, which has been commercially operated successfully for several years. It is interesting to note that Chin and Eckert (1976) observed no greater efficiency levels even using packed carbon electrodes.

One of the advantages of electrolytic cyanide oxidation is that it can also treat complexed cyanides of nickel and iron, for the latter of which chlorination is unsatisfactory. Where metals are present, these may deposit on the cathode (offering some potential recovery opportunities) and/or form metal hydroxides. Difficulties can be experienced with some spent cyanide metal-stripping solutions, e.g. those containing *m*-dinitrobenzene sulphonate, but generally the method offers advantages over chlorination in terms of cost for the concentrated liquors. The operation is not fast, as is shown by the representative curve of Fig. 5.14 (J. E. Entwistle,

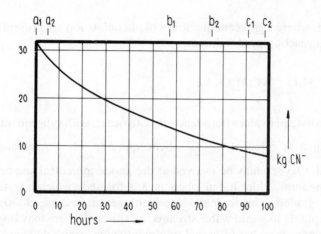

Fig. 5.14 — Rate of anodic destruction of cyanide using the Entwistle cell. a_1–a_2: 0.9 mol per Faraday. b_1–b_2: 0.5 mol per Faraday. c_1–c_2: 0.3 mol per Faraday. Overall: 0.7 mol per Faraday.

personal communication, 1975), and is not readily adapted to continuous treatment of flowing wastes. However, a significant improvement can be obtained when chloride is present, probably simply from the formation of chlorine or hypochlorite, thus to provide an alternative route for cyanate formation, with a faster reaction and improved current efficiency. There is, however, some tendency to form chlorate, ClO_3^- (Byrne *et al.* 1958, Ibl and Landolt 1968), which is an undesirable side reaction, and a more rapid deterioration of carbon anodes occurs when chloride is present.

When large amounts of chloride are present, even dilute solutions of cyanide can be electrochemically oxidised rapidly. Drogon and Pasek (1965) used retention times of $1\frac{1}{2}$ hours, but Hillis (1975) has demonstrated that only 1–3 minutes is necessary to treat up to 400 mg l^{-1} CN^- to undetectable residual concentrations. Hillis used a flow cell with a platinised titanium anode at a throughput rate of 273 l h^{-1}, a cell voltage of 9.8 V and a current density of 200 A m^{-2}, with a background medium of 3% sodium chloride. Of course, in some circumstances this high concentration of salt would be undesirable in a discharge.

Treatment of organic wastes

It may be supposed that anodic treatment of organic materials may present a means for lowering C.O.D. levels in wastes, and interest has been shown in a number of studies, although with only limited success and inadequate viability with respect to alternative treatment methods.

One obvious area for treatment is phenolic compounds. The mechanisms are not simple, because there are effects that depend on pH, the background medium, and the formation of degradation products that themselves are then involved in further reactions. Vermillion and Pearl (1964), following Hedenburg and Freiser (1953), consider that with phenol the primary reaction at low pH is

$$C_6H_5OH \rightarrow C_6H_5O^+ + H^+ + 2e$$

At high pH, where significant quantities of phenolate ion are present, a one-electron free radical reaction predominates:

$$C_6H_5O^- \rightarrow C_6H_5O\cdot + e$$

At intermediate pH values both reactions can occur, with subsequent decomposition products such as catechol, $\underset{}{\overset{OH}{\bigcirc}}^{OH}$, hydroquinone, $\underset{OH}{\overset{OH}{\bigcirc}}$, quinone, $\underset{O}{\overset{O}{\bigcirc}}$ and maleic acid. Oxygen may be evolved at the anode in neutral and acid solutions. In chloride medium, chlorinated phenols are formed, which are themselves only partially degraded. This is clearly disadvantageous, because chlorinated phenolics are unacceptable in waste water streams, so although there may be more operating efficiency the process has little real treatment value. Smith de Sucre and Watkinson (1981) have demonstrated that phenol can be oxidised virtually completely in the absence of chloride, the efficiency increasing with decreasing pH and initial phenol concentration, but total organic carbon residuals are still significant. It is also clear from the work of Surfleet (1969) that electrolytic treatment of phenols is uneconomic.

Some application of electrolysis to the treatment of kraft mill wastes has been reported, particularly in the Soviet literature. Das Gupta (1979) has used a carbon fibre based electrode cell to decolorise kraft mill bleach plant effluent, obtaining over 90% colour removal in approximately 6 minutes, with 50–60% C.O.D. removal. Oehr (1978) used a flow-through cell with lead dioxide anode and stainless steel cathode on caustic extraction and total mill effluents, finding that decolorisation was enhanced with increasing sodium chloride content and decreasing current density. There is a risk of the introduction of lead into the water when lead dioxide electrodes are used, however (A. B. Wheatland, personal communication, 1985).

Electrolytic flotation
In conventional flotation processes for phase separations, small gas bubbles are formed within the bulk of the liquid phase, usually by sudden pressure release. These small bubbles attach themselves to phase boundaries and lift suspended solids (or oil droplets) to the surface of the liquid phase. The generation of gas bubbles by passing a small electric current between suitable electrodes constitutes an alternative procedure, and commercial cells have been developed and used for waste water applications. The subject is outside the main theme of the present text, but is mentioned briefly as being an electrolytic operation.

Electrodes are normally inserted in a region below the flotation tank inlet distribution system, using a voltage sufficient to exceed hydrogen and oxygen overpotential values (5–20 V), and current densities of the order of 100–30 A m^{-2} may be employed; lower voltages are possible with higher solution conductivities, with consequent power savings. Increasing the current density generally reduces the flotation time, producing more gas bubbles, but is eventually wasteful through overproduction. There are some advantages in separating the cathode and anode zones, e.g. by a membrane, to utilise one electrode only; Matis (1980) for example found that although electrogenerated oxygen usually required the presence of a coagulating agent with an oil-in-water emulsion, the hydrogen at the cathode may be capable of breaking the emulsion on its own. Retention times necessary to achieve a given degree of separation increase with increasing initial concentration of material to be separated.

The method has been used for protein recovery from dairy wastes (Lewin and Forster 1974). It is also useful as an aid in emulsion breaking, as with latex wastes (Barrett 1976) and paint wastes (Backhurst and Matis 1981), but generally a coagulant is necessary as the primary agent, as with conventional dissolved air flotation processes. Indeed, in a comparative study between dissolved air flotation (see Chapter 7 for a discussion of this) and electroflotation for treatment of packing-house waste waters, Ramirez et al. (1976) found no difference in performance between the two.

Two criticisms can be levelled at electrolytic flotation: the relatively high cost of the equipment, and the limited life of the electrodes. There is, moreover, a small explosion risk associated with the generation of hydrogen and oxygen (or chlorine).

THE PRACTICAL REALITIES

It must be said that notwithstanding all the research and development that has been carried out on electrochemical waste water treatment, only limited serious plant application has resulted. One significant exception is for recovery of precious metals such as silver and gold, where conditions are particularly favourable, and there is a developing interest, again in the metal finishing industry, in the recovery of specific process solutions, such as spent etchants, giving metals such as copper as a byproduct. Nevertheless, as has already been mentioned earlier in this chapter, the apparent attractiveness of cost calculations does not necessarily result in the use of electrochemical techniques: indeed it has been known for companies practising electrochemical metal recovery to take the systems out of service, ostensibly because operational demands were inadequately offset by advantages gained.

Outside recovery, for waste water treatment alone, electrochemical methods have so far gained negligible acceptance by dischargers, with the main exception of application to destruction of concentrated cyanides. One possible reason may be that the capital investment in relation to benefits that may be gained over direct additions of chemical reagents is too high: but it seems anyway that it is another example where academic enthusiasm is outweighed by wider commercial considerations.

REFERENCES

Adamson, A. F., Lever, B. G., and Stones, W. F. (1963) *J. Appl. Chem.* **13** 483.
Agarwal, J. C., Buchon, A. M., Gieser, H. D., and Sparling, A. B. (1984) *Wat. Res.* **18** 237.

Apfelbach, R. D. (1979) *Galvanotechnik (Saulgau)* **70** 144.

Backhurst, J. R., and Matis, K. A. (1981) *J. Chem. Tech. Biotechnol.* **31** 431.

Backhurst, J. R., Coulson, J. M., Goodridge, F., Plimley, R. C., and Fleischmann, M. (1969) *J. Electrochem. Soc.* **116** 1600.

Barker, B. D., and Plunkett, B. A. (1976) *Trans. Inst. Metal Finish.* **54** 104.

Barrett, F. (1976) *Chem. Ind. (London)* 880.

Bennion, D. N., and Newman, J. (1972) *J. Appl. Electrochem.* **2** 113.

Bettley, A., Tyson, A., Cotgreave, S. A., and Hampson, N. A. (1981) *Surf. Technol.* **12** 15.

Bishop, P. L., and Breton, R. A. (1984) *Proc. 38th Ind. Waste Conf. Purdue Univ.* 473.

Bockris, J. O'M., and Reddy, A. K. N. (1970) *Modern Electrochemistry*, Vol. 2, Macdonald.

Byrne, J. T., Turnley, W. S., and Williams, A. K. (1958) *J. Electrochem. Soc.* **105** 607.

Cannell, J. F. (1980) *U.S. Patent* No. 4, 214, 964.

Chin, D. T., and Eckert, B. (1976) *Plat. Surf. Finish.* **63** 38.

Chu, A. K. P., Fleischmann, M., and Hills, G. J. (1974) *Appl. Electrochem.* **6** 323.

Crow, D. R. (1979) *Principles and Applications of Electrochemistry*, 2nd edn, Chapman and Hall.

Dart, M. C., Gentles, J. D., and Renton, D. G. (1963) *J. Appl. Chem.* **13** 55.

Das Gupta, S. (1979) *Pulp Paper Mag. Can.* **80** 68.

Das Gupta, S., and Fleet, B. (1976) *Nature* **263** 122.

Dobolyi, E. (1978) *Wat. Res.* **12** 1113.

Drogon, J., and Pasek, L. (1965) *Electroplat. Metal Finish.* **18** 310.

Easton, J. K. (1966) *Plating* **53** 1341.

Entwistle, J. E. (1976) *Effl. Wat. Treat. J.* **16** 123.

Føyn, F. (1959) *Proc. Int. Conf. Waste Disposal in the Marine Environment* 279.

Gabe, D. R., and Walsh, F. C. (1983) *J. Appl. Electrochem.* **13** 3.

General Electric Co. (1971) *U.S. Patent* No. 3, 788, 967.

Great Lakes Carbon Corporation (1968) *Brit. Patent* No. 1, 233, 444.

Hedenburg, J. F., and Freiser, H. (1953) *Anal. Chem.* **25** 1355.

Heger, K., Nowack, N., Pechtheiden, C., and Schloten, F. (1982) *Galvanotechnik (Saulgau)* **73** 1071.

Hillis, M. R. (1975) *Trans. Inst. Met. Finish.* **53** 65.

Hillis, M. R. (1979) *Trans. Inst. Metal Finish.* **57** 73.

Hillis, M. R. (1983) *Ion Exchange Membranes*, ed. Flett, D. S., Ellis Horwood.

Hillis, M. R. (1984) *Trans. Inst. Metal Finish.* **62** 21.

Holland, F. S. (1978) *Chem. Ind. (London)* 453.

Ibl, N., and Landolt, O. (1968) *Electrochem. Sci.* **45** 713.

Jansson, R. E., and Ashworth, G. A. (1977) *J. Appl. Electrochem.* **7** 309.

Jenke, D. R., and Diebold, F. E. (1984) *Wat. Res.* **18** 855.

Kammel, R., and Lieber, H. W. (1978) *U.S. Patent* No. 4, 123, 340.

Kramer, G. R., Buyers, A., and Brownlee, B. (1978) *Proc. 33rd Ind. Waste Conf. Purdue Univ.* 673.

Kuhn, A. T. (1971) *J. Appl. Chem. Biotechol.* **21** 29.

Lewin, D. C., and Forster, C. F. (1974) *Effl. Wat. Treat. J.* **14** 142.

Lopez-Cacicedo, C. (1975) *Trans. Inst. Metal Finish.* **53** 74.

Lopez-Cacicedo, C. (1976) *British Patent* No. 1, 423, 369.

Lopez-Cacicedo, C. (1981) *J. Separ. Proc. Technol.* **2** 34.

Mantell, C. L. (1960) *Electrochemical Engineering*, McGraw-Hill.

Marson, H. W. (1967) *J. Inst. Sewage Purif.* **66** 109.

Matis, K. A. (1980) *Wat. Pollut. Control* **79** 143.

Matlosz, M., and Newman, J. (1982) *Proc. Electrochem Soc.* **82** 53.

Meyer, W. R., Muraca, R. F., and Serfass, E. J. (1953) *Plating* **40** 1104.

Mohanta, S., and Das Gupta, S. (1982) *Proc. 29th Ont. Ind. Waste Conf.* 81, Ontario
    Water Resources Commission.

Müller, K.-J. (1983) *Galvanotechnik (Saulgau)*, **74** 902.

Newman, J. S. (1973) *Electrochemical Systems*, Prentice-Hall.

Nojiri, N., Tanaka, N., Sato, K., and Sakai, Y. (1980) *J. Wat. Pollut. Control Fed.* **52**
    1898.

Oehr, K. (1978) *J. Wat. Pollut. Control Fed.* **50** 286.

Onstott, E. I., Gregory, W. S., and Thode, E. F. (1973) *Environ. Sci. Technol.* **7**
    333.

Ontario Research Foundation (1969) *British Patent* No. 1, 252, 629.

Pinner, R. (1967) *Electroplat. Metal Finish.* **20** 208; 248; 280.

Pletcher, D. (1982) *Industrial Electrochemistry*, Chapman and Hall.

Ramirez, E., Johnson, D. L., and Clemens, O. A. (1976) *Proc. 31st Ind. Waste Conf.
    Purdue Univ.* 563.

Ramirez, E., Barber, L. K., and Clemens, O. A. (1977) *Proc. 32nd Ind. Waste Conf.
    Purdue Univ.* 183.

Robertson, P. M., and Ibl, N. (1977) *J. Appl. Electrochem.* **7** 323.

Robertson, P. M., Schwayer, F., and Ibl, N. (1975) *J. Electroanal. Chem.* **65** 883.

Robertson, P. M., Scholder, B., Theis, G., and Ibl, N. (1978) *Chem. Ind. (Lond.)*
    457.

Robertson, P. M., Lendolph, J., and Maurer, H. (1983) *Plat. Surf. Finish.* **70** 48.

Sawyer, L. B. and Day, R. J. (1963) *J. Electroanal. Chem.* **5** 195.

Schmidt, H., and Meinert, H. (1957) *Z. Anorg. Allgem. Chem.* **293** 214.

Smith de Sucre, V., and Watkinson, A. P. (1981) *Can. J. Chem. Eng.* **59** 52.

Sperry, L. B., and Caldwell, M. R. (1949) *Plating* **36** 343.

Surfleet, B. (1969) *The Electricity Council Research Centre (UK) Report
    ECRC/R204.*

Surfleet, B., and Crowle, V. A. (1972) *Trans. Inst. Metal Finish.* **50** 227.

Tarjanyi, M., Tonawanda, N., and Strier, M. P. (1971) *U.S. Patent* No. 3, 755, 114.

Tirrell, C. E. (1971) *U.S. Patent* No. 3, 761, 369.

Tison, R. P. (1981) *J. Electrochem. Soc.* **128** 317.

Tyson, A. G. (1984) *Plat. Surf. Finish.* **71** 44.

Van der Heiden, G., Raats, C. M. S., and Boon, H. F. (1978) *Chem. Ind. (London)*
    465.

Vermillion, F. I., Jr., and Pearl, I. A. (1964) *J. Electrochem. Soc.* **111** 1392.

Wabner, D., and Grambow, C. (1983) *Vom Wasser* **60** 181.

Williams, J. M. (1975) *U.S. Patent* No. 3, 859, 195.

6

Chemical and adsorptive precipitation and flocculation

THE SIGNIFICANCE OF PRECIPITATION IN POLLUTION ABATEMENT

If oxidation is one of the primary means available for eliminating pollution, with pH adjustment as a factor both in the processes and in the aquatic environment itself, then precipitation of contaminants and residues is another. Oxidation processes and pH adjustment themselves frequently generate insoluble materials, the removal of which is part of the overall treatment process. In other cases, oxidation (or reduction) may not be relevant, homogeneous chemical change either not being feasible or producing materials equally unacceptable to the environment. Selective precipitation and subsequent removal of the solids formed then constitutes the most effective treatment procedure.

The extent to which the formation of insolubles is a complete answer to pollution depends very much on their character. Phase change is not uncommon in the biological or other oxidative degradation of organic contaminants, including of course sewage, where not only are solids settled as a first treatment stage, but are formed as a consequence of oxidation, which therefore necessitates secondary sedimentation. An important consideration in any treatment scheme is therefore the ultimate disposal of the solids formed. Where these can be further oxidised by aerobic or anaerobic digestion, treatment can be nearly complete, but with inorganic and some industrial organic materials, the solids formed present an ultimate disposal problem. Nevertheless, for many industrial pollutants the only available 'treatment' is one of phase change, to remove the species physically from solution through formation of another product. The simple fact of ultimate indestructibility means that solving one pollution problem can generate another; but while this cannot be escaped, there is the possibility in principle of containment, so that water purification can be undertaken with some advance knowledge of the eventual consequences.

In this chapter the various types of precipitation, coagulation and flocculation processes are analysed in the context of their application in waste water treatment. The associated techniques of adsorption on to a prepared solid phase, such as activated carbon or ion exchange materials, are briefly mentioned where apposite, but they are broadly outside the defined scope of this book. For discussions on both fundamentals and applications of these subjects the reader is referred to reviews such as the following:

Activated carbon adsorption: Hassler (1963), Gloyna and Eckenfelder (1970), Culp and Culp (1971), Mattson and Mark (1971), Eckenfelder and Cecil (1972), Weber (1972), Cheremisinoff and Ellerbusch (1978), and McGuire and Suffet (1983).

Ion Exchange: Helfferich (1962), Arden (1968), Dorfner (1972), Weber (1972), and Kunin (1986).

PRECIPITATION

Fundamental concepts in relation to the precipitation process

In a simple ideal system, the equilibrium between an ionic solid and its ions in solution may be represented for a 1:1 electrolyte thus:

$$MX_{(solid)} \rightleftharpoons M^+_{(solution)} + X^-_{(solution)}$$

for which reaction the equilibrium expression is

$$K = a_{M^+} \cdot a_{X^-}/a_{MX}$$

Ideally the activity of the solid phase a_{MX} is fixed by the presence of excess pure solid, and it is conventional and also convenient to assign to this a value of unity; i.e. the pure solid is taken to be the thermodynamic standard state. Adopting this convention, the equilibrium expression for a 1:1 electrolyte simplifies to:

$$K^0_{sp} = a_{m^+} \cdot a_{X^-} = c_{M^+} \cdot c_{X^-} \cdot f^2_{\pm} \tag{6.1}$$

where the equilibrium constant K^0sp is referred to as the solubility product and $f_{\pm}$ is the mean ionic activity coefficient (of the participating ions). For very sparingly soluble salts in the absence of other electrolytes $f_{\pm}$ may be taken as unity, since the concentration of the dissolved ions tends to zero. However, this ideal situation is seldom, if ever, realised in practice, and in both natural and waste water streams there are invariably other ionic species present which may have quite marked and even extreme effects on the observed solubilities of precipitates. In view of the great importance of precipitation in waste water treatment processes, it is appropriate that we examine these effects in some detail. (For texts on the formation and properties of precipitates, see Nielsen 1964 and Walton 1967.)

Factors affecting the solubilities of precipitates

Ionic strength

In any solid/solution system such as that obtained when a sparingly soluble salt is in equilibrium with its ions in solution, any change in the activity coefficient of the dissolved solute species will result in a concomitant change in the solubility of the substance, as measured for example by analysis of the solution; i.e. the stoicheiometric solubility product K_{sp} will change. It should be noted, however, that the thermodynamic solubility product K_{sp}^0 will not change. Since in general it is the stoicheiometric concentration of the substance left in solution that is of practical interest in waste water treatment then the extent of the change in K_{sp} is clearly important.

In relatively dilute aqueous solutions (say < 0.1 M), activity coefficients for ionic species are less than unity and decrease with increasing ionic strength. Under these circumstances, the addition of inert electrolytes will tend to increase the solubilities of sparingly soluble salts, since activity coefficients and hence activities of the constituent ions of the salts will be decreased, and thus more solid will dissolve in order to re-establish the activities in accordance with, for example, equation (6.1).

Application of the extended Debye–Hückel law or the Davies equation to the equilibrium expression confirms this argument and quantitatively relates the actual solubility (as measured by K_{sp}) to the ionic strength via the thermodynamic solubility product K_{sp}^0 and the ionic charges. Thus for the general equilibrium

$$M_xX_{m(\text{solid})} \rightleftharpoons xM^{z_m+} + mX^{z_x-}$$

$$K_{sp}^0 = c_M^x \cdot c_X^m \cdot f_M^x \cdot f_X^m \qquad (6.2)$$

$$= K_{sp} \cdot f_M^x \cdot f_X^m$$

or

$$\log K_{sp}^0 = \log K_{sp} + x \log f_M + m \log f_X \qquad (6.3)$$

where f_M and f_X are the single ion activity coefficients for species M^{z_m+} and X^{z_x-} respectively.

The Davies equation relates the activity coefficient of an ion to the ionic strength by

$$\log f = -0.5Z^2 \left[\frac{I^{\frac{1}{2}}}{1+I^{\frac{1}{2}}} - 0.2I \right] \qquad (6.4)$$

where I is the ionic strength defined by

$$I = \frac{1}{2} \sum_i c_i z_i^2$$

The summation is for all ionic species present of concentration c_i and charge z_i.
Substituting (6.4) into (6.3) and rearranging we obtain the important result

$$\log K_{sp} = \log K_{sp}^0 + 0.5\left[\frac{I^{\frac{1}{2}}}{1+I^{\frac{1}{2}}} - 0.2I\right]\left[xz_M^2 + mz_X^2\right] \tag{6.5}$$

It is apparent from equation (6.5) that not only is the solubility of a salt dependent on the ionic strength, but this effect of added inert electrolyte is much more pronounced for those sparingly soluble salts which contain polyvalent ions (e.g. $CaSO_4$) because of the second power dependence on z, the ionic charge.

On the other hand, in relatively concentrated electrolytes it is possible that the solubility of a sparingly soluble salt may decrease compared to its value in more dilute solutions of electrolytes. This is sometimes referred to as a 'salting-out' effect, and is due to the retention of solvent in the hydration spheres of the electrolyte ions reducing the effective solvent concentration. This phenomenon is not restricted in its effects to the reduction of solubilities of ionic substances, but is also responsible for the reduced solubilities of gases and other non-ionic substances in the presence of added electrolytes. It is to be expected that highly hydrated ions (small size and/or high charge) will constitute the most effective 'salting-out' electrolytes, and in general this is observed to be the case.

The common ion effect

If an electrolyte is added to a solution of a sparingly soluble salt that is in equilibrium with undissolved solid, then if the added electrolyte contains an ion that is also a constituent of the sparingly soluble salt the solubility of the salt will be suppressed as the equilibrium is driven in the direction of undissolved solid. This 'common ion effect' is usually held to be a manifestation of the ubiquitous Le Chatelier's principle. It is important to note that in addition to exerting a common ion effect such an electrolyte will also increase the ionic strength of the solution, and hence tend to change the solubility as described in the section above. In dilute solutions of electrolytes, these two effects operate in opposite directions, although it is usually the common ion effect that has the greater influence on the observed solubility.

Complex formation

The precipitation of insoluble products can be markedly influenced by the co-presence of other species that are able to exert a complexing action on the ions concerned. A complex may result in a lower solubility, but more usually causes some inhibition of full precipitation. Well-known examples occur in the inhibiting effect of complexing agents on the precipitation of heavy metal ions as hydrous oxides, discussed in Chapter 2 (p. 77 et seq.).

Particle size

Although it is common experience and universally recognised that the rate of dissolution of small particles is greater than the rate of dissolution of larger particles of the same substance, it is perhaps not so widely realised that the ultimate solubility of a substance, the position of final equilibrium, may also be a function of particle size. This behaviour is not at variance with thermodynamic principles, but stems from the limited validity of the assumption made when formulating the solubility product expression (equation (6.1) for example) that the activity of the solid phase is constant. This assumption, whilst valid for macroscopic particles of the same well-defined crystal form, is not necessarily true for microscopic particles of colloidal dimensions where a large proportion of the atoms or ions of the solid phase are at the surface, and where in view of the large surface : volume ratio surface energy effects become important. It is also true and well recognised that even at the macroscopic level different crystal forms of the same compound have different solubilities. For example, calcite is less soluble ($\log K_{sp} = -8.42$) than aragonite ($\log K_{sp} = -8.22$), although both are calcium carbonate. In view of the energy involved in breaking bonds on sub-division of a solid, it is not surprising that the solubility of a pure solid is to some extent a function of particle size.

If the solid particles are dispersed in water, the energy that matters in comparing the behaviour of large and small particles is the difference between the lost lattice energy and the gain in hydration energy of the new surfaces. The picture that emerges is that in general small particles have higher surface energies (or interfacial energies when dispersed in a liquid), and hence are less stable and therefore more soluble than larger particles of the same material.

Quantitatively it can be shown that the increase in free energy ΔG resulting from the sub-division of a coarse material suspended in an aqueous solution into a finely divided solid of molar surface S is given by

$$\Delta G = 2/3 \gamma S \tag{6.6}$$

where γ is the mean surface energy of the solid/liquid interface. It follows therefore that the relationship between the solubility of the macroscopic (coarse) solid of negligible molar surface ($S \to 0$) and the microcrystalline solid is given by

$$\log K_{sp(s)} = \log K_{sp(s \to 0)} + \frac{2/3 \gamma S}{2.303 RT} \tag{6.7}$$

Schindler (1959) and Schindler *et al.* (1965) have verified an equation of this form for ZnO, $Cu(OH)_2$ and CuO.

The fact that very small particles are more soluble than larger particles has important practical implications in that over a period of time there is a tendency for larger particles to grow at the expense of smaller ones ('For I say unto you, that unto everyone that hath shall be given; and from him that hath not, even that he hath, shall be taken away . . .' St Luke 19: 26; alternatively, 'The rich get richer and the poor get poorer'!). This natural 'ageing' of finely precipitated solids, which is generally accelerated by increased temperature, leads to improved filterability and settling characteristics and results in lower concentrations of soluble material remaining in solution.

The solubility differences between different particle sizes are accentuated when precipitation occurs from solutions which are grossly oversaturated. In this case not only is there initial formation of a very finely divided precipitate, but the initial solid phase frequently has a highly disordered lattice and may be thermodynamically much less stable than the normal solid phase. In spite of this thermodynamic instability, such highly disordered structures may persist for a considerable time, constituting a metastable state which transforms to the thermodynamically stable state only slowly. Such solid phases have been called 'active' forms (Feitknecht and Schindler 1963), and are often considerably more soluble than one might be led to expect from a consideration of solubility product data derived from measurements involving normal solid phases.

In view of the number of factors which can affect the solubility of a precipitate and the extremely complex, variable and ill-defined nature of many waste water streams, it is apparent that tabulated values of solubility products are only useful as a guide to possible treatment feasibilities, and no more. They do not in any way provide a means of making precise quantitative predictions; indeed, the tabulated values themselves sometimes vary widely, with orders of magnitude differences not unknown. The situation is also frequently complicated by the fact that the precipitates are not always or necessarily of a stoicheiometrically defined nature, and may incorporate occluded material. A significant example is in the precipitation of metal hydrous oxides, where the products by no means correspond to a known hydroxide theoretical composition.

FLOCCULATION AND COAGULATION OF COLLOIDS

Introduction
Particles suspended in water, whether generated precipitatively as part of a treatment process or present as a natural constituent of a waste stream, can vary in size over many orders of magnitude. The larger particles with dimension of 10^{-2} cm or more are, in general, readily separable from the aqueous phase by settlement or conventional filtration. Particles of colloidal dimensions ($<$ ca. 0.5×10^{-4} cm), however, frequently form more stable dispersions which do not settle and which are not amenable to conventional filtration processes. Such systems require preliminary treatment in order to destabilise the colloidal particles and induce particle growth or aggregation before separation of the solid phase is attempted. In a limited number of cases, simple ageing will be effective, but it is often necessary to add reagents to encourage coagulation or flocculation. The specific properties required of coagulating or flocculating agents will depend on the nature of both the colloidal dispersed phase and the liquid dispersing medium, and in order to make decisions regarding reagent choice it is necessary to understand something of the fundamental nature of the colloidal state.

The nature of colloids
For present purposes we shall restrict our consideration of colloidal systems to those in which the dispersion medium is water, or at least an aqueous solution. The discussions which follow can be usefully augmented by reference to other publications, such as those from Mysels (1959), Faust and Hunter (1967), Stumm and O'Melia (1968), Gloyna and Eckenfelder (1970), Eckenfelder and Cecil (1972), O'Melia (1972), Ives (1978), and Stumm and Morgan (1981).

The definition of what constitutes an aqueous colloidal system, referred to as a 'sol', is not clear-cut, but in general it is recognised to be where the suspended particles have at least one dimension between 10^{-7} cm and 10^{-4} cm. This covers the range between relatively small molecules and what the analytical chemist would regard as an ultrafine precipitate. The particles are not removed by passing through a conventional filter. Generally, colloidal particles are not individually easily visible under conventional microscopes, but they are sufficiently large compared to the wavelength of light to scatter visible radiation, and by virtue of this property are often visible under the ultramicroscope in much the same way as dust particles, normally invisible to the naked eye, are picked out in a shaft of sunlight. Sols can be classified into two broad sub-divisions: 'hydrophobic', which are dispersions of an essentially insoluble solid phase in water, and 'hydrophilic', which are true solutions of molecules having colloidal dimensions. Solutions of many polymers and other macromolecules fall into this latter category.

This division of aqueous colloidal systems is widely accepted, but it is important to realise that the so-called hydrophobic sols also have some affinity for the dispersion medium, otherwise no wetting of the particles would occur and no dispersion would be possible. Thus metal oxide or hydrous oxide colloidal particles are classed as hydrophobic even though they interact quite strongly with water molecules and the particle surface is extensively hydrated. From a thermodynamic standpoint an important distinction between hydrophobic and hydrophilic sols is that the latter are in principle thermodynamically stable (although polyelectrolyte solutions, as examples of hydrophilic sols, do degrade as a consequence of depolymerisation): they are true solutions and will spontaneously form when the two components are brought into contact. Hydrophobic sols on the other hand are thermodynamically unstable by virtue of the very large interfacial energy previously referred to in the discussion of the effects of particle size on solubility. Nevertheless, for reasons which will become apparent, these thermodynamically unstable hydrophobic sols may persist as metastable states for considerable periods of time, resisting the energetically favourable processes of separation into two distinct macroscopic phases with minimum interfacial area. One of the objects of treatment is to destabilise such systems in order to achieve separation.

In general, the stability and persistence of hydrophobic sols arises principally from the fact that the colloidal particles carry an electrical charge. The charge on the dispersed particles may be positive or negative and is confirmed experimentally by observation of the movement of the particles under the influence of an electric field (electrophoresis). In the colloidal dispersions encountered in water and waste water streams the particles usually carry a negative primary charge, although the magnitude and even the sign of the charge is dependent on conditions (notably the pH of the dispersing medium). Since under any particular set of conditions the charge on all the particles will have the same sign, it is tempting to ascribe the observed stability of such sols to simple coulombic repulsion between the particles preventing aggregation. Unfortunately this viewpoint is rather too simplistic, and although coulombic forces arising from the surface charge do play a crucial role in determining the stability of hydrophobic sols, the nature of the interactions is somewhat more complex.

Origin of surface charge

The electrical charge present at the surface of dispersed particles may arise as a result of one or more processes. Some of the more important mechanisms for charge generation are as follows.

Chemical reactions at the surface

Many solid surfaces contain functional groups exposed to the liquid phase which may undergo chemical reaction. Of prime importance in aqueous dispersions are association and dissociation reactions involving H^+ or OH^- ions. For example, the surface of a silica particle which contains free silanol groups can become either positively or negatively charged as a result of reactions such as:

$$\overset{\displaystyle H}{\underset{\displaystyle}{\diagup}} \quad \overset{\displaystyle H}{\underset{\displaystyle}{\diagup}}$$

$$-Si-O \quad + \; H^+ \; \rightleftharpoons \; -Si-O^+$$

$$\diagdown H$$

or

$$\overset{\displaystyle H}{\underset{\displaystyle}{\diagup}}$$

$$-Si-O \quad + \; OH^- \; \rightleftharpoons \; -Si-O^- + H_2O$$

The total nett surface charge, σ_0, is defined by

$$\sigma_0 = F(\Gamma_{H^+} - \Gamma_{OH^-})$$

where Γ_{H^+} and Γ_{OH^-} are the numbers of moles of H^+ and OH^- adsorbed per cm², and F is the Faraday constant.

Similar reactions occur with other oxides, hydrated oxides and hydroxides. The difference $(\Gamma_{H^+} - \Gamma_{OH^-})$ and hence σ_0 can be measured by pH-potentiometric methods. It is clear that both Γ_{H^+} and Γ_{OH^-} will be functions of pH (Γ_{H^+} decreasing and Γ_{OH^-} increasing with pH) and it therefore follows that at a sufficiently low pH colloidal particles of oxides or hydroxides will be positively charged, the charge decreasing as the pH is increased until at high pH the particles will have a nett negative charge. At some intermediate pH, which is characteristic of the colloidal material, the particles will have a nett surface charge of zero, i.e. $\Gamma_{H^+} = \Gamma_{OH^-}$. The pH at which this occurs, often termed the isoelectric pH (pH_{iso}), depends significantly on the electronegativity of the central atom. For example, silica has a greater affinity for OH^- ions than for H^+ ions, and the point of zero charge occurs at approximately pH 2, whereas $\alpha - Fe_2O_3$ (haematite) is relatively basic, with $pH_{iso} = 8.5$.

Organic colloidal material containing ionised or potentially ionisable groups (e.g. $-COOH$, $-NH_4^+$) may also develop surface charges by a similar mechanism to the above, the magnitude and sign of which will again be related to the pH of the surrounding medium. Where the organic substance is ampholytic, containing potentially negative and potentially positive groups (e.g. proteinaceous material) then again there will exist a pH where the nett charge of the particle is zero. This isoelectric pH or 'point of zero charge' is an important operational parameter in treatment regimes, and it is also discussed from this standpoint in Chapter 2.

Non-stoicheiometric adsorption of constituent ions

A typical example of surface charge development by this mechanism is afforded by metal sulphide precipitates. Colloidal suspensions of copper sulphide, for example, may become positively charged by the uptake of Cu^{2+} ions, or negatively charged by adsorption of S^{2-} ions. An important point in this type of adsorption is that, once adsorbed, the Cu^{2+} or S^{2-} ions are indistinguishable from the ions constituting the solid phase, i.e. they occupy normal lattice sites and form part of the crystal structure.

Isomorphous replacement

In naturally occurring silicate minerals, isomorphous replacement is extremely common. Frequently ions such as Al^{3+} will replace some of the Si atoms (formally in the $+4$ oxidation state) without changing the crystal structure, thus leading to a charge imbalance. Since these isomorphous replacements are not confined to the surface of the particles but are also found within the bulk of the solid phase, the resultant charge is not so sensitive (at least in the short term) to changing ionic concentrations in the dispersion medium. Particles of this type are found for example where natural storm-water runoff enters waste streams or in waste water arising from mineral-processing operations.

Adsorption of surfactant ions

A wide variety of surface-active substances may adsorb on to colloid particles. If strong specific chemical bonding occurs then the adsorbed surfactant ions may be considered to form an integral part of the surface structure, and hence are properly contributors to the primary surface charge σ_0. If on the other hand the adsorption is fairly weak, reversible and non-specific, it may be more realistic to consider the adsorbed species to constitute part of the diffuse double layer structure (discussed in the next section).

The electrical double layer

The nature of the double layer

From the experimentally observed fact that the colloid particles are charged although the colloidal suspension as a whole is not, it follows inescapably that the charge on the surface of the colloid particle must be balanced by an equal and opposite charge in the dispersion medium. This counteracting charge is not distributed uniformly throughout the dispersion medium but, as may be expected, is concentrated in a mobile layer near the solid/liquid interface. The charged surface of the particle and the oppositely charged layer of 'counter-ions' adjacent to it constitute an 'electrical double layer', a term introduced by Helmholtz in 1879. It is clear that if there is an equivalent number of counter-ions to the particle primary charge contained in a compact layer close to the surfaces of the particles, then the combination of particle plus counter-ions is electrically neutral and hence at large distances no nett repulsive forces between particles would be expected. Of course, if the two particles approached each other so closely that the individual double layers overlapped then coulombic repulsion would occur.

This simple idea due to Helmholtz was subsequently extended by Gouy and Chapman and later by Stern (1924). The currently accepted model (see Lyklema 1978) is shown diagrammatically in Fig. 6.1. Stern envisaged the region of the liquid

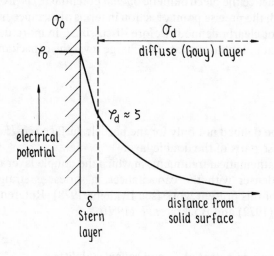

Fig.6.1 — Electrical potentials in the electrical double layer.

near the interface to be divided into two parts. Immediately adjacent to the solid surface is a layer of ions more or less specifically adsorbed which can be referred to as the Stern layer. Outside this compact layer of ions there is a second, more diffuse layer which is often referred to as the Gouy layer, in which the excess of counter-ions of charge opposite to that on the colloid particles over those of the same charge ('co-ions') decreases. Ions in this diffuse layer are considered to be subject only to non-specific coulombic forces of attraction or repulsion, whereas ions in the Stern layer, in addition to being under the influence of electrostatic forces, may also be intimately involved in specific chemical interactions with the surface (e.g. covalent, or hydro-gen-bonded interactions).

It is possible, and in fact often occurs, for these specific chemical forces to be larger than the electrostatic ones; in such cases it is feasible, for example, for anions to be adsorbed on to an already negatively charged surface. Of direct relevance to water treatment processes is the fact that many positively charged metal hydroxy complexes specifically adsorb on to negatively charged particles so strongly that the negative surface charges of the colloidal particles are over-compensated and charge reversal occurs. This will be discussed in more detail later in the chapter, when direct applications and problems associated with over-dosage are examined.

Electrical potentials associated with double layers

The electrical potential (ψ_x) at any point at distance x from the surface of the solid is defined such that the product $e.\psi_x$ is the reversible isothermal electrical work required to bring a single elementary charge e from infinity (or from a reference point) to the point x. If we consider our reference point of zero potential to be the bulk solution, then we can define potentials associated with the various parts of the double layer: for example ψ_0, the potential at the solid surface; ψ_d, the potential of the diffuse layer; and so on. Some of these are illustrated in Fig. 6.1. In the context of water treatment practice the electrokinetic or zeta potential (ζ) is probably the most important, although the precise point of action in terms of distance from the surface of the particle is not clearly defined. Before discussing ζ in more detail, it may be noted in passing that potential is related to charge σ via the capacitance C such that

$$C = \frac{\partial \sigma}{\partial \psi}$$

Capacitances can be defined not only for the particle surface ($\partial\sigma_0/\partial\psi_0$) but also for the Stern and diffuse parts of the double layer.

The detailed mathematical treatment, in which the double layer is regarded as a parallel plate condenser with the capacitances in series, is straightforward but outside the scope of this present work (see Lyklema 1978). Reference may also be made to Sparnaay (1972) and Bockris *et al.* (1980).

The electrokinetic or zeta potential, ζ, and colloid stability

In the presence of an electric field, charged colloid particles are observed to move, a phenomenon known as electrophoresis. During electrophoresis colloid particles, even so-called hydrophobic ones, retain a thin layer of liquid and counter-ions which adhere to the solid surface and move with the particle. There is therefore a hydrodynamic boundary or 'shear plane' between the stationary bulk liquid phase and the liquid constrained to move with the particle. It is the potential of this shear plane which we identify as the zeta potential, and which is measured experimentally by electrophoretic methods. Some authors equate the zeta potential with the potential at the inner limit of the diffuse layer (or the outer limit of the Stern layer), i.e. with ψ_d. Identifying the shear plane with the outer boundary of the Stern layer is not necessarily justifiable, but in the absence of any certain knowledge of the location of the hydrodynamic shear plane it can be regarded as a 'best guess' (Lyklema 1978).

As a parameter in considerations of colloid stability ζ has decided advantages over ψ_0, the surface potential. Unlike ψ_0, ζ is, at least in principle, determinable by direct experiment, whereas for colloid particles ψ_0 is not. In addition, the correlation between ζ and colloid stability is good and as an operational parameter in process design is extremely useful. If it is the case that the principal reason for hydrophobic colloid stability is the prevention of inter particle contact by coulombic repulsion, then it is apparent that any reduction in the charge on the particles will reduce this repulsion and may allow the weaker attractive forces to induce aggregation of the particles. In this context it is important to note that it is ζ or the associated electrokinetic charge σ_ζ which appear to be the critical parameters, rather than ψ_0 or

σ_0. Qualitatively it is easy to see that if one can increase the number of counter-ions within the shear plane then a reduction of ζ and σ_ζ will occur. This may be achieved by increasing the concentration of indifferent electrolyte present, i.e. by having more counter-ions available. Since it is the charge carried by the counter-ions which is important, it is apparent that polycharged counter-ions will be more effective than singly charged species. Thus for colloid particles in aqueous dispersions, which in general are negatively charged, it is to be expected, and indeed is observed, that Ca^{2+} is more effective in inducing coagulation than Na^+, and that trivalent ions such as Al^{3+} or Fe^{3+} are even more effective.

This qualitative assessment is somewhat oversimplistic even for non-specifically adsorbing electrolytes, since in addition to the reduction in charge densities concomitant with the adsorption of more counter-ions close to the particle surface, a reduction in the thickness of the diffuse part of the double layer also occurs with important effects on interparticle interaction energies; and Ishibashi (1980) has demonstrated that coagulation is possible even at high ζ-potential values, suggesting that adsorption and bridging may be more important than ζ-potential in some cases. Stumm and O'Melia as long ago as 1968 also warned that double layer models are inadequate to describe natural systems. Nevertheless, at an unsophisticated level, the charge neutralisation ideas (which are the basis of the empirical Schülze–Hardy rule) are useful. This rule simply states that the effectiveness of counter-ions in inducing coagulation is approximately proportional to the sixth power of the counter-ion charge; so, for example, the relative concentrations of $Na^+ : Ca^{2+} : Al^{3+}$ required to produce aggregation of a negative lyophobic sol would be in the ratio of approximately $700 : 60 : 1$.

DLVO theory
Between 1935 and 1948 Derjaguin and Landau (1941) in the USSR and Verwey and Overbeek (1948) in the Netherlands independently developed a quantitative theory of the stability of lyophobic colloids. This theory, now commonly referred to as the 'DLVO theory', evaluates the interaction energy of two colloid particles approaching one another. Whilst the detailed theoretical arguments are out of place in a general text such as this, a brief résumé of the main features of the model is appropriate.

The first assumption made is that the sole attractive force operating between the two particles is the van der Waals' attraction, and that repulsion is due solely to the coulombic interaction of the double layers. A further assumption is that these forces are completely independent and hence can be evaluated separately. The total potential energy for the system is then simply given by the algebraic sum of the attractive and repulsive energies

$$V_T = V_A + V_R$$

It can be shown that whereas V_R decreases exponentially with separation distance, V_A, the attractive component, varies approximately as the reciprocal of the square of the separation distance. The implication of this is that both at very small and very large distances the attractive force is greater than the repulsive force, but that at intermediate distances repulsion may predominate, giving rise to a potential energy maximum which must be overcome if particle contact and hence aggregation is to occur. This is illustrated in Fig. 6.2.

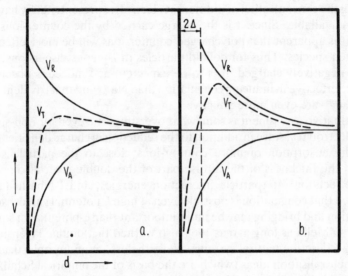

Fig. 6.2 — Potential energy distributions in the electrical double layer showing the effect of an adsorbed layer of non-ionic material on the repulsion between colloidal particles. a. Potential energy–interparticle distance curves. b. Potential energy–interparticle distance curves with adsorbed material, showing increase in repulsion. V_R = repulsion potential, V_A = attraction potential, V_T = total potential. (Reproduced from Ives 1978.)

In addition to the exponential decrease of V_R with distance, V_R also depends strongly on ψ_d, the potential at the outer boundary of the Stern layer (NB, $\psi_d \simeq \zeta$). The effect of added indifferent electrolyte is therefore to reduce V_R by reducing ψ_d (ζ) and hence reducing or even entirely eliminating the potential energy maximum in the V_T vs. distance curve. Fig. 6.3 illustrates the effect of added indifferent 1:1 electrolyte on the total interaction energy.

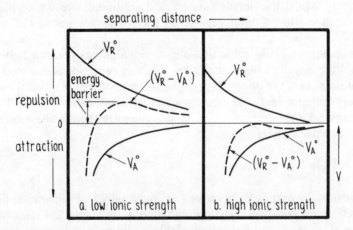

Fig. 6.3 — The effect of added indifferent 1:1 electrolytes on total interaction energy in the electrical double layer. V_R = repulsion energy, V_A = attraction energy. (Reproduced from Weber 1972.)

Useful as it is in accounting for stability of hydrophobic sols under laboratory conditions, the DLVO model is of limited quantitative applicability in waste water treatment processes, for the simple reason that in the majority of situations encountered the assumptions on which the theory is founded are not valid. In particular, specific adsorption effects are of critical importance in real processes and treatment reagents seldom behave as simple indifferent electrolytes. Qualitatively, however, the predictions of the DLVO theory are in accord with observation.

Specific adsorption
The importance of specific adsorption effects is further illustrated in Fig. 6.4, where typical coagulation curves (residual turbidity vs. coagulant concentration) are plotted for a number of coagulants (O'Melia 1972). Although these curves refer to a natural rather than to a waste water, the principles are the same, and it is instructive to consider the curves individually.

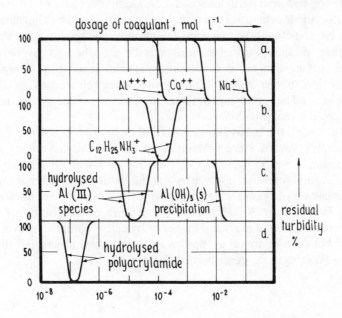

Fig. 6.4 — Coagulation curves for various coagulants (after O'Melia in Weber 1972).

The curves for Na^+, Ca^{2+} and (at low pH) Al^{3+} in Fig. 6.4(a) show the expected form as expected from the empirical Schülze–Hardy rule and as predicted by DLVO theory. Here the coagulants are behaving as simple indifferent electrolytes and non-specific coulombic interactions predominate. In Fig. 6.4(b) the dodecyl ammonium cation $C_{12}H_{25}NH_3^+$ is seen to be anomalous in two respects. Firstly, for a monovalent electrolyte it is far more effective at low concentrations than is predicted from DLVO theory or as expected from a comparison with, say, Na^+ in Fig. 6.4(a).

Secondly, at higher concentrations, restabilisation of the colloid is observed, a feature not shown by Na$^+$, and thus there is a fairly narrow optimum concentration region where coagulation occurs. The reason for this behaviour is to be found in the surface-active properties of long-chain amines of this type. The amine molecule is only soluble in water because of the hydrophilic nature of the amine (or ammonium) end group, which will be extensively solvated. The solvation energy of the hydrocarbon chain is not favourable, and more energy has to be expended to force water molecules apart to accommodate it than can be recovered by subsequent favourable interactions with water molecules. Consequently, if it were possible for the hydrocarbon end of the molecule to 'escape' from the aqueous phase whilst still allowing the amine group to project into the aqueous phase, then this would be an energetically favourable process. Such molecules therefore tend to congregate at water/solid (or water/hydrophobic liquid phase) interfaces with the molecules orientated in such a way that the hydrophobic part of the structure is next to the surface, with the hydrophilic end group projecting into the aqueous phase. The adsorption of such molecules on colloidal particles arises therefore not because of any particular affinity of the molecule for the solid but rather by default, since the alternative of staying in solution is energetically so much less favourable. The tendency for this to occur is so strong that sufficient surfactant ions may be absorbed to 'over-compensate' the negative surface charge of the colloid particle, leading to charge reversal and restabilisation of the colloid. This is sometimes referred to as 'superequivalent adsorption', and has important practical implications, since overdosing of coagulants of this type is a real possibility.

In Fig. 6.4(c), the behaviour of Al(III) at higher pH is shown, presenting an even more complex pattern. Here aluminium is present as hydroxymetal complexes rather than the simple aquo ion $Al(H_2O)_6^{3+}$ as in Fig. 6.4(a). Successive substitution of hydroxy groups for water molecules (or the equivalent and mechanistically more likely ionisation of bound water protons) leads initially to monomeric ions such as $Al(OH)(H_2O)_5^{2+}$ or $Al(OH)_2(H_2O)_4^+$. As the concentration of Al(III) and/or OH$^-$ increases, then dimers and higher polynuclear complexes form by elimination of water between hydroxyl groups on adjacent metal atoms and the formation of Al $-$ O $-$ Al bridges, as described in Chapter 2:

$$(H_2O)_5Al^{2+} \underline{\quad\quad} O\!\mid\!H \quad H\,O \mid\!\underline{\quad\quad} {}^{2+}Al(H_2O)_5$$

$$[(H_2O)_5Al \underline{\quad\quad} O \underline{\quad\quad} Al(H_2O)_5]^{4+}$$

Eventual precipitation of $Al(OH)_3$ or some similar product occurs as the polymers build in size and the solubility product is exceeded. The small hydroxy complexes (monomers, dimers) are reasonably soluble and are strongly adsorbed on to colloid particles. Thus aluminium salts (and also Fe(III) salts) under conditions where hydrolysis is occurring are extremely effective coagulants. However, because of specific, non-coulombic chemical binding of these hydrolysed species, over-dosing can lead to charge reversal and restabilisation, as indicated in Fig. 6.4(c).

At still higher concentrations coagulation again ensues but for quite a different reason. In this second destabilisation region, the precipitating aluminium hydroxide enmeshes or traps the colloid particles within its floc structure. Removal of colloidal material by this mechanism is termed 'sweep flocculation'. In water treatment practice it is normal to use concentrations of Al(III) or Fe(III) such that sweep flocs are generated, rather than to attempt to induce coagulation by adsorptive destabilisation with its attendant problems of overdosing. Colloid particles themselves can serve as nuclei for precipitate formation, so precipitation increases with increasing concentration of colloidal particles to be removed (Packham 1965), all other factors being constant. This can result in an inverse relationship between an optimum coagulant dose and concentration of material to be precipitated. Nevertheless, the likelihood of overdosing in the treatment of waste waters with large concentrations of suspended solids is less than in the case of the much less turbid raw waters or potable waters.

Interparticle bridging
Colloid destabilisation by polymers

So far we have discussed two principal mechanisms by which colloid particles may be destabilised and aggregation induced, namely reduction of the ζ potential (either by specific adsorption of counter-ions or by the non-specific effect of indifferent electrolytes) or enmeshment of particles within a sweep floc. In recent years there have been significant developments in the use of high molecular weight synthetic polymers as flocculants in the water and waste water treatment industries. These substances are generally termed polyelectrolytes, inasmuch as they have a number of ionised or potentially ionisable functional groups within their structures (although non-ionic polymers are also used). These polymeric materials are often highly effective at low doses (see Fig. 6.4(d)), and it is apparent that both cationic and anionic polymers may destabilise the negatively charged colloid particles usually encountered in aqueous systems. This being so, it is apparent that non-coulombic forces must be important, and clearly so in the case of anionic polyelectrolytes, where adsorption of the polymer must occur in spite of charge repulsion. It is noteworthy, however, that flocculation of colloid particles by polyelectrolytes of the same charge type almost always also requires the presence of other dissolved electrolytes in order to effect the same initial destabilisation and promote adsorption of the polymer on to the particle surface (see Gregory 1978). This will be referred to again later in the chapter.

A general mechanism for the destabilisation of colloids by large polymeric molecules has been developed by a number of workers (Ruehrwein and Ward 1952, Michaels 1954, La Mer 1966, 1967, La Mer and Healy 1968) which involves the bridging of colloid particles by the polymer chain, hence forming larger structural units that are readily separated from the aqueous dispersing medium. In order for a polymer to be effective in this role it must have certain minimum structural features which can be identified as follows. In order to be able to bind simultaneously to two or more colloid particles the polymer must have functional groups (binding sites) present that are sufficiently separated to allow interaction to occur at particle–particle distances, such that overlap of the diffuse double layers of the particles is not significant. If this is not the case, then coulombic repulsions between the double layers will not allow the second particle to approach within capture distance of the polymer bound to the first particle (see Fig. 6.5, in which binding sites are labelled B). It follows from this that extended-chain linear polymers will be more effective

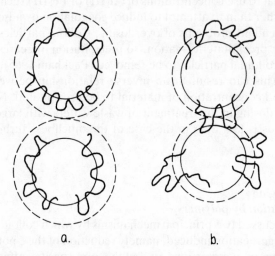

a.   b.

Fig. 6.5 — Representation of failure of adsorbed polyelectrolyte to bridge across colloid particles because binding sites are inadequate to extend beyond double layer.

than less extended or cross-linked polymers of similar molecular weights, and also that any compression of the double layer which can be induced (by added electrolytes for example) should enhance bridging by allowing closer particle approach (Fig.

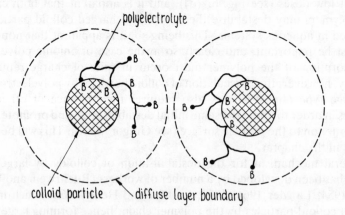

Fig. 6.6 — The effect of ionic strength on bridging efficiency of polyelectrolytes. a. Low ionic strength (bridging prevented by electrostatic repulsion). b. High ionic strength (bridging across repulsion distance). (Reproduced from Ives 1978.)

6.6). This may be the reason why the presence of electrolytes is necessary if anionic polyelectrolytes are used to destabilise and flocculate negative colloids, although the marked improvement in performance when certain divalent cations are present (e.g. Ca^{2+}) suggests that these ions exert a specific intermediate bridging role in addition to any non-specific double layer compression.

Polymer conformation

The conformation of polymer chains in solution is determined by a number of factors, and is clearly of great significance in view of the proposed bridging mechanism for colloid flocculation. In the absence of strong solvent interactions a non-ionic linear polymer can be statistically expected to have a random coil conformation and it is highly unlikely that any significant concentration of spatially extended structures will be present. Thermodynamically, the decrease in configurational entropy in going from a randomly coiled to a straight (extended) chain structure is very large; therefore ΔG, the associated free energy change, is large and positive (i.e. unfavourable) since in

$$\Delta G = \Delta H - T\Delta S$$

the enthalpy change ΔH is close to zero for this process, because no chemical bonds are made or broken on going from a random coil to a straight chain conformation —they are merely rotated. The spontaneous extended-to-coiled transformation is therefore principally an entropy-driven process.

If, however, there are present along the polymer chain functional groups possessing the same electrical charge (e.g. $-COO^-$), then the coulombic repulsion between the groups will tend to impose an extended conformation on the polymer as these segments containing the charged groups attempt to maximise their separation. The actual conformation adopted by a particular polymer will therefore reflect the balance of the coulombic and entropic (statistical) driving forces, and can to a certain extent be controlled by the number and nature of functional groups incorporated into the polymer chain. Additional complications are introduced by the presence of electrolytes in solution, since these ions will in effect 'screen' the charged groups from each other and hence favour a random coil configuration. Tan and Schneider (1975) studied the effect of dye binding to anionic polymers in the presence of electrolytes, and concluded that the transition between the extended conformation obtaining at low ionic strength to the compact configuration existing at high ionic strength occurred in the range 0.1–0.5 M NaCl.

It is therefore apparent that the effects of indifferent electrolytes on the bridging of colloid particles by polymers are manifold and may be quite complex; for whereas on the one hand electrolytes may assist bridging by reducing double layer repulsion and allowing closer approach, on the other high concentrations of electrolytes reduce the extension of the polymer and may have an adverse effect on bridging. The optimum conditions are rarely, if ever, predictable in waste water treatment practice, and must be determined by prior experiment for each particular case.

Stabilisation of colloids by polyelectrolytes

So far the emphasis in this review has been on those properties of polyelectrolyte molecules which assist the destabilisation and flocculation of colloidal material. Polymer adsorption may, however, also increase colloid stability by increasing the coulombic repulsion between particles. This can arise if the adsorbed polyelectrolyte carries the same charge as the primary surface charge of the colloidal particles or, in the case of polyelectrolytes of opposite charge, the initial lowering of the zeta potential and destabilisation may be followed by superequivalent adsorption, charge reversal, and concomitant restabilisation of the colloidal particles. Even in the case of non-ionic polymers, interparticle coulombic forces may be affected simply by virtue of the fact that a strongly adsorbed polymer may well physically displace counter-ions from the Stern layer, thus increasing the diameter of the diffuse layer. The larger diffuse layers will generate repulsive interactions between particles at greater distances and hence lead to increased stability.

It is apparent that polyelectrolytes, low molecular weight surfactants and high molecular weight polymers present something of a two-edged sword as a weapon to deal with colloidal dispersions. It follows from the discussions above that either destabilisation or restabilisation can occur on introduction of a polyelectrolyte into a colloidal system. Critical in deciding the efficacy of treatment is the concentration of polymer employed in relation to the colloid concentration. The dose/response curves for polymer addition show a definite stoicheiometric dependence on colloid concentration; this behaviour is contrary to that for colloids in the presence of indifferent electrolytes, where specific adsorption is excluded and where the concentration of electrolyte required to induce coagulation is almost independent of colloidal concentration. If we accept the bridging theory as a model for colloid destabilisation by polymers, then it is readily apparent that in order for a bridge to be formed between two adjacent particles there must be (a) a polymer molecule bound to one particle and (b) a surface binding site vacancy on the other. This may appear to be trivially obvious, but if we extend this simple idea to include all the colloid particles present, then we arrive at an important result. If λ denotes the fraction of surface sites occupied by polymer and hence $(1 - \lambda)$ the fraction of unbound sites, then optimum bridging will occur when the product $\lambda \times (1 - \lambda)$ is a maximum, i.e. when $\lambda = 0.5$ (50% coverage). Since polymers are strongly adsorbed, it is reasonable to assume that all of the added polymer is taken up at the surface of colloid particles, and it follows that the optimum concentration of polymer required for flocculation may be very small (frequently less than 1 mg l^{-1}). Over-dosing is thus a real danger, particularly in solutions of relatively low turbidity, such as those encountered in tertiary 'polishing' stages or in potable water treatment, although it is usually less significant in primary treatment of waste waters.

Particle transport
Aggregation mechanisms

The destabilisation of colloid particles either by indifferent electrolytes or specifically adsorbed surfactants or polymers is really only the first step in the final aggregation that is required to generate a separable bulk solid phase. Unless the destabilised particles come into contact or, in the case of polymer-treated systems, come within 'grappling' range of the adsorbed polymer chains, then the formation of occur. Separation of the treatment process into the two steps of destabilisation and aggregation is somewhat artificial, since in many cases the two processes will occur simultaneously; nevertheless a consideration of the separate processes is useful in that it enables an analysis of important operational parameters to be undertaken. We have already considered the factors affecting the destabilisation process and we will now turn our attention to particle transport problems.

The rate of aggregation of thermodynamically unstable hydrophobic colloids is is dependent ultimately on two factors, viz. the frequency of collisions between particles, and the effectiveness of these collisions in causing particle–particle adhesion. The collision effectiveness is related to the particle destabilisation that has already been discussed. The frequency of collisions is a particle transport phenomenon and is to a large extent independent of the degree of destabilisation.

In an unstirred quiescent dispersion there are two principal transport mechanisms by which particle–particle contacts occur:

(1) Collisions may occur by virtue of the thermally induced motions of the particles, i.e. 'Brownian diffusion'. Flocculation arising as the result of such contacts is termed 'perikinetic flocculation'.
(2) Differential settling under gravity may also result in contacts between particles in heterodisperse systems, where because of the different settling rates of particles of different sizes, faster moving particles may overtake (and in doing so may collide with) particles that are moving slowly.

In stirred dispersions where velocity gradients are induced within the fluid there is a third mechanism, aggregation from which is referred to as 'orthokinetic flocculation'. Of course in real systems contributions from all three mechanisms may be expected, although one or other usually predominates, depending on such parameters as particle size, number, degree of heterogeneity and stirring rate.

Perikinetic flocculation
Smoluchowski (1917) first developed a quantitative theory for the rates of perikinetic flocculation, in which he showed that for monodisperse systems (i.e. only one type and size of particle), the rate of coagulation is independent of particle size, inversely proportional to the viscosity of the dispersing fluid, and proportional to the square of the particle concentration. Specifically,

$$-\frac{\partial N}{\partial t} = \frac{4}{3}\alpha \cdot \frac{kT}{\eta} \cdot N^2 \tag{6.8}$$

where α is a collision efficiency factor $(1 > \alpha > 0)$ and is related to the extent of destabilisation of the particles; k is the Boltzmann constant $(1.38 \times 10^{-16}$ erg deg$^{-1})$; N is the total concentration of particles remaining in suspension at time t (number of particles cm^{-3}); and η is the fluid viscosity (for water $\eta = 0.01$ poise). This relationship is a simple second order rate equation and integration leads to the result

$$t_{\frac{1}{2}} = 3\eta/4kT\alpha N_0 \tag{6.9}$$

where N_0 is the original colloid concentration and $t_{\frac{1}{2}}$ is the time to reduce it by 50%.

Substituting values for the constants and assuming $\alpha = 1$ (i.e. the colloid is completely destabilised and every collision results in adhesion), then at 20°C

$$t_{\frac{1}{2}} = 1.8 \times 10^{11}/N_0 \text{ seconds}$$

Hence dilute dispersions (low turbidity) will only coagulate slowly if perikinetic flocculation is the sole mechanism. Of course, if α is much less than unity, implying poor destabilisation, then even slower aggregation times are predicted.

Orthokinetic flocculation

In many cases it is observed experimentally that stirring or agitation significantly accelerates the aggregation of a destabilised colloid. The spatial variation in fluid velocity throughout the dispersion induced by stirring clearly presents opportunities for particle–particle contact in addition to those occurring as a result of Brownian diffusion. If we again assume a uniform particle size (monodisperse system) then the agglomeration rate as a result of this so-called orthokinetic flocculation is given by the second order rate equation:

$$\frac{\partial N}{\partial t} = \tfrac{2}{3}\alpha G d^3 N^2 \qquad\qquad (6.10)$$

where G is the mean velocity gradient and d is the particle diameter; α is again the collision efficiency factor.

It is of interest to compare the relative rates of perikinetic and orthokinetic flocculation. From equations (6.8) and (6.10) we obtain the result

$$\frac{(-\partial N/\partial t) \text{ ortho}}{(-\partial N/\partial t) \text{ peri}} = \eta \frac{G d^3}{2kT} \qquad\qquad (6.11)$$

For particles of diameter ca. 10^{-4} cm the ratio of the orthokinetic to perikinetic flocculation rates is approximately unity with velocity gradients of about to 10 s^{-1} (about the rate at which one might stir a cup of tea!). For larger particles, orthokinetic flocculation becomes increasingly important because of the dependence on d^3, whereas stirring has little effect on the coagulation of particles much smaller than 1 μm. In particular, viruses ($d < 10^{-6}$ cm) cannot be aggregated until they are adsorbed on or enmeshed in larger particles.

Of important practical significance in the context of waste water treatment, it should be noted that frequently the flocs produced by polymer bridging are of comparatively low mechanical strength and may be broken down by excessive shear forces. It is therefore possible, under conditions of high turbulence or too rapid stirring, that floc breakdown and colloid restabilisation can occur.

In real systems the initial stages of coagulation would be perikinetic, but a transition to orthokinetic flocculation would occur with increasing floc growth, and ultimately sedimentation would become significant.

Differential settling

In an ideal monodisperse system clearly no collisions will occur as the result of settling, since all particles will be falling through the medium with identical velocities and the rate constant for coagulation will be zero. True monodisperse systems are, however, seldom encountered outside the research laboratory, and any heterogeneity that is present in, or can be introduced into, the system will allow some contribution from this mechanism.

Heterodisperse systems

The second order rate constants given above for perikinetic and orthokinetic flocculation were derived for and strictly apply only to monodisperse systems. In heterodisperse systems, both particle size and size distribution become important. As we have already stated, the vast majority of real systems with which one has to deal are heterogeneous in respect of particle sizes. Inevitably, the theoretical treatment of such systems is more complex; but the important result of such calculations is that in almost all cases increased heterogeneity significantly increases aggregation rates and hence improves process performance. In Table 6.1 the second

Table 6.1 — Collision rate constants for monodisperse and heterodisperse systems

Transport mechanism	Rate constant (monodisperse)	Rate constant (heterodisperse), particle size d_1 and d_2
Brownian (perikinetic)	$k = \dfrac{4kT}{3\eta}$	$k = \dfrac{2kT\,(d_1 + d_2)^2}{3\eta\,d_1 d_2}$
Laminar shear (orthokinetic)	$k = \dfrac{2Gd^3}{3}$	$k = \dfrac{(d_1 + s_2)^3}{6}$
Differential settling	$k = 0$	$k = \dfrac{\pi g(\rho - 1)(d_1 + d_2)^3(d_1 - d_2)}{72v}$

Rate $= k.N_{d_1}.N_{d_2}$ (or $k\,N^2$ if $d_1 = d_2$).
G = mean velocity gradient (s^{-1}).
η = absolute viscosity (poise = g s^{-1} cm^{-1}).
v = kinematic viscosity (cm^2 s^{-1}).
g = acceleration due to gravity (981 cm s^{-2}).
k = Boltzmann constant (1.38×10^{-16} erg deg^{-1}).

order rate constants for collisions between particles is compared for a monodisperse and a simple heterodisperse system in which particles of two different sizes are assumed to exist.

O'Melia (1978) considers that in the past insufficient attention has been given to the enhancement of coagulation rates by heterogeneity, and gives the example of the striking improvement in aggregation rates for viruses (particle size approximately

10^{-6} cm) that can be achieved by introducing heterogeneity through addition of bentonite (particle size 10^{-4} cm). Using the information given in Table 6.1, O'Melia calculated that in the absence of bentonite a suspension containing 10^4 viruses cm^{-3} would require 200 days before the concentration is reduced by half as a result of perikinetic flocculation (the most significant mechanism for particles of this size). Addition of bentonite at a concentration of 10 mg l^{-1} (7.35×10^6 particles cm^{-3}, assuming a particle size of 1 μm and density 2.6 g cm^{-3}), results in a reduction of virus concentration to only 2.6 particles cm^{-3} after a contact period of only one hour. The principal aggregation mechanism is still perikinetic ($k_p = 3.12 \times 10^{-10}$ cm^3 s^{-1}; $k_0 = 1.72 \times 10^{-12}$ cm^3 s^{-1}) and so stirring has little effect.

The use of clay particles such as bentonite or the introduction of alum flocs simply as a means of increasing the hereogeneity of a colloidal dispersion has significant effects on aggregation rates as a result of the increased likelihood of particle contacts. This is in addition to any improvement in process performance which accrues from the increased collision efficiency (destabilisation) achieved as a result of adsorptive, enmeshment, or electrokinetic effects of such additives. This is particularly important in potable water treatment or in tertiary treatment of waste water where turbidities are low. In the primary stages of waste water treatment, a sufficient number of large particles are probably present to allow effective coagulation of flocculation of small particles by orthokinetic processes, provided the particles are adequately destabilised.

The influence of surfactants

Surfactants may interact with particulate or ionic species to produce effects that may be used to assist in the separation of species from solution. Surface-active agents, as their name implies, exhibit particular characteristics at interfaces, notably by causing a reduction in the surface tension with respect to the solvent water through being adsorbed preferentially at the surface. Synthetic detergent solutions demonstrate their collector properties for cleansing through micellar envelopment.

Three main classes of detergent surfactant can be recognised:

(1) anionic, these being in the majority of commercial formulations, including soaps, RCOO$^-$Na$^+$ (e.g. sodium oleate); sulphated fatty alcohols, ROSO$_3^-$Na$^+$ (e.g. sodium dodecyl sulphate); sulphonated hydrocarbons, RSO$_3^-$Na$^+$ (e.g. sodium lauryl sulphonate), which may also include counter-ion groups as in the salts of aminobenzene sulphonic acid; sulphonated fatty acid condensates, RCOR'SO$_3$Na$^+$ (e.g. oleyl methyltaurine); and sulphonated dibasic esters, RCO$_2$·CH$_2$·CH(CO$_2$R)SO$_3^-$Na$^+$ (e.g. sodium dioctyl monosulphosuccinate);

(2) cationic, including quaternary ammonium salts, R'R''R'''R''''N$^+$X$^-$, and pyridinium salts, RN$^+$(C$_6$H$_5$)X$^-$, where X$^-$ is a halide ion;

(3) non-ionic, which are not dissociated, and which include, for example, ethylene oxide–alcohol (or phenol) condensates, R(OCH$_2$CH$_2$)OH, and the naturally occurring saponins and digitonins.

The particular feature of all surfactants that is of importance is their combined oleophilic and hydrophilic properties, which facilitates the micellar entrapment of oily or greasy materials immiscible with water. Here we are concerned with their modes of interaction with ions or solids, which can be classified into groups. These groups define the ways in which the surfactants can be used, which is always in conjunction with gasification (normally fine bubble aeration).

Foam fractionation or separation

An ion or sometimes a colloidal particle may react with a surfactant to form a soluble or micellar product that retains surface-active properties, and which will therefore be absorbed on the surface of bubbles ascending through the liquid to form a foam that can be removed, and with it the species formed. This is particularly effective in removing metal solutes from very dilute solutions. The sublates probably involve simple electrostatic association between the ion or colloid and the surfactant charged group.

Ion separation or flotation

When the collector surfactant and the ion or colloid react to from an insoluble product, that product can also be brought to the liquid surface by fine bubble aeration. Here, as in foam separation, stoicheiometric or greater quantities of surfactant are required with respect to the material being removed.

Precipitate flotation

This third grouping is different in that only relatively small or trace amounts of collector are required, being governed only by the surfactants's own ability to form a stable foam at the liquid surface on aeration. This technique is particularly applied where coagulants such as hydrolysed Fe(III) and Al(III) salts have been used as coagulants for colloids, and is the normal form of flotation technique employed in waste water treatment.

On the other hand, it should be recognised that surfactants can introduce difficulties when sedimentation operations are involved, by causing some of the material to float, and so complicating the separation engineering. Jola (1975) investigated the effect of surfactants on the sedimentation of nickel, copper and chromium hydrous oxides in the co-presence of neutral chlorides and a polyelectrolyte. He found that sodium lauryl sulphonate (anionic) and polyethylene oxide (nonionic) surfactants had no effect on sedimentation rates at 50–100 mg l^{-1} concentrations, but that with N-cetylpyridinium chloride (cationic) some flotation occurred. An excess of calcium caused further deterioration with nickel and copper, but improved the chromic hydrous oxide settlement. This Jola interpreted in terms of adsorption related to charge characteristics on the metal hydrous oxide precipitates.

For full discussions of separation techniques with surfactants, see Fuerstenan (1962), Rubin and Gaden (1962), Sebba (1962), Lemlich (1972), and Grieves (1975).

COMMON COAGULANTS AND FLOCCULANTS

The reagents most widely used in waste water treatment are polyelectrolytes, polyvalent cations such as those of aluminium and iron, and calcium hydroxide. In the case of the metal salts, it is usually the hydrolysis products that are the active species, as discussed in this chapter and in Chapter 2.

Polyvalent metal cations

It is clear from the earlier discussions that polyvalent cations such as Al(III) or Fe(III) may fulfil one or more of several possible roles in waste water treatment:

(1) They may act as specific precipitants for anions such as phosphate, with which they form sparingly soluble salts.

(2) They may act as colloid destabilisers, reducing the zeta potential and inducing coagulation either by virtue of their status as 'indifferent electrolytes', or, as is usually the case in waste water treatment, because of the strong specific adsorption capacities of their hydrolysis products.

(3) They may behave as sweep flocculants, whereby colloidal material is removed by enmeshment and/or adsorption within a hydroxide floc generated *in situ*.

(4) Since polyvalent cations are invariably strongly hydrated, they are useful in higher concentrations as 'salting-out' agents (in protein recovery, for example).

Sometimes polyvalent metal ions may be used interchangeably, in which case reagent choice is largely governed by economic considerations and convenience of storage and application. In other circumstances, the particular properties of a water or waste water stream may dictate the choice of reagent or combination of reagents. Rarely is it possible to predict the most suitable reagents and conditions without recourse to laboratory testing and field trials.

Polyelectrolytes and non-ionic polymers

The water-soluble polymers used in waste water treatment are generally classified as cationic, anionic or non-ionic, according to the nature of the functional groups present along the polymer chain. Predominant among polymer types are those based on polyacrylamide or derivatives thereof. Polyacrylamide itself is synthesised by catalytic polymerisation of acrylamide monomer thus:

$$CH_2=CH \quad \quad \quad \ldots \quad \left[CH_2-CH \right] \quad \ldots$$
$$\underset{NH_2}{\overset{|}{\underset{|}{\overset{C=O}{|}}}} \quad \xrightarrow{\text{catalyst}} \quad \underset{NH_2}{\overset{|}{\underset{|}{\overset{C=O}{|}}}}$$

The degree of polymerisation, i.e. chain length, can be controlled to a certain extent by altering the reaction conditions, and is particularly sensitive to the amount of catalyst used, low catalyst loadings favouring high molecular weights and vice versa. Polyacrylamide is clearly non-ionic in the pH range usually encountered in waste water streams, but modifications of the polymer are possible yielding anionic and cationic types. For example, partial hydrolysis of the amide groups will result in some of them being converted to carboxylate groups to provide an anionic polymer, the degree of anionicity being controlled by the extent of hydrolysis. The same result is achieved by copolymerising acrylamide with sodium acrylate:

$$\text{CH}_2=\text{CH} \quad + \quad \text{CH}_2=\text{CH} \quad \rightarrow \quad \ldots \quad \left[\text{CH}_2-\text{CH}-\text{CH}_2-\text{CH}\right] \quad \ldots$$

$$\text{COO}^- \qquad\qquad \text{C}=\text{O} \qquad\qquad\qquad\qquad \text{COO}^- \qquad \text{C}=\text{O}$$

$$\text{NH}_2 \qquad\qquad\qquad\qquad\qquad\qquad\qquad\qquad \text{NH}_2$$

In this case the total charge on the polymer can be controlled by varying the proportion of acrylate to acrylamide monomers in the starting reactants. In an analogous fashion, cationic polymers can be generated by copolymerising acrylamide with a monomer containing a positively charged quaternary ammonium group, or by chemically modifying a previously prepared polymer. The possibilities of structure variants along these lines are enormous. An exhaustive account of all the polymer modifications which are available is out of place in this present text, and interested readers are referred to the text edited by Schwoyer (1981), which gives an introduction to the manufacture of polyelectrolytes as well as a comprehensive account of their applications in both potable water and waste water treatment. A review of the patent literature on the subject is given by Gutcho (1977), and reference may also be made to Davidson (1980).

An important consideration in choosing a polyelectrolyte for evaluation is whether the desired function is that of a coagulant (colloid destabilisation by charge neutralisation) or that of a flocculant (interparticle bridging). These two functions, although not necessarily mutually exclusive, certainly cannot both be optimised within the same polymer molecule. Coagulants depend on a high charge density for their effective action and those polyelectrolytes used in waste water treatment are almost exclusively strongly cationic polymers of relatively low molecular weight. On the other hand, for reasons discussed earlier, bridging flocculants are almost invariably high molecular weight linear polymers, which may be cationic, anionic or non-ionic. An essential part of the bridging mechanism is the binding of the polymer to the colloidal particles via the functional groups on the polymer molecule, and it is therefore apparent that some control of specificity of action is attainable by selecting polymers with different functional groups. In this context it is notable that wide variations in effectiveness occur even between polymers of the same charge type. The pH of the waste water is also extremely important either as a control parameter affecting the performance of a particular polyelectrolyte or as a factor to be considered when selecting a polyelectrolyte for a particular application. For example, the pK of a carboxylic acid group attached to a polyacrylamide chain is approximately 6, so at pH $= 6$, 50% of the groups are dissociated; at pH $= 8$, approximately 99% would be dissociated; and at pH values less than 4, the polymer would be essentially non-ionic, since less than 1% of the carboxyl groups would be in the anionic ($-\text{COO}^-$) form. The effect of this behaviour is that throughout the pH ranges commonly encountered in waste streams, the anionicity and binding properties of acrylamide/acrylate copolymers will be strongly pH-dependent. In general, the higher the charge density, the narrower the pH band within which a polyelectrolyte is effective. On the other hand, if anionicity is conferred on a polymer by the presence of sulphonic acid ($-\text{SO}_3\text{H}$) groups, then the anionicity is essentially pH-independent, since the pK of the sulphonic acid group is low and it is virtually completely dissociated throughout the pH range encountered in waste water treatment practice. Similar pH insensitivity occurs with cationic polymers, where quaternary ammonium groups give essentially pH-independent behaviour, but amine groups confer pH dependence on the cationic nature of polyelectrolytes in which they occur. As well as synthetic polymers, natural or modified natural materials can be and are used (see Davidson 1980 for a description of substances available). One interesting product, chitosan, is produced from chitin, a natural carbohydrate polymer found in the shells of crustacea, being itself a linear polymer of 2-deoxy-2-acetyl-amine glucose (analogous to glucose in chemical structure). When chitin is deacetylated with strong alkalis the product is chitosan, a cationic polyelectrolyte (Peniston and Johnson 1977). It has found particular use in the treatment of food wastes (see later).

The low molecular weight polymers, being simply charge neutralisers, act to form insoluble flocs, and have the advantage over other coagulants such as ferric and aluminium salts in not increasing the salts' level. However, their effectiveness with a given waste water always has to be compared with the metal salts, and they are often relatively uneconomic, given even the extra sludge production from formation of the metal hydrous oxides. Quite commonly, the ferric or aluminium salts are used firstly to effect coagulation and then a high molecular weight polyelectrolyte is added to flocculate the coagulated solids. In general, the higher the molecular weight the larger the floc, although this may not always prove a practical advantage from bulking of the precipitate finally obtained. If the solids to be flocculated are very highly dispersed, either low molecular weight polymers or metal salts are usually needed; high molecular weight polymers are more effective with larger particles.

Polyelectrolytes can also be added to sludges as conditioning agents in a post-treatment operation, where they again improve de-watering characteristics (see, for example, Novak and O'Brien's (1975) study of the effects of a range of polymer species on chemical sludges).

In view of the wide variation in properties encountered in the large range of polyelectrolytes currently available it is usually impossible to select the most suitable polymer for a particular application without extensive trials ranging from simple laboratory jar tests (Jackson and Sheiham 1981) to full-scale plant evaluation. In this respect the manufacturers of the various polyelectrolyte formulations are almost invariably willing to assist in product evaluation for specific applications. Practical aspects of dosing are considered later.

Other flocculant aids
Apart from metal salts and polymers, other materials have been employed, primarily as floc structure improvement aids. For example, activated silica, prepared by partial neutralisation of sodium silicate with acids, has been promoted to assist in binding suspended solids, such as colloidal sulphur. Like alumina and activated carbon, also used, it probably acts as an adsorbent to some extent.

So-called 'weighting agents' are sometimes used to improve the sedimentation or filtration characteristics of suspended solids. Clay materials such as bentonite, kaolin, montmorillinite, and diatomaceous earths, are commonly added, but these rarely provide much benefit as coagulants. They are added as slurries to give final concentrations of hundreds to thousands of milligrams per litre, according to the nature and quantities of suspended solids being treated. In a converse manner, suspensions of bentonite separated as a slurry during reclamation of foundry sand can be flocculated with polyelectrolytes, beneficially promoted by the addition of calcium sulphate (A. B. Wheatland, personal communication, 1985). This latter phenomenon may be associated with the ion exchange properties of bentonite and related materials, where the sodium form is converted to the calcium form by addition of calcium sulphate.

APPLICATIONS

Chemical treatment of sewage

In recent years, a substantial interest has developed in chemical methods for the treatment of sewage. This has evolved partly to cater for those circumstances where conventional biological treatment plants are overloaded or subject to seasonal peaks, and partly where specific discharge requirements apply, e.g. strict control of phosphorus levels. The advantages of chemical processing, sometimes allied to other forms of advanced treatment such as sand filtration and activated carbon adsorption, are that it can be started and stopped without acclimatisation, is cheaper in capital costs and less space-demanding than biological treatment, is more easily controlled, and is better able to cope with toxic inputs; and, with lime treatments, there is the possibility for ammonia removal by air stripping at the high pH employed. It suffers generally from imposing a higher operating cost, and for this reason is only used as an adjunct or in special circumstances, such as overload relief at holiday resorts, or in coastal districts for marine discharge. (It should be noted moreover that in the UK, at least, biological oxidation is necessary as a precursor to activated carbon treatment to produce high quality effluent. This is because UK sewage is stronger than, for example, in the USA, and low molecular weight organic matter not removed by chemical treatment is also not removed by activated carbon.)

A vast literature exists on the chemical and physico-chemical treatment of sewage, and in a text such as this it is impossible to provide more than a basic outline. Minton and Carlson (1976) have given a critical review of lime treatments, and Jenkins *et al.* (1977) have summarised conclusions of previous studies.

The principal reagents used are lime, either as CaO or $Ca(OH)_2$, aluminium sulphate or chlorohydrate, and ferric sulphate. Polyelectrolytes are also employed, mostly to enhance the performance of the inorganic salts and to improve sludge quality. (Polyelectrolytes are used, as are inorganic salts, to condition normal biological treatment sludges, but this is a different application.) Lime has been extensively investigated, mainly as a means of coagulating raw sewage to aid in settlement, and reduce both suspended solids and B.O.D. levels below those otherwise achievable. This precedes conventional biological treatment, but can be used as a preliminary step for a complete physico-chemical treatment operation. Reagent addition may also be made after biological aeration, where lime consumption is then less.

Where lime is used, normally at pH values between 10 and 11.5, some recovery can be achieved by carbonation with carbon dioxide: the pH is lowered to 9, where calcium carbonate has minimum solubility and is thus removable with the primary sludge for subsequent calcining and lime recovery. If after the chemical treatment activated carbon adsorption is to be applied as a polishing stage—a common practice where the treatment is totally physico-chemical—then a second-stage carbonation may be used to bring the pH to 7, which is more suitable for carbon adsorption, is normally more acceptable as a discharge value, and which causes less scaling of equipment. Clough (1977a) has, however, pointed out that there are cost penalties in this, not least in the recovery equipment, and that if a lower discharge standard is acceptable less lime is needed, and so less carbon dioxide. Mineral acid addition or single-stage carbonation may be used if calcium carbonate recovery is not practised, and pH adjustment is not necessary if the treatment following lime dosing is a completely mixed activated sludge plant, where the carbon dioxide produced by bio-oxidation serves the purpose. The requirement for pH adjustment is avoided altogether if aluminium or iron salts are used for coagulation; these are more expensive materials, but generally produce a better quality effluent.

A large number of pilot plant trials have been carried out, including major ones in the UK at Coleshill (see Clough 1977a, b) and at Davyhulme, where chemical treatment was applied before and after activated sludge treatment (see Jenkins *et al.* 1977); other UK work has been reported by Cooper and Thomas (1974), Cooper (1975), and Banks (1975). Stamberg *et al.* (1971), Jenkins and Lee (1977) and Leentvaar *et al.* (1978) have described studies in the USA, while Rebhun and his co-workers (1974, 1975) have applied lime and ferric chloride coagulation to the relatively stronger municipal wastes arising in Israel. Leentvaar *et al.* applied treatment to raw sewage, and noted the improvements shown in Table 6.2, making

Table 6.2 — Chemical treatment of raw sewage (reproduced from Leentvaar *et al.* 1978, Pergamon Journals Ltd)

Coagulant	Dose level	pH	Percentage removal T.O.C.		
			Soluble	Suspended	Total
$FeCl_3$	61 mg l^{-1} Fe	5.3	27	93	54
$Al(SO_4)_3$	47 mg l^{-1} Al	5.9	24	94	58
$Ca(OH)_2$	520 mg l^{-1} $Ca(OH)_2$	11.0	17	90	44

the comment that the results were not as good as some other workers had observed. Ulmgren (1975) reported on the operation of some 50 full-scale plants in Sweden, where phosphate removal was the prime objective (see later section) but noting improvements to B.O.D. as well.

Stones (1981) has described the results of using aluminium sulphate with and without lime on both domestic sewage and industrial waste waters in the UK. Progressive improvements in P.V. (permanganate value) were obtained with increasing concentrations of salt up to 500 mg l^{-1} $Al_2(SO_4)_3.18H_2O$ for industrial waste waters, but better and more dramatic P.V. reductions (approximately 60%) were achieved on domestic sewage when the dose level reached 300 mg l^{-1}. When lime was also used, a drop in performance occurred at pH values around 9.5, but improved again as the pH was increased to 10–12. However, Stones concluded that optimum precipitation with aluminium sulphate occurs at pH 5.0 (achieved by co-addition of sulphuric acid) —see also Table 6.2. The active species promoting coagulation appears to be, as may be expected, the positively charged hydroxy metal complexes; this is in spite of the fact that the isoelectric pH for domestic sewage is approximately 3.0, the largely unhydrolysed Al^{3+} cation being ineffective here.

Sea water substantially improves the lime clarification of primary sewage because of its magnesium content, and where chemical treatments are applied in coastal regions, the introduction of small percentage flows of sea water at the point of lime addition offers advantages: where high pH (>11) treatment is used, 1–2% sea water is sufficient, although where the pH is less than 10.5 some 10% or more may be needed for beneficial action (Ferguson and Vråle 1984). Experience in Norway has shown that the treatment is worthwhile if no more than 75% of soluble B.O.D. removal is required. Vråle (1978) has reported that suspended solids (and phosphorus) removals may be greater than 90%, with practical lime dosages in the region of 250–300 mg l^{-1}.

Workers at the National Institute of Water Research in South Africa have over several years conducted extensive studies on combined biological/physico-chemical treatment operations for sewage with the primary aim of producing re-usable water, mainly at Windhoek. Their results have been summarised in a series of papers (Van Vuuren *et al.* 1970, 1975, Cillié and Van Vuuren 1978), and represent a significant step in the full-scale practical application of these techniques.

Many of the physico-chemical treatment plants use activated carbon, normally after a pre-filtration stage, as final polishing treatment. Extensive studies have been made, notably in the USA, of the criteria and performances (see, e.g., Gloyna and Eckenfelder 1970), but it is apparent that the relative expense of the material inhibits its widespread use beyond those situations where a very high quality effluent is required. Operating problems can also develop, for example from biological action in the bed and generation of reduced sulphur compounds such as H_2S and $(CH_3)_2S$.

One disadvantage of using inorganic coagulants in sewage or any other treatment where biological alternatives are possible is the increase in sludge quantities, and the fact that the metal content largely renders the sludge unsuitable for distribution on land. However, polyelectrolyte flocculants can reduce the dose levels needed for iron and aluminium salts, as well as improving the sludge filterability. The reasons for this can be traced back to the coagulation mechanisms discussed earlier. Although the hydrolysis products of polyvalent metal cations are effective at low doses in coagulating colloidal particles to form larger agglomerates, these are not so readily settled as those generated by large doses, which give sweep flocculation—but also result in high sludge volumes. High molecular weight polyelectrolytes can, however, provide a bridging of the smaller microflocs, so it is then feasible to use smaller quantities of inorganic coagulants.

Food processing wastes
Most of the liquid effluent resulting from food processing is biodegradable and differs from normal domestic sewage in its treatability usually only by virtue of its high B.O.D., which may reach many thousands of milligrams per litre compared

with a typical 300 mg l^{-1} in raw municipal waste water. Consequently many treatment processes applied to food processing waste waters are simply variants of the usual sewage purification methods taking into account the increased B.O.D. and suspended solids' loadings. The use of polyvalent metal cations and polyelectrolytes to coagulate and flocculate colloids and suspended solids, improve effluent clarity and condition sludges follows similar patterns to those already discussed.

Some application examples will illustrate the possibilities. The processing of poultry generates waste waters from washings that contain very high B.O.D. level, fats, grease, various solids and blood. Separation of the blood for recovery and re-sale is an important step to minimise pollution treatment, and screening of the gross solids must be applied before removal of the suspended solids, fats and grease. Here aluminium sulphate is highly effective as a coagulating/flocculating aid, particularly when used with dissolved air flotation (an example study has been reported by Woodward *et al.* 1972). Wastes from slaughterhouses, bacon processing and ice cream manufacturing can be similarly treated to remove solids, fats and a substantial part of the B.O.D.. Al(III) or Fe(III) concentrations (as metals) can be in the range 50–150 mg l^{-1}, but overdosing should be avoided to obviate undue sludge gene-ration. A high molecular weight polyelectrolyte can be introduced advantageously at 5–10 mg l^{-1} concentration to aid flocculation.

Leentvaar *et al.* (1979), in a comparative study of the alternatives of ferric chloride, aluminium sulphate and lime for coagulation of beet sugar waste after pre-sedimentation, have also emphasised the need to avoid overdosing: they recommend for this application 50–100 mg l^{-1} Fe, or 30–60 mg l^{-1} Al or 1500–2000 mg l^{-1} Ca(OH)$_2$. With any possibilities for the growth of sulphate-reducing bacteria there is the risk of hydrogen sulphide generation, as for example if aeration is inadequate, and so as chloride salt may be preferred to a sulphate one.

An interesting variation in the mode of use of chemical flocculation has been demonstrated by Beer and Gibbs (1975). Because of the high strength of piggery effluent, biological treatment by activated sludge methods is unable to achieve Royal Commission standards (20 mg l^{-1} B.O.D., 30 mg l^{-1} suspended solids). However, by using ferric chloride and lime in a tertiary treatment stage, the standards can be realised. The merit of this approach, of course, is that the chemical demand is then much less than with chemical pre-treatment.

Fish processing wastes are also treatable in the same manner. With aluminium salts a pH of 5 is a suitable coagulation condition, and B.O.D. reductions of the order of 60–90% can be achieved, with total solids' reduction of 30–90%, depending on waste input characteristics. The polyelectrolyte chitosan produced from the natural product chitin (discussed earlier) has been used for the treatment of a variety of wastes, and particularly from fish processing (see Bough 1976). For example, shrimp processing discharges, treated with 10 mg l^{-1} chitosan and 5 mg l^{-1} anionic polyelectrolyte, showed C.O.D. and suspended solids' reductions of approximately 75% and 95% respectively after settlement for an hour. With 100 mg l^{-1} chitosan levels and dissolved air flotation, the C.O.D. and solids' reductions were 90% and 98% respectively (which demonstrates the use of polyelectrolytes as sweep coagu-lants, as discussed earlier in this chapter). Other applications of chitosan have been found in poultry processing, egg breaking and meat packing, where C.O.D. and suspended solids' reduction of the order of 80% and 90% may be achieved.

With food processing wastes there is always the possibility that by-product recovery (proteins, fats, carbohydrates) may be feasible: this may then impose restrictions on the choice of coagulant or flocculant reagent. Large amounts of metal coagulants would clearly be inappropriate for protein recovery if the sludge were to be incorporated into an animal feed without further processing to remove added metal. One chemical dosing technique that has been applied to the recovery of fats and protein from meat processing waste waters uses lignosulphonic acid (LSA) as a coagulant at low pH (< 3), with removal of the precipitated material by dissolved air flotation (see Tønseth and Berridge 1968 and Hopwood and Rosen 1972). Reductions in B.O.D. and suspended solids of over 80% and 90% respectively are possible from slaughterhouse waste water, the further advantage being that the sludge has a potential use as an animal feed component: LSA appears to exhibit no toxic effects when included up to a concentration of 4% in the diet of growing pigs. Another flocculant that has been examined for protein recovery is sodium alginate, a material obtained from brown seaweed. In application to meat processing wastes Russell *et al.* (1984) used a pH of 3.5–4.5, obtaining a product they claim could be used as an animal feed supplement, feasible because alginates are non-toxic.

Phosphates can also be used as a precipitant for proteins (Perlmann and Hermann 1938). Spinelli and Koury (1970) examined the reactions of various condensed phosphates with soluble fish proteins, and demonstrated the formation of insoluble complexes in acid conditions. Finley *et al.* (1973) also used condensed phosphates at low pH to treat the washings from wheat flour doughs, and found that nearly 90% of the protein could be extracted; a similar result was obtained by Cooper and Denmead (1979), using sodium hexametaphosphate at pH 3.5 on slaughterhouse waste water.

The applicability of all recovery methods rests on how readily usable the product is, or whether further processing is necessary: theoretical viability calculated from pilot studies has to be measured against the commercial practicalities, not the least of which may be a varying price of normal supplies against the cost of the recovered material. Installation of a recovery plant at extra expense over normal waste treatment can prove to be a costly error if the normal feedstock price falls, as it has in the past with protein supplies.

Textile and wool wastes

Textile effluents can vary widely in content, according to the nature of the textile fibre material and whether dyeing wastes are included. Textile material wastes themselves may include biodegradable species, but many, such as those from cotton processes, include biologically resistant ones (also discussed briefly in Chapter 2, p. 64). Dyes, while not necessarily generating a heavy biological load, are usually objectionable in receiving watercourses simply on account of their colour, although it should be mentioned that some are biologically 'hard' as well.

In the practice of treating textile waste waters, apart from oxidation processes (p. 113 *et seq.*, pp. 121, 135), the most general chemical method is by addition of a coagulating/flocculating agent, usually with simultaneous pH adjustment to form a floc, followed by aeration and flotation of the solids. Common coagulants are ferrous sulphate (at pH 9) ferric chloride or sulphate (at pH 6–9), and aluminium sulphate (at pH 6), although lime has also been used, and has been claimed by Shelley *et al.*

(1976) to be superior to aluminium sulphate. This general form of treatment not only removes most of the colloidal and suspended material, but also provides a means of reducing colour levels, and can constitute a first-stage operation before subsequent biological treatment(s). Levels of Fe and Al may be in the region of 20–100 mg l^{-1} in the effluent stream, which must be balanced to a maximum degree to smooth out concentration peaks. Broadly speaking, acidic, chromic and substantive dyestuffs are precipitated readily but basic types are less satisfactorily treated. Load and flow balancing is of particular importance with dyehouse waste waters, which usually arise as batches of used dye are discharged from the vats. These may be subjected to oxidation, e.g. with chlorine or hydrogen peroxide but this rarely provides full colour removal, and adsorptive treatment with ferrous or ferric sulphate formed into flocs with alkali can be a supplementary operation. Stahr *et al.* (1980) favour the use of alum together with a cationic polymer. Magnesium carbonate at a pH > 10.7 has also been shown to be effective for vat, disperse and sulphur dyes, where an advantage is that most of the $MgCO_3$ is in principle recoverable (Judkins and Hornsby 1978). Polymeric flocculants have been used on their own in some instances, the advantage being a significant reduction in the quantity of sludge produced. For example, Blank *et al.* (1976) refer to the use of polymers with water-soluble anionic dyes, while Crowe *et al.* (1977) applied cationic polymers with disperse dyes.

Scouring wastes, e.g. from wool, contain substantial amounts of grease and alkali, with emulsions of the two. One treatment method involves adding calcium chloride, which precipitates soap and carbonate that can be subsequently floated. Acid cracking at a pH of less than 4 is also commonly used, and provides a means for recovering wool grease (see Anderson 1965). However, Christoe (1977) has pointed out that where non-ionic surfactants are present the treatment is less effective, and the clarified phase still contains substantial quantities of B.O.D. and grease. He made a study of ferrous sulphate, ferric chloride and sulphuric acid as alternative coagulants, and found that aluminium salts appear to be the most effective, at a pH of 5, used together with polyelectrolyte flocculant, preferably, though not necessarily, cationic. The efficiency of the treatment depends on the composition of the waste liquor, and C.O.D. removals were observed between 47% and 93%. Christoe also noted that a small amount of biological pre-treatment is beneficial in both improving the coagulation performance and dewatering properties of the sludge produced.

Pulp and paper wastes
The pulp and paper industry generates substantial pollution-loaded waste waters, not only of a biodegradable nature, but also with a high content of non-biodegradable material, principally lignin. In addition, the effluents are usually strongly coloured, particularly from kraft mills, and biological treatment does not remove this. A preferred procedure is thus to segregate the non-biodegradable and coloured wastes for chemical processing while the degradable wastes go to conventional biological treatment.

Waste water from pulp bleaching gives rise to the main source of pollution, of over 60% of the C.O.D. (up to 5000 mg l^{-1}) and some 60% of the colour (5000–20 000 mg l^{-1} Pt units in the caustic extract). Oxidation techniques, carbon adsorption, ion exchange and reverse osmosis have been used as treatment procedures, but so-called massive lime dosing (Berger *et al.* 1964) has been applied on several plants in the USA. It involves addition of lime to pH > 11 to produce a precipitate that adsorbs the colour (some 90%); this is then clarified and dewatered, used in re-caustication steps and eventually calcined to effect recovery for re-use. Alternative coagulant materials have been examined, including aluminium and ferric chloride or sulphate, but although, for example, aluminium sulphate gives equally good removal, the sludge produced is difficult to de-water, and recovery opportunities are limited or expensive, making the process relatively less attractive. Olthof and Eckenfelder (1975), for example, report on the relative colour and C.O.D. removals, shown in Table 6.3. The iron and aluminium salts were used in the

Table 6.3 — Relative efficiencies of coagulants for pulp and paper waste waters (reproduced from Olthof and Eckenfelder 1975, Pergamon Journals Ltd)

Reagent	Colour removal %	C.O.D. removal %
250–500 mg l^{-1} Fe(III) sulphate	85–92	50–60
250–400 mg l^{-1} Al sulphate	91–92	44–53
1000–1500 mg l^{-1} Ca(OH)$_2$	85–92	40–58

Residual turbidity was < 10 APHA units.

pH range 3.5–4.5, although it was acknowledged that pH 8.5–9.5 may have advantages (poor results were obtained in the neutral region). These results were essentially confirmed by Nasr and MacDonald (1978), whose studies on kraft mill caustic extraction effluent showed 98%–99% colour removals with 23 mg l^{-1} Al (at pH 5) and Fe (pH 3.3). Goos *et al.* (1974), however, found that for both organic carbon and colour removal from board mill effluents ferric salts were superior to lime, with lime being more effective than aluminium salts. These workers noted little benefit from the addition of polyelectrolytes. Ferric sulphate coagulation has also been shown (O'Shaughnessy and Blanc 1978) to provide good colour removal on waste waters from a corrugated box manufacturing plant when operated at pH 3.2–3.5 and followed by granular activated carbon adsorption, providing some possibilities for water recycling.

Cationic polyamines have beeen used in laboratory studies by Kisla and McKelvey (1978) to precipitate colour bodies, with promising results, but the high concentrations of coagulant (550–800 mg l^{-1}) raise questions of economic viability—although sludge volumes would be less than with the metal salt precipitants, an advantage found by Roberts and Lin (1983) in application of cationic surfactants for colour removal from sulphite waste waters.

Fellmongering and tanning wastes

Fellmongering, involving the de-hairing and conditioning of animal skins prior to tanning, generates a heavy polluting load in waste waters that are usually highly alkaline, containing hundreds of milligrams per litre of sulphide and several thousands of milligrams per litre B.O.D. Older forms of treatment consisted of adding ferrous salts to precipitate the sulphide as ferrous sulphide, which was then settled out before pH adjustment, leading on to biological treatment. Nowadays, as described in Chapter 3, the sulphide is normally oxidised by catalytic aeration after preliminary screening and before subsequent treatments. Although these latter are normally biological, some work has indicated that chemical coagulation is capable of reducing colloidal and suspended solids and B.O.D. values to levels where the discharge can be accepted directly into a municipal sewage treatment plant. This offers the advantage to the factory that it need not install its own biological treatment facility.

Studies carried out at two fellmongerers (Mattock and Bryan 1983, 1984) showed that addition of either ferric or aluminium salts to the catalytically aerated but unsettled effluent gave improvements in C.O.D. and suspended solids' levels. The pH was adjusted and maintained at selected values, and 500 mg l^{-1} Fe$_2$(SO$_4$)$_3$ or 500 mg l^{-1} AlCl$_3$·6H$_2$O were introduced, followed by anionic polyelectrolyte (1–10 mg l^{-1} according to type). After settlement the C.O.D. and suspended solids were determined and compared to those values pertaining with settlement only (on untreated oxidised samples, flocculant alone had negligible effect). The results showed significant improvements in both C.O.D. and suspended solids' levels, although they were not highly reproducible: some data are given in Table 6.4.

Table 6.4 —The effect of coagulating agents on aerated fellmongering waste water

	C.O.D. % change		S.S. % change	
Coagulant	Tank A	Tank B	Tank A	Tank B
500 mg l^{-1} Fe$_2$(SO$_4$)$_3$, 10 mg l^{-1} Magnafloc 156, pH = 8	33	15	66	63
500 mg l^{-1} AlCl$_3$.6H$_2$O, pH = 6	65	89	91	95
H$_2$SO$_4$, pH = 6	23	40	64	87
H$_2$SO$_4$, pH = 3	53	83	82	97

Settlement was good, and floc sizes were large. It is apparent that acid alone was excellent as a treatment, but the relative benefits of Fe(III), Al(III) and acid need to be assessed on the basis of reagent and sludge disposal costs; the availability of waste acid would, of course, be a help, but alkali would be needed to neutralise before discharge.

Polyelectrolyte additions alone do not in general improve C.O.D. levels, but Bitcover *et al.* (1980) have found that anionic polymers can effect 80–85% reductions in suspended solids levels by flocculating paddle vat wastes, while for hide processing wastes some 90% improvements are possible; but the requirements are specific, and careful laboratory testing is necessary.

The sulphide problem does not arise with tanning operations, where the skins are received after initial fellmongering, and treatments may be totally biological after initial screening, pH adjustment (to precipitate Cr(III)), and settlement. Bishop (1978) has, however, reported that coagulation and flocculation of a secondary tannery waste after screening and neutralisation and before clarification can give a discharge quality broadly acceptable to municipal sewage systems, showing 98% reduction in Cr to 1 mg l^{-1}, and suspended solids, B.O.D.$_5$ and C.O.D. reductions of 92%, 75% and 75% respectively.

Paint wastes in aqueous streams

The increasing use of electrophoretic paints, which consist of electrically charged paint bodies dispersed in water that deposit when a suitable polarity is applied to the item to be painted, has generated an aqueous waste problem. In electropainting technology much of the paint waste is recovered by ultrafiltration, but a significant residue is always discharged, and this degrades on dilution to give rise to tacky masses that clog pipes and watercourses. An effective technique for treatment is by destabilising the paint particles by charge neutralisation and adsorption on fresh ferric or aluminium hydrous oxide precipitates, together with flocculation by very high molecular weight polyelectrolytes. The behaviour of untreated paint is controlled by the pH; anodic (negatively charged) paints become tacky in acid conditions, while cathodic paints destabilise in alkaline ones. One method, not very satisfactory, for treating anodic paints is to add lime slurry to a high pH, but ferric sulphate has proved very satisfactory in practice (Briscoe and Mattock 1972), using a controlled pH of 8–9. Aluminium salts work better at a pH of 6–7, and use of sulphate salts ensures good removal simultaneously of co-present lead and chromium. Eilbeck and Mattock (1983) studied the treatment to determine optimum operating process parameters, and established that for both cathodic and anionic paints a desirable dose level is 1 mg l^{-1} Fe(III) or Al(III) per 10 mg l^{-1} paint solids. Alkali addition to provide the optimum pH, together with polyelectrolyte, are essential to provide flocs that can be floated or settled efficiently. There is some tendency for both flotation and settlement, but it is possible to orientate towards one or the other. For settlement, Fe(III) is the preferred coagulant in conjunction with Ca(OH)$_2$ addition to pH 8 and a low (0.25–0.5 mg l^{-1}) concentration of high molecular weight polyelectrolyte (such as Allied Colloids (UK) Magnafloc 1011). For flotation aluminium sulphate can be used with NaOH or Na$_2$CO$_3$ addition to pH 7, and with a higher (1–2 mg l^{-1}) polyelectrolyte level. These criteria were found to apply to epoxy cathodic, acrylic anodic and polybutadiene anodic types of paint.

Paint spray booth effluents also present a problem, since these paints are solvent-based, and when carried away by a water mist, used as a safety procedure in spray booths, they separate and form sticky masses. It is common practice to recirculate spray booth water after some flotation in a lagoon or after aeration, but denaturation is necessary. Various proprietary chemicals that are essentially alkalis are often used, but an odour problem develops after a time, and then the entire water load has to be disposed of. Eilbeck and Mattock (1983) found that ferric chloride dosing with calcium hydroxide to achieve a pH of 8 effects good denaturation with thermosetting acrylic and stoving alkyd paints, but that relatively high concentrations (3 g l^{-1}) of surfactant were also needed. A more effective method of treatment is provided by a blend of a primary aminoalcohol, a polyether polyamine and a water-soluble salt of an amphoteric metal (Gabel and Seitz 1977), and this type of product is available commercially (Nalfloc (USA)).

Latex and rubber plant wastes
Latex wastes can be coagulated by pH adjustment to acid conditions, but flocculation is improved by the addition of polyelectrolytes. Groves and Lundgren (1973), for example, used a cationic polymer to flocculate suspended carbon black, oil, latex and rubber solids before applying dissolved air flotation to separate the solids; suspended solids levels were reduced from approximately 1100 mg l^{-1} down to an average 16 mg l^{-1}. Inorganic salts can also be used for carbon black wastes, with subsequent settlement of the coagulated solids.

Removal of heavy metals
Heavy metal ions present in waste waters, typically derived from electroplating or other metal finishing facilities, are frequently removed by precipitation as their hydrous oxides at high pH (see Chapter 2). The resultant solids after suitable ageing are usually settled, with the addition of polyelectrolyte flocculant aids. In some circumstances, heavy metal salts, notably Fe(III), are added, to produce flocs which have functions additional to those of a coagulant or sweep flocculant. It is well established that freshly precipitated iron(III) hydrous oxide has an extensive adsorptive capacity for transition metal ions and some of their complexes, and so in addition to being an effective coagulant it helps to reduce the concentration of soluble material present in the waste water.

Not only are the more common heavy metals co-precipitated when adjustment is made to an appropriate pH, but others, not so readily precipitable by pH adjustment alone, can be removed by treatment with hydrous oxide flocs. For example, metals such as mercury, antimony and vanadium (Hannah *et al.* 1977), lead (Hartinger 1973) and molybdenum, which requires a pH of only 3 for substantial removal of low concentrations from solution (Le Gendre and Runnells 1975), have all been successfully treated. Fe(III) or Al(III) hydrous oxide flocs, when used for adsorption of heavy metal ions, can also be treated with a surfactant such as sodium lauryl sulphate to impart hydrophobicity, and thus permit flotation of the solids (Thackston *et al.* 1980, McIntyre *et al.* 1983). The principal advantage here is a smaller sludge volume.

Precipitation of Fe(III) hydrous oxide flocs can sometimes exert a complex-breaking effect, as for example has been demonstrated for metal–phosphate complexes (Gleisberg 1976). However, action is usually more pronounced when Fe(II), Fe(III) or Al(III) salts are added to provide competitive complexing in the solution phase rather than as a co-precipitant: this is reviewed in Chapter 2 (pp. 81, 83).

Although most of the adsorptive precipitation studies and plant applications have used iron or aluminium hydrous oxides as the co-precipitating material, others can be employed. Hydrated manganese oxides are also effective for example (Morgan and Stumm 1964), and although the relative cost is probably too high to justify general application, their effectiveness in removing mercury has been demonstrated by

Lockwood and Chen (1973) and Thanabalasingam and Pickering (1985). Another approach that has apparently been used successfully in plant practice has been reported by Okuda *et al.* (1975), who describe the removal of zinc, cadmium and lead from waste waters by adding ferrous sulphate, neutralising and then oxidising so that magnetic ferrites are formed which act as co-precipitants. It was claimed that at one plant, zinc was reduced from 18 to 0.016 mg l^{-1}, cadmium from 240 to 0.008 mg l^{-1}, and lead from 475 to 0.01 mg l^{-1}.

Direct adsorption, as for example on activated carbon, is also capable of removing traces of heavy metal species (Linstedt *et al.* 1971), and the technique has been used on plant installations where it has followed as a final stage after conventional high pH lime treatment, settlement, sand filtration and neutralisation. Total residual metal levels for zinc, cadmium, chromium, nickel, copper and iron mixtures of less than 0.5 mg l^{-1} have thereby been achieved (Mattock 1970). pH has a significant influence on adsorption efficiency, and competitive adsorption occurs with metal ion mixtures. Netzer and Hughes (1984), for example, observed an adsorption sequence Pb > Cu > Co for these metals. Huang and Wu (1975) have studied chromium adsorption, finding that efficiency of removal increases with decreasing metal concentration at pH values less than 2. Adsorption of Cr(VI) is at a maximum in the pH range 5–6, Cr(III) being adsorbed less; since some reduction of Cr(VI) to Cr(III) occurs below pH 6 it is therefore preferable to operate in the region of pH 7, which is consistent with plant experience. With arsenic, Huang and Wu (1984) observed maximum adsorption at pH 4–5, while with cyano iron complexes, pH 3 is optimum, Fe(CN)$_6^{3-}$ being more strongly adsorbed than Fe(CN)$_6^{4-}$ (Saito 1984). Vanadium is also adsorbed on activated carbon (Kunz *et al.* 1976), and the reducing properties of the material assist in the removal of metals such as palladium, gold and mercury (see Chapter 4).

Other natural materials have been proposed as adsorbents, including clinoptilite (Semmens and Martin 1980), vermiculite (Keramida and Etzel 1982) and peat moss, which act as ion exchangers and can be regarded as disposable or regenerable. Peat moss has been studied by Coupal and Lalancette (1976), who removed less than 1 mg l^{-1} quantities of Hg, Cd, Zn, Cu, Fe, Ni, Cr, Ag, Pb and Sb, together with various organics, oil, detergents and dyes, by operation in the pH range 3.0–8.5. Smith *et al.* (1977) treated peat moss with sulphuric acid to develop ion exchange properties, with the opportunity for regeneration. Fly ash can also act as an adsorbent—see, for example, the study of Panday *et al.* (1985) on copper removal. Even shredded motor car tyres have been used (Rowley *et al.* 1984)!

Natural polyelectrolytes, such as alginic acid or polygalacturonic acid, PGA (obtained from pectin), form sparingly soluble heavy metal complexes, and the behaviour of these materials has been studied by Jellinek and Sangal (1972). Cationic species, e.g. of Cu, Cd, Zn and Ni, are removed from solution to low residual levels, and the adsorbent can be regenerated with hydrochloric acid. Where a metal is complexed in anionic form, e.g. Au, Pt, Pd, Rh, a polyethylene immine is proposed as initial additive before PGA addition to form a 'sandwich' structure in the final product.

Of course, commercial synthetic ion exchange resins can be used directly as metal removers. The efficacy of such treatment can be high provided no complexing agents forming electrically neutral species are present, and cation exchangers can be used.

Where anionic complexes are formed, these may be adsorbed on anion resins directly, but if some complex splitting occurs metal ions may then either leak through or be precipitated on to the resin.

Ion exchange is most commonly applied for purification to recover water, but ion exchange resins can be used as specific agents for the removal of traces of metal ions following conventional chemical treatment, which can be more cost-effective than total ion exchange as an effluent treatment procedure. Weakly acidic cation resins when used in the sodium form will remove many heavy metals, the adsorption sequence usually being $Cu > Pb > Fe(II) > Zn > Ni > Cd > Ca > Mg$. So-called chelating resins, more recently introduced, have the further advantage that they do not adsorb the alkaline earth metals that are often co-present in substantial amounts in treated metal-bearing effluents, although they are more expensive. Selective ion exchange is best applied after conventionally treated waste waters have been adjusted to have a pH in the range 5–8, where adsorption is most effective: pre-treatment by sand filtration after settlement is necessary, and other interfering agents such as surfactants should desirably be removed if present. Good performances can be achieved, down to levels of 0.02–0.05 mg l^{-1} metal concentration for copper, lead and cadmium.

It should be appreciated that adsorption on to agents such as carbon, peat moss and ion exchange resins as a means of removal of heavy metals should be generally regarded as a supplement to conventional precipitation procedures, and should thus only be used as a 'polishing' device to remove trace metal quantities. If used as a primary treatment above a few milligrams per litre the procedure is not only likely to be uneconomic but, certainly in the case of carbon, may also be less efficient than the usual methods. Of course, with ion exchange resins, there is always the alternative approach of using them as a means of recovering metals, but this then constitutes a totally different approach, and must itself be subject to careful design analysis: mixtures of metals, although readily adsorbed on to ion exchange resins, are not necessarily easily separated thereafter, and prior stream separation before individual stream treatments is usually preferable.

Other novel approaches to trace metal's removal are reviewed in Thompson (1986).

PRECIPITATION OF ANIONS AND CATIONS

The fundamental principles governing the precipitation process and the factors affecting the solubilities of substances have already been discussed. In the treatment of waste water, precipitative methods can be used not only to remove particular anions or cations, but may be employed to lower the concentrations of total dissolved solids. For example, in the removal of, say, Cu(II) from solution as the hydrous oxide by increasing the pH the following two reactions (idealised) might be considered:

(a) $CuSO_{4(aq)} + 2NaOH_{(aq)} \rightarrow Cu(OH)_2 \downarrow + Na_2SO_{4(aq)}$

(b) $CuSO_4 + Ca(OH)_2 \rightarrow Cu(OH)_2 \downarrow + CaSO_4$

Either of these reactions would reduce the concentration of Cu(II) in solution, but (b) could also reduce the total concentration of dissolved solids if the solubility product of $CaSO_4$ is exceeded, whereas (a) would not. The reduction of total dissolved solids is of course at the expense of an increase in sludge solids. Choice of reagent may not therefore be as arbitrary as may at first be supposed.

Phosphate removal

The removal of phosphate from aqueous waste streams, whether of domestic or industrial origin, consists of conversion of the soluble phosphates to an insoluble solid phase, followed by removal of the solid phase from the aqueous waste stream by settlement, filtration or flotation. Much attention has been directed to the problems arising from discharge of phosphates into rivers and lakes, where in the presence of nitrogen algal growth can be enhanced, giving rise to eutrophication. As with the treatment of sewage by chemical means, a substantial literature exists on phosphate removal, and it is possible here only to outline the key aspects and point to important work in the field.

Conventional biological treatment processes (e.g. trickling filters, activated sludge) are limited in the extent of phosphate removal that can be accomplished by the phosphate uptake of the micro-organisms during the growth phase. The carbon:phosphorus ratio in domestic waste waters is about 70:1, whereas in the cell mass in activated sludge it is 100–110:1, and insufficient carbon is therefore present in domestic wastes to produce enough organisms to incorporate all the available phosphorus into the cell mass. As a consequence, much of the soluble phosphate will pass through biological treatment processes. In these circumstances, if phosphate removal is a priority, it is apparent that chemical treatment has an important role in enhancing or sometimes even replacing conventional biological treatment.

Typically, raw domestic waste water has a total phosphorus concentration of approximately 10 mg P l^{-1}, the principal forms being orthophosphate (5 mg P l^{-1}), pyrophosphate (1 mg P l^{-1}) and tripolyphosphate (3 mg P l^{-1}), together with smaller amounts of organic phosphates. Industrial waste waters may contain much more than this, and the speciation may be quite different, although the principles of treatment are the same. Orthophosphate (PO_4^{3-}), pyrophosphate ($P_2O_7^{4-}$) and tripolyphosphate ($P_3O_{10}^{5-}$) are all protonated to a lesser or greater degree depending on the pH of the solution (see the discussion in Chapter 2, p. 60). At pH 7, for example, the predominant species are $H_2PO_4^-$, HPO_4^{2-}, $H_2P_2O_7^{2-}$, $HP_2O_7^{3-}$, $H_2P_3O_{10}^{3-}$ and $HP_3O_{10}^{4-}$, all of which will form complexes with metal ions, as will the parent anions. Some of these complexes are soluble and others very insoluble. The distribution of phosphorus between all the various species will be dependent on the relative concentrations of phosphate and metal ions and particularly will be affected by changing pH. In principle, knowing the formation constants for all the metal and proton complexes, it is possible to compute a complete species distribution, and indeed this is an elegant computational exercise into which some workers have been tempted (e.g. De Boice and Thomas 1975, Kavanaugh et al. 1978, Arvin and Petersen 1980). The problem with this approach is two-fold. Firstly, there is a lack of accurate formation constant data obtained under the conditions of particular waste streams, and secondly, with so many species there is frequently no unique solution to the equations. Often several possible models, sometimes quite dissimilar, are mathematically acceptable, without any of them necessarily being a true reflection of physical reality. DeBoice and Thomas for example conclude on the basis of such an exercise that 'calcium phosphate precipitation in wastewater may be adequately described by an equilibrium model that considers precipitation of $Ca_3(PO_4)_2$ and $CaCO_3$', whereas in fact the insoluble calcium phosphate phase is not $Ca_3(PO_4)_2$ but hydroxyapatite, $Ca_{10}(PO_4)_6(OH)_2$. We do not wish to imply that all such calculations are futile but merely sound a note of caution. Certainly equilibrium calculations have a useful role in identifying major species and indicating possible limitations of particular treatment regimes, but it is well to remember that waste water streams are frequently not in a state of equilibrium. This is particularly true when precipitation processes are occurring and kinetic considerations become of vital importance. As has been stressed by Diamadopoulos and Benedek (1984), many intermediate complexes are formed in phosphate precipitation, and equilibrium descriptions based on single compounds are not adequate: a range of variables, including Al : P ratio, OH^- : Al ratio, and the presence and concentrations of other ions (notably sulphate, which promotes phosphate removal, especially at lower pH values) all have a significant effect.

The three most common reagents used for phosphate precipitation are aluminium sulphate or chlorohydrate, ferric chloride (or sulphate) and calcium hydroxide, which are used either alone, in combination, or in conjunction with a polyelectrolyte to improve the settling characteristics of the resultant precipitate. In Fig. 6.7 the

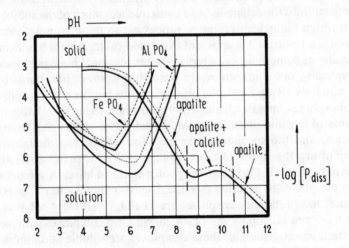

Fig. 7 — Equilibrium solubility diagrams for the $Fe-PO_4$, $Al-PO_4$ and $Ca-PO_4$ systems as a function of pH. [P diss] = dissolved phosphorus concentration (reproduced from Jenkins *et al.* (1971), Pergamon Journals Ltd).

solubilities of the thermodynamically stable Al(III), Fe(III) and Ca(II) phosphate species are plotted as a function of pH, from which it is clear that pH is an extremely important parameter in deciding the ultimate efficiency of these reagents. In the case of Al(III) calculation indicates and experiment confirms that the most favourable pH for phosphate removal is approximately 5.5–6.5, whereas the optimum for Fe(III) is about one pH unit lower. Both Al(III) and Fe(III) react with orthophosphate with a 1:1 stoicheiometry, but invariably reagent doses in excess of this stoicheiometric requirement are needed to achieve low phosphate residuals. This is not surprising, since not only will there be competition from OH^- ions for the metal ions (in fact initial precipitates are usually mixed hydroxo–phosphate complexes) but the tendency for the metal ions and their hydrolysis products to adsorb on to particulate matter will be great. This is no disadvantage, since destabilisation of colloidal particles is assisted and settlement properties may be improved.

Early work on the application of both aluminium and iron(III) salts for phosphate removal from domestic waste water has been well reviewed by Jenkins et al. (1971). More recently Lin and Carlson (1975) have advocated the addition of aluminium sulphate to the aeration tank in the activated sludge process. Addition to the tank near the exit is preferred to addition at the inflow end for the following reasons. Firstly, significant hydrolysis of condensed phosphates to orthophosphate occurs during the biological treatment. This is more readily precipitated and leads to improved total phosphorus removal. Secondly, the extreme turbulence in aeration tanks imposes shear forces on the flocs causing breakdown of the floc structure and subsequent poor settlement. The precipitation of aluminium phosphates and hydroxyphosphates occurs very rapidly and there is consequently sufficient time between the waste leaving the aeration tanks and entering the settlement tanks for reaction to occur. There is an additional argument that addition of reagent at the end of the aeration tank will result in lower doses being required, since some of the phosphate will have already been removed by the growing micro-organisms. The dose levels of metal ion adopted for sewage treatment has usually been of the order of 10–30 mg l^{-1} Al to achieve 90% removal of total phosphorus, but Fe(III) quantities are usually higher (by weight), and some workers have employed twice these levels. Opinions differ on the relative efficiencies of aluminium and iron salts, and it is clear that the conditions and modes of addition (whether before, during or after biological treatment) have influence, so it is not possible to generalise. (Reference can also be made to the review by Minton and Carlson (1972), which examines the relative efficiencies of the two reagents.)

The situation with calcium hydroxide is quite different from that for aluminium or iron(III) salts. In some ways, there are fewer problems with the use of calcium hydroxide, since the precipitation is usually conducted at high pH (> 10.5) and the stoicheiometry is less significant, the pH required usually dictating the amount of calcium hydroxide added rather than the phosphate concentration. On the other hand the chemistry is more complex, since there are several insoluble phases, many of indeterminate composition. Under these high pH conditions thermodynamically the most stable of these, in the absence of fluoride, is hydroxyapatite, $Ca_{10}(PO_4)_6(OH)_2$, although because of kinetic restraints on both nucleation and crystal growth the initial precipitates are frequently amorphous, of variable stoicheiometry, and considerably more soluble than predicted from equilibrium calculations. Under laboratory conditions seeding of the solution with hydroxyapatite dramatically improves the crystallisation rate and crystal form, and there is an obvious case to be made for recycling of precipitated solids in plant scale operations. Between pH 9 and 10.5 in the presence of carbonate the precipitation of calcite $(CaCO_3)$ may compete with hydroxyapatite precipitation. In general, lime has been applied at pH levels above 10.5, and Stamberg et al. (1971) found it necessary to use values as high as 11.5 to reduce the total phosphorus level (as P) down to the order of

0.3 mg l^{-1}. Jenkins and his co-workers, however, have shown that it is possible to achieve 2 mg l^{-1} with orthophosphates by lime addition in only slightly alkaline conditions in the pH region 8–9 (Jenkins *et al.* 1972, Ferguson *et al.* 1973). Whereas at very high pH the presence of magnesium in the water assists in reducing residual turbidity and hence phosphate levels, in the lower pH regions it tends to increase calcium phosphate residuals, and Jenkins *et al.* (1972) suggest that for those treatment regimes the maximum magnesium that can be tolerated is about 10^{-3} M, or 100 mg l^{-1} expressed as $MgCO_3$. Moreover, precipitate recycle is best avoided in these circumstances, because there is a tendency for precipitate re-dissolution. While operation at the lower pH is much more attractive economically, it is less efficient, and its use either for treatment of simple phosphate solutions or for sewage is dependent on the discharge qualities sought.

So far we have tacitly assumed that all phosphate forms are removed by the reagents we have been discussing. This unfortunately is not an entirely valid assumption, for whereas orthophosphate is effectively precipitated, condensed phosphates such as pyrophosphate and tripolyphosphate form much more soluble complexes with metal ions. Jørgensen *et al.* (1973) found with raw sewage that although orthophosphate could be removed down to < 0.05 mg l^{-1} (as P), polyphosphate and organic phosphate residuals were of the order of 0.2–2 mg l^{-1} (as P), unless polyelectrolyte was also used to improve flocculation and hence final filterability, and lime appeared to be better than iron and aluminium salts. Removal of condensed and organic phosphates appears to occur more by adsorption on to particulate matter such as aluminium and iron hydroxyphosphates or lime dispersions rather than by direct precipitation. Adsorption on to other particles in the waste water is also possible via specific interactions. Non-specific coulombic interactions are of course precluded, since under normal treatment conditions the colloidal particles and the condensed phosphates are likely to have the same negative charge sign, particularly at higher pH values. Under these circumstances, colloid restabilisation can occur, resulting in poor settlement and high total phosphate levels in the final effluent. Like Jørgensen *et al.*, Bancsi *et al.* (1975) have demonstrated that marked improvements can be achieved using polyelectrolytes in conjunction with aluminium sulphate, cationic polymers being particularly effective. Typical dose levels may be 2 mg l^{-1} polymer (Allied Colloids (UK) Percol 728) and 150 mg l^{-1} $Al_2(SO_4)_3.16H_2O$, corresponding to an Al : P molar ratio of 3 : 2.

Adequate residence times must be allowed for crystal growth and adsorption processes to occur to a maximum degree. It is normal to allow residence times in CFSTR units of the order of 10 minutes to achieve a pseudo-steady state condition; but processes can continue for much longer than this before true equilibrium is attained. On the other hand, further times are allowed in subsequent operations such as settlement (less so with flotation, but 'apparently without disadvantage—see Merrill and Jorden 1975), and ageing can occur through a complete treatment plant.

Fluoride removal
Fluoride occurs in a wide range of industrial effluents, the principal sources being waste streams from aluminium and steel production, metal finishing and electroplating, glass and ceramic processing, and phosphate rock processing in the fertiliser industry.

The most commonly used treatment for the removal of relatively high levels of fluoride (>200 mg l^{-1}) is precipitation as the calcium salt using lime both as the source of calcium ions and as a means of pH adjustment. The solubility of calcium fluoride in water is approximately 16 mg l^{-1} at ambient temperatures, corresponding to a fluoride concentration of about 8 mg l^{-1}, which thus represents the theoretical limit for fluoride removal using stoicheiometric quantities of calcium ions. This, however, is rarely realised with use of lime alone unless the fluoride concentrations are quite low and a very high pH is acceptable. Czukräsz and Hompasz (1981), for example, report that using a four-fold stoicheiometric excess of lime gave a residual fluoride level of 30–55 mg l^{-1} when used on etching wastes from electric light bulb manufacture. To improve precipitation, advantage can be taken of the mass action effect of adding calcium chloride with simultaneous pH adjustment with lime. In the treatment of a fluoride containing waste from a TV tube manufacturing plant it was found that by adding a five-fold excess of calcium as calcium chloride over that stoicheiometrically required to form CaF_2, with simultaneous lime addition to maintain a pH of 8.5, input fluoride levels of about 100 mg l^{-1} could be reduced to 10 mg l^{-1} (Mattock 1970); the retention time used in a continuous flow reactor was 30 minutes. Rohrer (1974) used $Ca(OH)_2 : CaCl_2$ weight ratios of 1:1 to 1:2 in the pH region 5.7–8.0, obtaining residues of 15–20 mg l^{-1} F^-. However, with lime alone at high pH, much longer times are often needed (Zabban and Jewett 1967), and Rohrer (1971) reported that with a TV tube manufacturing waste, use of $Ca(OH)_2$ at a pH $>$ 12 and a retention time of 1–6 days gave residual fluoride of 2 mg l^{-1}. The simultaneous use of sodium orthophosphate to precipitate the lead here almost certainly contributed to the residual fluoride being less than theoretical. Miller (1974) points to the addition of sodium hexametaphosphate with alum or aluminium sulphate to lime-treated fluoride wastes as being effective in reducing residual fluoride levels from 16 mg l^{-1} to 2–5 mg l^{-1} (according to dose), suggested as being due to formation of an insoluble product approximating in composition to $NaPO_3AlF_3$. The recommended alum:sodium hexametaphosphate weight ratio is 4:1. Zabban and Helwick (1975) also recommend the use of phosphate to form a less soluble fluorapatite species, $Ca_{10}(PO_4)_6F_2$.

The long reaction times required are a reflection of the kinetic restraints on nucleation and crystal growth, and it is noteworthy in this context that in continuous treatment processes recirculation of precipitated calcium fluoride to the reaction tank, maintaining as high a concentration of suspended solids as practicable, leads to enhanced performance.

In order to reduce final fluoride concentrations still further alternative precipitants are required. For this aluminium sulphate has proved effective, apparently fulfilling the dual role of precipitant and coagulant and resulting in fluoride removals down to approximately 1 mg l^{-1} (Zabban and Helwick 1975). This level of fluoride corresponds to the concentration deliberately introduced into potable water supplies in those areas where fluoridation of drinking water is a normal part of water treatment as a prophylaxis for dental caries. In fluoride removal processes, pH control is yet again an important control parameter. The optimum pH ranges for precipitation of calcium fluoride have been shown to be 8–9 or above pH 12 (Paulson 1977). For the process using aluminium sulphate the optimum pH range is 6.5–7.5, since effective coagulation and settlement of the co-precipitated aluminium hydrous oxide and fluoride is important to achieve a good quality effluent clarity and low fluoride residual. The amount of aluminium sulphate used is also important, the quantity required increasing as the desired residual fluoride level decreases. For example, Paulson (1977) quotes that with a solution containing 15 mg l^{-1} F^-, to obtain a residual of 4 mg l^{-1} it is necessary to use about 30 g aluminium sulphate hydrate per gram of fluoride, while to obtain 1.5 mg l^{-1} residual this requirement increases to about 90 g aluminium salt. Link and Rabosky (1976), using 26 mg l^{-1} ferrous sulphate as coagulant in a secondary treatment stage at pH 8, achieved 96–98% fluoride removals down to levels of 6–7 mg l^{-1} F^-.

Activated alumina can also be used, bypassing the fluoride-containing waste water through an adsorbent bed at a low flow rate (Zabban and Helwick 1975): 167 l min^{-1} per litre bed volume can reduce F^- levels to less than 1.5 mg l^{-1}, with a definite exhaustion breakpoint in the cycle, but at four times this rate there is continuous breakthrough, increasing as the run proceeds. The alumina can be regenerated with aluminium sulphate or acid, but this produces further effluent.

When complexing agents are co-present, e.g. boron (forming BF_4^-), the fluoride removal by precipitation is generally less effective. Acid hydrolysis with heating can be used for fluoborates (see Chapter 2, p. 67), but the co-presence of lead with the addition of phosphate can also improve results with calcium precipitation.

References to other work can be found in a brief literature review given by Link and Rabosky (1976).

Sulphate removal

In cases where the final effluent from an industrial operation is discharged into a sewer, limits may be imposed on the concentration of sulphate. The reasons for imposing limits on what is essentially a non-toxic species are, firstly, that concrete used in the fabrication of sewer pipes is liable to destructive attack by sulphate as a result of the formation of alumino–sulphate complexes, which swell and crack concrete made from high alumina cements, and secondly, where anaerobic conditions arise, sulphate is converted to sulphide; and the evolution of highly toxic hydrogen sulphide gas can present hazards to personnel. Generated sulphide also dissolves in moisture on exposed surfaces and is oxidised by sulphur-oxidising bacteria to sulphuric acid, which then attacks the fabric of the sewer and iron material in the manholes. For conventional concrete sulphate (as SO_3) limits of 300 mg l^{-1} are often sought, although with the so-called sulphate-resistant varieties, higher tolerances are permissible (up to 1000 mg l^{-1} as SO_3). This limitation can pose a real problem where large concentrations of sulphate are present in a discharge.

In the laboratory, the reagent used to precipitate sulphate is normally barium chloride; the solubility product of barium sulphate is extremely small ($K_{sp} \approx 10^{-10}$) and sulphate residuals of 1 mg l^{-1} may be achieved. However, this reagent is precluded for waste water treatment operations not only on the grounds of cost but particularly also because of the toxicity of soluble barium salts. In view of this the usual reagents employed are calcium based, e.g. $Ca(OH)_2$ for acids, or calcium chloride, although here it is not possible to reduce sulphate levels (as SO_3) much below about 1200 mg l^{-1} because of the solubility of calcium sulphate (2000–2200 mg l^{-1} at room temperature in the absence of other salts). Some benefit might be expected from the use of large excesses of Ca^{2+}, possibly by augmentation of lime treatment where acids are being neutralised, through a mass action effect, but calcium sulphate shows a marked tendency to supersaturate, and in practice little improvement is secured. Moreover, such a procedure can create an unacceptably high total salts' level in the discharging water. Consideration of the solubility product of gypsum (the stable form of calcium sulphate in water below 42°C) would indicate that sulphate residuals of about 500 mg l^{-1} should be theoretically achievable, but in practice this also is not possible, supersaturation being commonly observed. Where sulphate is present as a consequence of the treatment or even the manufacturing process, consideration may need to be given to using an alternative reagent, e.g. hydrochloric acid—although this of course is more expensive. Alleviation can be obtained by separating the sulphate-containing stream(s) for individual treatment before dilution with other wastes—a not unusual device where difficult discharge concentration limits apply.

The presence of aluminium salts either added or adventitiously present can, however, result in a reduction in residual soluble sulphate, through precipitation of a solid phase reported as having the empirical formula [$3CaO.Al_2O_3.3CaSO_4.31H_2O$] (Kelly 1960). In an investigation of the use of aluminium salts with lime in wool scouring wastes, Christoe (1976) identified formation of an insoluble complex at pH 10, containing Al, SO_4 and Ca in the ratio 2:3:6, which corresponds to Kelly's characterisation. Christoe was able to reduce the level of SO_4 in a typical waste from 1820 mg l^{-1} to 350 mg l^{-1} by such a treatment, although he also implies that lower residuals should be possible by adjusting the Al and Ca levels to maximise precipitation. The process is clearly limited by the cost of the aluminium salt, e.g. aluminium chlorohydrate, $Al_2(OH)_5Cl$; the amounts of chlorohydrate and lime necessary to remove 1000 mg l^{-1} SO_4, for example, would be 1800 mg l^{-1} and 3800 mg l^{-1} respectively.

Other anions

Other anions commonly encountered in waste water streams and which are subject to strict discharge limits are chromate, cyanide and sulphide. In principle, any of these could be removed by precipitative processes, since they all form insoluble salts with some metal cations, e.g. barium and lead chromates, ferric ferrocyanide, and most heavy metal sulphides. The deliberate addition of toxic metal ions such as lead or barium for the precipitation of chromate is not an appropriate treatment, and the preferred procedure for chromate removal is via chemical reduction to chromium(III) as discussed in Chapter 4, followed by precipitation as the hydrous oxide (Chapter 2). If heavy metal ions such as lead are already present in the waste stream then precipitation of insoluble chromates will occur and could be exploited, although this may lead to sludge disposal problems.

Vanadium removal from industrial waste waters has been studied by Kunz *et al.* (1976), who point out that only V(IV) and V(V) are likely to be soluble; as the pH increases, cationic forms become anionic, and V(IV) is oxidised to V(V). Simple pH adjustment with sodium hydroxide is ineffective, but Kunz *et al.* found that a ten-fold excess of Fe(II) over stoicheiometric levels, calculated on FeV_4O_9 or $Fe(VO_3)_2$, precipitated insoluble species satisfactorily in the pH range 8–9 with soluble V residuals of the order of 2 mg l^{-1}. At lower and higher pH values, solubilisation occurred.

The preferred method for cyanide removal is nowadays by oxidative destruction (Chapter 3), but where Fe(II) and Fe(III) cyanide complexes are present in a waste stream, this is not always effective. It is not uncommon in these circumstances for other heavy metals to be present, and precipitation of highly insoluble metal cyanide complexes then occurs anyway: the same may also occur with the somewhat oxidation-resistant $Ni(CN)_4^{2-}$ complex. It must be borne in mind, however, that the sludge formed has potentially toxic properties.

Sulphide may also be removed oxidatively by conversion to sulphur or sulphate (Chapter 3) or by precipitation as an insoluble metal salt (e.g. FeS). Since iron salts are frequently used in water treatment regimes and are relatively non-toxic, the objections raised to the precipitative process for chromate removal do not apply here. A vitally important parameter in sulphide precipitation processes is the pH. Sulphide ions, S^{2-}, are readily protonated to form HS^- ($pK_a > 14$) and H_2S ($pK = 7$) and the concentration of free sulphide is therefore very low except in alkaline solution. Consequently, only the most insoluble metal sulphides will precipitate from acidic solutions (e.g. HgS, $K_{sp} = 10^{-52}$; CuS, $K_{sp} = 10^{-36}$; CdS, $K_{sp} = 10^{-28}$). Iron(II) sulphide for which the solubility product is only about 10^{-18} will not precipitate from acid solutions, although precipitation from neutral and alkaline ones occurs readily.

Removal of heavy metal ions
The most common method for the removal of heavy metals from waste water streams is via pH adjustment leading to precipitation of the hydrous oxides, as described in Chapter 2. Whilst normally this is a successful approach there are circumstances where it cannot give satisfactory results if it is the sole treatment method. Because of the amphoteric nature of a number of metal hydrous oxides, the optimum pH for precipitation is not the same for every metal, as is shown in Figs. 2.8, 2.9 and 2.10 in Chapter 2. Where very strict discharge conditions apply there may be the need for stream segregation and individual treatments (with concomitant extra capital expense), or post-treatment stages involving selective cation exchange for removal of residual dissolved metals.

Simple pH adjustment of metals in anionic forms, e.g. chromate, molybdate and vanadate, may not effect adequate metal precipitation, and other techniques have to be employed (see previous sections, and Chapter 4). In other cases, the metal hydrous oxides may have too high a residual solubility to rely on their precipitation as a sole treatment method: examples are silver and lead, for which strict discharge limits are often applied. For silver, chloride precipitation is often effective (provided no complexing agents are present), and even with high pH treatment alone, where other metals are being precipitated there is usually some co-adsorption that reduces the residual silver levels; otherwise reduction to the metallic state can be used, as it can for other noble metals (see Chapter 4). Lead is precipitated reasonably well by sodium carbonate in the pH range 7–8, and soluble residuals of 1–2 mg l^{-1} have been observed in plant practice (Mattock 1970). Lime is much better than caustic soda as a precipitant for lead and can be used as the treatment method where other metals are also being precipitated. A combination of sodium hydroxide and carbonate is also effective.

Another co-precipitant is sodium orthophosphate, which forms insoluble basic phosphates at pH values around 9 with most heavy metals, to leave residuals often lower in concentration than by using pH adjustment alone, e.g. for tin and cadmium (see also Chapter 2). Treatment may be relatively less effective if substantial quantities of calcium are present, presumably from the competing effect of insoluble calcium phosphate formation, so lime should not generally be used as the pH adjustment reagent in these cases.

A significant problem can arise, however, when complexing agents are present, as already reviewed in Chapter 2. Attention has therefore been directed to searching for precipitants that can compete effectively with, for example, EDTA, NTA, NH_3 and other complexing agents that bind heavy metal ions strongly.

Use of sulphides

One possible candidate for effective competition with complexing agents is sulphide, since heavy metal sulphides are many orders of magnitude less soluble than the analogous hydrous oxides, and sulphide will precipitate metal ions such as copper, zinc and cadmium from most of their complexes. There are, however, obvious practical difficulties associated with the use of either H_2S or soluble sulphides (such as sodium sulphide) as precipitating reagents, insofar as they are highly toxic (H_2S is as toxic as HCN) and excess reagent cannot be tolerated (sulphide is itself subject to strict discharge limitations). Under these circumstances it is necessary to have either extremely efficient and reliable monitoring and control equipment or to incorporate subsequent treatment stages to eliminate excess sulphide. Nevertheless, sodium sulphide has been used to remove mercury (Bianchi and Palmese 1982), and for the same metal Findlay and MacLean (1981) have suggested the use of polysulphides, prepared by saturating an alkaline solution of sodium sulphide with sulphur, with which a typical chloralkali plant waste gave $< 1 \mu g l^{-1} Hg^0$ and 20–$40 \mu g l^{-1}$ total Hg in the effluent. However, excess S^{2-} is undesirable, because of the formation of soluble HgS_2^{2-}, and these authors recommend the use of, for example, Fe(III) to precipitate excess S^{2-}. Brown *et al.* (1979) indeed suggest the use of naturally occurring iron sulphide minerals, such as pyrrhotite (FeS) and pyrites (FeS_2) as mercury removal agents for chloralkali wastes.

Ferrous sulphide in fact provides an alternative means of precipitating most heavy metals in the presence of complexing agents, and has been recommended by Korchin (1979), but as the freshly prepared slurry rather than as the mineral (which offers a much smaller surface area for precipitation). On one plant receiving Cd, Cr, Cu, Pb, Ni and Zn, lime neutralisation to pH 8.5 was applied, followed by FeS slurry and a polymer before the waste passed to a clarifier. Extremely low metal levels were achieved in most cases (although Cu and Cr were 1 mg l^{-1}) but very long retention times were used (19 hours for FeS in the sludge blanket).

The principle is based on the fact that FeS is highly insoluble, with a solubility product of approximately 10^{-18}; hence the free sulphide, S^{2-}, in equilibrium with solid FeS is only about 10^{-9} M. If one assumes a pK for HS^- of approximately 14 then it is readily calculable that the total soluble sulphide $[HS^-] + [S^{2-}]$ under the operating conditions (pH 9) is less than 5 mg l^{-1}. Nevertheless, this low concentration is sufficient to precipitate as sulphides those metals encountered in waste streams which have solubility products for their sulphides lower than that for FeS (see Table 6.5). The nett reaction in which solid FeS acts as a large buffer reserve of S^{2-} is simply

$$M(OH)_2 + FeS \rightarrow MS + Fe(OH)_2$$

Chromates are also reduced to Cr(III):

$$CrO_4^{2-} + FeS + 4H_2O \rightarrow Fe(OH)_2 + Cr(OH)_3 + S^0 + 2OH^-$$

a procedure used by Aldrich (1984) and Higgins and Sater (1984). An alternative to FeS is MnS, which, although more expensive, is also highly effective in removing metals from soluble complexes.

Table 6.5 —Solubility products of heavy metal sulphides

Metal sulphide	$-\log K_{sp}$ (approximate)
Fes	18
NiS	19
CoS	20
ZnS	22
CdS	28
PbS	28
CuS	36
HgS	52

The problem with sulphide precipitation techniques, however, is that the sludge produced is not only highly toxic, but potentially liable to create serious problems if allowed to become acidic. We consider that the risks outweigh the technical advantages.

Xanthate complexes

Another approach, also effective in precipitating heavy metals from their soluble complexes, involves the use of starch xanthates (Wing and Doane 1977, 1978). Starch that has been treated with cross-linking agents such as epichlorhydrin, $POCl_3$, sodium trimetaphosphate, formaldehyde and others is xanthated with carbon disulphide in alkaline conditions, and reacted to form an insoluble alkali metal–magnesium starch xanthate. This type of product gives extremely insoluble heavy metal products in the presence of a cationic polymer, and has shown effectiveness in competing with complexing materials such as NH_3, EDTA and tartrate, particularly after preliminary acidification of the metal solution to pH 3 to break the complex (Wing 1981). Nevertheless, even here the long-term stability of the sludges produced must be queried; and the application would appear best as a polishing one to remove final traces of metals.

Other sulphur-containing precipitants

Some large molecular weight sulphur compounds can form highly insoluble metal complexes, where the sulphur atom acts as an electron donor to bind the metal ion. Examples include diethyldithiocarbamate (Oda *et al.* 1976, Toma 1976, Yamaguchi *et al.* 1976), mercaptobenzothiazole (Ichiki *et al.* 1976), and the sodium salt of *s*-trimercaptotriazine,

(Nakamura *et al.* 1973), marketed commercially by Degussa (W. Germany). This last chemical can be applied for precipitation of heavy metal ions such as Cu, Cd, Ag, Hg and Pb in the presence of complexing agents such as NH_3 and NTA in the pH range 7–10. The reaction can apparently be followed by a redox measuring system, although for complete precipitation an excess of reagent is needed, plus a retention time of approximately 15 minutes. With stronger complexing agents such as EDTA it is necessary to add Fe(III) as a competitive stripping agent, which is, of course, similar to the procedure described in Chapter 2 using Fe(II); and it would therefore seem that in these cases there is little advantage in using what is a relatively expensive reagent.

A somewhat simpler alternative is thiourea, $(NH_2)_2CS$, which has been proposed by several claimants, e.g. Szolnoki and Poloczek (1977) and Patron *et al.* (1978), but this material is environmentally unattractive itself. Indeed many sulphur-containing compounds must to some extent be suspect because of the risks associated with possible subsequent degradative reactions that may take place. They are also often nitrification inhibitors, and can result in unacceptable ammonia levels in discharges from sewage treatment works if present in significant amounts.

Foam separation and ion flotation

Foam fractionation, the principles of which were discussed earlier in this chapter, has been used to remove traces of metal ions down to fractions of one $mg\,l^{-1}$: Huang and Talbot (1973), for example, studied the removal of copper, cadmium and lead by the technique, and Chou and Okamoto (1976) applied chelating surfactants for cadmium. The surfactant:metal ratio is important, metal removal increasing until a critical surfactant concentration is reached; after this, further surfactant decreases the degree of removal, due to excess surfactant over that required for complex formation competing for bubble formation in the separation process. Similarly, too high a metal ion concentration, above the critical collector concentration (Rubin and Gaden 1962), results in reduced removal efficiency. The optimum pH range is normally in the region 8–10, but does depend on the surfactant as well as the metal.

Ion flotation, also mentioned earlier in this chapter, is another approach that could be used for removing traces of metal ions. Space does not permit a full discussion of the practical aspects of these surfactant-based techniques, and the reader is referred to the various publications cited for further information.

Summary

There are obviously numerous options available for precipitative and related removal methods for metal ions. The choice in any given situation must depend on a number of factors: cost, the final metal concentrations sought, the presence or otherwise of complexing agents (and their type), and whether recovery value is relevant; and any one of the methods described in this section would only seriously be applied in preference to simple pH adjustment where special circumstances prevail.

Patterson (1975) has given a comparative review covering a range of metals, and his text is recommended as a starting point. As an indication of the variabilities

Table 6.6 — Comparative efficiencies of various treatment techniques for mercury removal

Source	Treatment	Lower concentration performance $\mu g\,l^{-1}$
1, 2, 3	Sulphide precipitation	0.1–60
1	Al(III) flocculation	1–10
1	Fe(III) flocculation	0.5–5
1	Ion exchange	1–5
1	Activated carbon	0.25–20
4	Na borohydride + filtration	10–100
5	Electrolysis	<5

Sources:
1. Patterson (1975).
2. Brown *et al*. (1979).
3. Findlay and MacLean (1981).
4. De Angelis *et al*. (1978).
5. Matlosz and Newman (1982).

between performances at lower concentration levels (where the method used becomes particularly important) Table 6.6 gives some data for mercury—although it must be pointed out that not all of the methods quoted have been applied in plant practice.

PRACTICAL ASPECTS

Dosing techniques

The fact that many coagulation flocculation procedures involve addition of coagulating agent, e.g. ferrous or ferric sulphate, or aluminium sulphate, with pH adjustment to develop the metal hydrous oxide floc, both of which should desirably be carried out simultaneously, brings up the question of dosing control. Automatic pH control is readily achieved (see discussion in Chapter 2), but automatic control of coagulant dosing, in the full sense of dosing optimum amounts related to the waste input quantities, is not. Likewise, the addition of polyelectrolyte(s), a common secondary operation carried out as a next reaction stage, is not easily controlled in relation to input of solids to be flocculated. The same remarks generally apply to the addition of precipitating agents, and in what follows it can be assumed that these are included within the term 'coagulants'. Exceptions in principle apply where specific sensors such as ion-selective electrodes are available. There are practical and cost limitations on the in-line use of these devices, however (see Chapter 1), and they seem hardly to have been applied to automatic dosing control. Fluoride and sulphide ion-selective electrodes could, for example, be used, but they do not meet the principle that for control it is better to monitor the reagent rather than the species being removed (see Chapter 1).

In practice it is usually necessary, with continuously flowing discharges, to arrange for a quantity of reagent sufficient to treat the highest mass input rate (concentration × flow rate) that is likely to arise. This calls for a regulation of discharges when, as is often the case, periodic batch discharges of relatively strong wastes occur as a superimposition on a continuously flowing stream; the concentrates have to be retained and fed carefully into the continuous stream over a maximum period between batch discharges.

Automated control of flocculant dosing is possible, using commercially available turbidimetric equipment. A suspended solids monitor has recently been described (Gregory and Nelson 1984, Brown et al. 1985), available from Rank Bros. (UK), in which the flowing suspension is carried in a transparent flexible tube fitting in a perspex block housing two fibre-optic probes. These carry the incident and transmitted light, the latter being monitored by a photodiode that provides an electrical signal suitable, after amplification, to give an indication of the degree of aggregation of particles. It must be noted that it is important with all photometric equipment that the optical surfaces be kept clean, or at least that fouling compensators be incorporated; and this, plus the high cost of the equipment relative to other control systems, has meant that only limited plant application has been found so far in waste water treatment.

Overdosing of coagulant rarely gives rise to treatment problems beyond creating larger quantities of sludge to be removed (at least when sweep flocculation is being used), but overdosing of polyelectrolyte, even with sweep flocculation, can be as unsatisfactory as underdosing in some cases. Polyelectrolyte underdosing usually results in poor flocculation ; as the quantities added are increased, it is often found that a critical point is reached at which the flocculation quite dramatically improves. Beyond this, further polyelectrolyte addition tends to increase the size of flocs, until the stage is reached where they are too large to agglomerate; this often occurs when the flocs are 5–20 mm across. At this point polyelectrolyte addition has become self-defeating, since the solids may then neither settle nor float readily. In passing, it can be mentioned that one useful technique for polyelectrolyte dosing is to do it in two stages: a first so-called sacrificial dose is applied with reasonably vigorous agitation to start the flocculating action, and then a second dose is introduced downstream with the mildest possible blending to develop the flocs. Not only are good flocs formed in this way but less polyelectrolyte is usually needed than in single-dose operations. An economy measure for dosing both coagulants and polyelectrolyte flocculants is to link reagents' feed with flow of the waste water, as for example by making reagent pumps start and stop with waste water transfer pumps.

Blending of the reagent with the waste water must be carried out with efficient mixing, of course, to ensure that as the coagulant generates precipitate it is thoroughly dispersed to provide efficient utilisation; this applies for solutions as well as for lime slurries, dosing techniques for which are discussed in Chapter 2. The arguments given in Chapter 2 concerning the general undesirability of dosing too strong a reagent solution apply, and adequate coagulation residence time must be ensured. So-called 'flash mixing' can be used as an initial stage—O'Melia (1969) considers it essential—but must be followed by an ageing time in a second mixed reactor; in general, this double treatment is rarely necessary in waste water treatment practice. For most treatments with coagulation reagents, a total residence time of 10–20 minutes is suitable; less than this may not provide adequate coagulation ageing before the next stage, which is often flocculation. A separate reactor can be provided for flocculation, and is sometimes the ageing stage for coagulation as well; generally, however, it is better to keep the operations separate. Flocculation development times are usually only of the order of a few minutes, but polyelectrolytes must be introduced with efficient dispersion—although vigorous stirring must be avoided.

A common coagulation dosage rate for many waste waters provides in the effluent stream a concentration of 100–500 mg l^{-1} of the reagent, the amount depending on the nature of the waste and on its concentration. Experimental trials frequently have to be carried out to determine the most effective reagent dosage. With polyelectrolytes used in sweep flocculation, concentrations used in the effluent stream likewise depend on the nature of the waste, but importantly also on the type of polyelectrolyte: the range 1–10 mg l^{-1} is usual. It is highly desirable to carry out trials to determine the best polyelectrolyte and then to optimise the concentration to be employed in the waste stream. Laboratory tests have to be carried out with an eye to the plant conditions that will arise, so for coagulation tests it is necessary to arrange for virtually simultaneous pH adjustment with addition of coagulant to the stirred sample. In tests with polyelectrolytes, it is important to use freshly coagulated samples (aged ones can exhibit different flocculation characteristics), and to add the appropriate amount of polyelectrolyte with only the minimum degree of agitation-

—just sufficient to blend the two solutions, more increasing the risk that the formed flocs will break down.

Most of the polyelectrolytes used for waste water treatments (as opposed to those used in other applications) are available as solids, and have to be distributed as gel 'solutions' for use. This process demands some care and the correct equipment. A common practice is to add the appropriate amount of polyelectrolyte to a funnel connected to an eductor into which water is introduced. This sucks the solid into the venturi where it is wetted evenly before it passes into the dissolution tank underneath, itself containing water in agitation. (Air mixing is particularly helpful here.) Casual addition of the polyelectrolyte solid to the water results in 'clumping' of the solid to form jelly-like aggregates that may take several days to disperse even with continuous mixing; and poorly dispersed polyelectrolyte solution is virtually useless as a flocculating agent. A further point to note is that polyelectrolyte solutions age and lose some of their effectiveness after a period of 1–2 weeks, due to depolymerisation. Too vigorous an agitation can also cause molecular breakdown during solution preparation. Reagent solution strengths are usually in the region 0.05–0.2%, being lower with higher molecular weight materials. The limitation is due to increasing viscosities with increasing solution strengths: dissolution can, however, sometimes be helped, particularly with anionic and non-ionic types, by addition of sodium chloride (5–10% of weight of polymer used), since this reduces the solution viscosity. In application it is best to introduce as dilute a solution as possible (say 0.05%), since this ensures efficient distribution and hence effective utilisation. It is better to apply final polyelectrolyte dosing after pumping or other agitation to avoid floc breakdown, and with settlement systems this is commonly done either in a gently agitated flocculation vessel immediately prior to solids removal, or directly at the inlet to a settlement tank. Dissolved air flotation does not seem to disturb flocs unduly, and polyelectrolyte may be added just after the aeration stage.

Removal of flocculated or precipitated solids

The approach to flocculation or precipitation in terms of reagents used and their practical application cannot be divorced from the technique that is to be employed for removal of the solids formed. The two principal methods used are either flotation or sedimentation; direct filtration is rarely used as a primary technique in waste water treatment, although polishing, such as by sand filtration after preliminary treatment, is applied when a high quality discharge is needed, sometimes followed by activated carbon adsorption, e.g. if water re-cycling is intended. It should also be remembered that with some organic wastes, e.g. sewage, the flocculating agent can be of major significance in aiding de-watering of settled sludge. Because of these aspects, consideration must always be given to the flocculation/precipitation procedure and its place in the total treatment scheme.

To cite an example, consider the treatment of electrophoretic paint wastes, where ferric or aluminium sulphate is added with simultaneous pH adjustments by alkali addition. Lime slurry with ferric sulphate produces a floc that tends to settle, while sodium hydroxide or carbonate as alkalis produce a much lighter material which can be subjected to dissolved air treatment for flotation. The choice must depend to some extent also on the nature of the starting material; those electrophoretic paints containing lead may tend to produce a settleable floc with sodium carbonate addition, for example. There are indeed pros and cons for both approaches. Thus if dissolved air flotation is not applied consistently, with removal of the floated material soon after aeration treatment, sedimentation also will occur. On the other hand, with some flocs, again in electrophoretic paint treatment as an example, there can be a tendency for solids to float as well as settle, which makes settlement tank design more difficult. However, with some materials there is less difficulty in choice. With food wastes, e.g. from pig and poultry processing, the grease present inevitably favours flotation. Where surfactants are present, the flocs also provide a means of adsorbing these, so this plus the fact that aeration treatment also concentrates them in the surface points to the use of dissolved air flotation again.

It is not appropriate here to review the design and engineering features of flotation and settlement systems (some discussions are given in Chapters 2 and 7). However, mention can be made in passing to some aspects of consequence. Where it is applicable, dissolved air flotation (or even simple flotation if this is feasible) gives a sludge that usually contains substantially less water than that from settlement — the solids content often being higher by a factor of 5–10, e.g. 10% w/v solids as opposed to 1–2%. This has its advantages when sludge dewatering has to be carried out. Flotation systems often take up less space than conventional settlement ones (although with inclined tube and plate settlement devices this is not necessarily the case). On the other hand, dissolved air flotation plant is more costly to install and operate, and demands a higher degree of maintenance than settlement tanks, which are simple and relatively uncomplicated.

Slurry de-watering and sludge disposal

Although this subject also is broadly beyond the scope of this text, mention must be made of the influence that coagulation and flocculation reagents can have on the ease with which slurries can be de-watered. In the case of biologically formed sludges, inorganic and/or polyelectrolyte filter aids are common, and, as has already been indicated throughout this chapter and Chapter 2, use of these reagents in the chemical treatment of industrial wastes has to be assessed in terms of the sludges they produce as well as the pollution reductions they achieve. As discussed in Chapter 2, for example, lime usually confers easier de-watering on metal hydrous oxide sludges than does sodium hydroxide when used as a precipitant; and the value of polyelctrolytes is not confined to promoting flocculation for better removal of supended solids, but is apparent in the better de-watering properties induced in the separated slurries.

Ultimately, slurries or de-watered sludges have to be disposed of. Although biologically formed sludges can be either digested or directly transferred to land, residues containing inorganic or toxic materials demand care in their disposal, and increasingly strict legislation is developing to prevent the accumulation of such hazardous wastes. Indiscriminate dumping, or even poorly stabilised sludges, can give rise to re-pollution by leaching action of rain or ground water. Although incineration has been applied to the disposal of particularly intractable wastes (again not without dispute in some cases), this is generally highly uneconomic with the majority of sludge wastes, and landfill or sea disposal are usually the approaches adopted. Some interest has accordingly developed in ways by which the potential risks from such operations can be minimised, and several chemical fixation processes have been developed. One example is the 'Sealosafe' process, developed in the UK, which involves the addition of polymerising agents to the sludge, resulting in the formation of what is claimed to be an impervious solid after about three days. Another, the 'Chemfix' method developed in the USA, utilises a reaction between inorganic soluble silicates and setting agents to produce a pseudo-mineral solid matrix (Fenn-Smith 1974, Connor *et al.* 1977). Low-grade cements can be blended with sludges to form non-load-bearing materials that may have value for practical uses, e.g. in tiles (Tuznik and Kieszkowski 1972). Trials have also been carried out in the USA for brick making using sludge starting materials.

COAGULATION REAGENTS

The choice of a coagulation reagent is governed firstly by its effectiveness, and secondly by its cost relative to alternatives. Other considerations do enter, however. Iron salts generally offer advantages over aluminium ones when pH adjustment is made to produce a floc, in that floc formation is less sensitive to pH with iron salts, owing to the amphotericity of aluminium; and Fe(III) is also often superior to Fe(II) because it has a wider pH range of action. On the other hand, for maximum effectiveness it is sometimes necessary to control the pH fairly tightly anyway (as previous discussion in this chapter has indicated). The importance of the method to be used for removing solids on the choice of alkali has already been stressed, and this also has some bearing on the coagulant: aluminium hydrous oxide flocs are usually lighter than ferric ones.

Some of the more important physical properties of commonly used reagents are given below.

Ferric sulphate, $Fe_2(SO_4)_3$

Although available as a crystalline hydrate, this material is normally supplied for plant use as a solution in sulphuric acid, and commercially available material may contain as much as 600 mg l^{-1} $Fe_2(SO_4)_3$ in 0.5% w/w H_2SO_4, with 115 g l^{-1} Fe(III), the specific gravity at 15°C being just over 1.5. (Different commercial solutions have different characteristics, so it is not possible to generalise; there is also a ferrous iron content.) Because of the acid nature of the solution, precautions similar to those used for sulphuric acid (see Chapter 2) have to be observed in storage and handling, and dosing devices have to be selected for corrosion resistance accordingly.

Ferrous sulphate, $FeSO_4$

A common form of commercial supply of this material is as wet crystals of copperas, $FeSO_4 \cdot 7H_2O$, often derived as a by-product from sulphuric acid pickling of steel. Because of this it normally contains some excess sulphuric acid (often about 2% w/w), and so storage and handling of both the crystals and of aqueous solutions prepared from it should be as for sulphuric acid (see Chapter 2). The bulk density of the solid material is approximately 570 kg m^{-3}. For use it is convenient to make up a 25% w/v solution, for which the specific gravity is 1.13 at 15°C, although strengths of up to 500 g l^{-1} can be realised (specific gravity at 15°C = 1.24). Because of the acid content, the pH is, for example, approximately 3.2 for the 25% strength; too low a storage concentration is not desirable because the reduced acidity (and hence increased pH) renders the solution liable to hydrolysis.

Ferric chloride, $FeCl_3$

Ferric chloride, although available in hydrated crystal form ($FeCl_3 \cdot 6H_2O$), is frequently supplied commercially as a concentrated solution, typical strengths of $FeCl_3$ being 38–40% w/w, for which the specific gravity at 15°C is approximately 1.45. Since free hydrochloric acid is present, handling and storage precautions need to be the same as for hydrochloric acid (see Chapter 2).

Aluminium sulphate, $Al_2(SO_4)_3$

Various hydrated crystalline forms of aluminium sulphate are commercially available, e.g. $Al_2(SO_4)_3 \cdot 14H_2O$ in granulated or ground condition, having a bulk density of approximately 910 kg m^{-3}; and $Al_2(SO_4)_3 \cdot 21H_2O$ (7.4% Al) in slab form, having a bulk density in the loosely tipped conditions of approximately 700 kg m^{-3}. Iron is a common contaminant in the commercial materials, and this has an effect on the solubility–temperature curve of the $21H_2O$ material, reducing the solubility increase with temperature. The practical limits of solubilities are such that solutions of up to about 55% w/w of $21H_2O$ and 49% w/w of the $14H_2O$ forms can be prepared, but it is better to operate at less than this strength for convenience: the freezing point is at a minimum (-16°C) in the concentration region 4.3% w/w Al; above this strength the crystallisation temperature increases sharply. Excess sulphuric acid is usually present in the commercial materials, giving a pH in the region 2.0–2.5. The specific gravity of a typical 8% Al_2O_3 strength solution is approximately 1.32 at 15°C.

Commercial aluminium sulphate is often referred to as 'alum', but this is a misnomer: the term should be reserved for the correct compositions, viz. the double sulphates of aluminium with potassium or sodium, e.g. $K_2SO_4 \cdot Al_2(SO_4)_3 \cdot 24H_2O$.

'Polyaluminium chloride', $Al(OH)_xCl_y$

This commercially available and partly hydrolysed solution, where x is in the range 1.35–1.65, and $y = 3 - x$, typically has an Al content of 5.3% w/w, a freezing point of 13°C, a specific gravity of 1.2 at 20°C, and a viscosity of 5 cp at 20°C, with the usual acid character (pH 2.3–2.9) due to the presence of hydrochloric acid (demanding appropriate handling and storage precautions). Although this material is stable in these concentrations, it degrades by hydrolysis when diluted, so that for example a dilution to about one-third strength will give a solution that is stable for only 1–2 weeks.

REFERENCES

Aldrich, J. R. (1984) *Met. Finish.* **82** 51.
Anderson, C. A. (1965) *Text. J. Aust.* **40** 11.
Arden, T. V. (1968) *Water Purification by Ion Exchange*, Butterworths.
Arvin, E., and Petersen, G. (1980) *Progr. Wat. Technol.* **12** 283.

Bancsi, J. J., Benedek, A., and Hamielec, A. E. (1975) *Wat. Technol.* **7** 369.

Banks, N. (1975) *Wat. Pollut. Control* **74** 312.

Beer, W. J., and Gibbs, D. F. (1975) *Wat. Res.* **9** 1047.

Berger, H. F., Gehm, H. W., and Herbet, A. J. (1964) *U.S. Patent* No. 3, 129, 464.

Bianchi, A., and Palmese, N. (1982) *Chim. Ind. (Milan)* **64** 71.

Bishop, P. L. (1978) *Proc. 33rd Ind. Waste Conf. Purdue Univ.* 64.

Bitcover, E. H., Cooper, J. E., and Bailey, D. G. (1980) *J. Am. Chem. Leather Assoc.* **75** 108.

Blank, H. U., Keller, W., and Kuhne, G. (1976) *Brit. Patent* No. 1, 490, 691.

Bockris, J. O'M., Conway, B. E., and Yeager, E. (1980) *Comprehensive Treatise of Electrochemistry. Vol. 1. The Double Layer*, Plenum Press.

Bough, W. A. (1976) *Process Biochem.* **11** 13.

Briscoe, R. V., and Mattock, G. (1972) *Trans. Inst. Metal Finish.* **50** 199.

Brown, G. M., Gregory, J., Jackson, P. J., Nelson, D. W., and Tomlinson, E. J. (1985) *Instrumentation and Control of Water and Wastewater Treatment and Transport Systems*, ed. Drake, R. A. R., Pergamon.

Brown, J. R., Bancroft, M., Fyfe, W. S., and MacLean, A. A. N. (1979) *Environ. Sci. Technol.* **13** 1142.

Cheremisinoff, P. N., and Ellerbusch., F. (eds.) (1978) *Carbon Adsorption Handbook*, Ann Arbor Science.

Chou, E. J., and Okamoto, Y. (1976) *J. Wat. Pollut. Control Fed.* **48** 2747.

Christoe, J. (1976) *J. Wat. Pollut. Control Fed.* **48** 2804.

Christoe, J. (1977) *J. Wat. Pollut. Control Fed.* **49** 848.

Cillié, G. G., and Van Vuuren, L. R. J. (1978) *New Processes of Waste Water Treatment and Recovery*, ed. Mattock, G., Ellis Horwood.

Clough, F. (1977a) *Chem. Ind. (London)* 811.

Clough, F. (1977b) *Wat. Pollut. Control* **76** 10.

Connor, J. R., Zanadzki, E. A., and Polsky, R. J. (1977) *U.S. Patent* No. 4, 072, 320.

Cooper, P. F. (1975) *Wat. Pollut. Control* **74** 303.

Cooper, P. F., and Thomas, E. V. (1974) *Wat. Pollut. Control* **73** 505.

Cooper, R. N., and Denmead, C. F. (1979) *J. Wat. Pollut. Control Fed.* **51** 1017.

Coupal, B., and Lalancette, J. M. (1976) *Wat. Res.* **10** 1071.

Crowe, T., O'Melia, C. R., and Little, L. (1977) *Proc. 32nd Ind. Waste Conf. Purdue Univ.* 655.

Culp, R. L., and Culp, G. L. (1971) *Advanced Waste Water Treatment*, Van Nostrand Reinhold.

Czukräsz, G., and Hompasz, G. (1981) *Galvanotechnik (Saulgau)* **72** 832.

Davidson, R. L. (1980) *Handbook of Water Soluble Gums and Resins*, McGraw-Hill.

De Angelis, P., Morris, A. R., and MacMillan, A. L. (1978) *U.S. Patent* No. 4, 098, 697.

DeBoice, J. N., and Thomas, J. F. (1975) *J. Wat. Pollut. Control Fed.* **47** 2246.

Derjaguin, B. V., and Landau, L. (1941) *Acta Phys. Chem. U.S.S.R.* **14** 633.

Diamadopoulos, E., and Benedek, A. (1984) *J. Wat. Pollut. Control Fed.* **56** 1165.

Dorfner, A. (1972) *Ion Exchangers. Properties and Applications*, Ann Arbor Science.

Eckenfelder, W. W., Jr., and Cecil, L. K. (eds.) (1972) *Application of New Concepts of Physical-Chemical Waste Water Treatment. Progress in Water Technology. Vol. I*, Pergamon.

Eilbeck, W. J., and Mattock, G. (1983) *Chem. Ind.* (*London*) 551.

Faust, S. D., and Hunter, J. V. (eds.) (1967) *Principles and Applications of Water Chemistry. Proc. 4th Rudolfs Conf.*, Wiley.

Feitknecht, W., and Schindler, P. (1963) *Solubility Constants of Metal Oxides, Metal Hydroxides and Metal Hydroxide Salts in Aqueous Solution*, Butterworths.

Fenn-Smith, C. W. K. (1974) *Anticorrosion* **21** 17.

Ferguson, J. F., and Vråle, L. (1984) *J. Wat. Pollut. Control Fed.* **56** 355.

Ferguson, J. F., Jenkins, D., and Eastman, J. (1973) *J. Wat. Pollut. Control Fed.* **45** 620.

Findlay, D. M., and MacLean, R. A. N. (1981) *Environ. Sci. Technol.* **15** 1388.

Finley, J. W., Gauger, M. A., and Fellers, D. A. (1973) *Cereal Chem.* **50** 465.

Fuerstenan, D. W. (ed.) (1962) *Froth Flotation*, American Institute of Mining.

Gabel, R. K., and Seitz, P. L. (1977) *U.S. Patent* No. 4, 055, 495.

Gleisberg, D. (1976) *Galvanotechnik* (*Saulgau*) **67** 346.

Gloyna, E. F., and Eckenfelder, W. W., Jr. (1970) *Water Quality Improvement by Physical and Chemical Processes*, Univ. Texas Press.

Goos, G., Hwang, C. P., and Davis, E. (1974) *A.I.Ch.E. Symp. Ser. No.* 144 213.

Gregory, J. (1978) *The Scientific Basis of Flocculation*, ed. Ives, K. J., Sijthoff and Noordhoff.

Gregory, J., and Nelson, D. W. (1984) *Solid–Liquid Separation*, ed. Gregory, J., Ellis Horwood.

Grieves, R. B. (1975) *Chem. Eng. J.* **9** 93.

Groves, S. E., and Lundgren, H. E. (1973) *Proc. 28th Ind. Waste Conf. Purdue Univ.* 99.

Gutcho, S. (1977) *Waste Treatment with Polyelectrolytes and Other Flocculants*, Noyes Data Corp.

Hannah, S. A., Johns, M., and Cohen, J. M. (1977) *J. Wat. Pollut. Control Fed.* **49** 2297.

Hartinger, L. (1973) *Metalloberfläche* (*München*) **27** 157.

Hassler, J. W. (1963) *Activated Carbon*, Chemical Publishing Co.

Helfferich, F. (1962) *Ion Exchange*, McGraw-Hill.

Higgins, T. E., and Sater, V. E. (1984) *Environ. Progr.* **3** 12.

Hopwood, A. P., and Rosen, G. (1972) *Process Biochem.* **7** (3) 15.

Huang, C. F., and Wu, P. L. K. (1984) *J. Wat. Pollut. Control Fed.* **56** 233.

Huang, C. P., and Wu, M. H. (1975) *J. Wat. Pollut. Control Fed.* **47** 2437.

Huang, R. C. H., and Talbot, F. D. (1973) *Can. J. Chem. Eng.* **51** 709.

Ichiki, M., Ogawa, N., and Ishu, M. (1976) *Jap. Patent* No. 76, 00, 764.

Ishibashi, T. (1980) *J. Am. Water Works Assoc.* **72** 514.

Ives, K. J. (ed.) (1978) *The Scientific Basis of Flocculation*, Sijthoff and Noordhoff.

Jackson, P., and Sheiham, I. (1981) *Water Research Topics. Vol. I.*, ed. Lamont, I. M., Ellis Horwood.

Jellinek, H. H. G., and Sangal, S. P. (1972) *Wat. Res.* **6** 305.

Jenkins, D., and Lee, F. M. (1977) *Progr. Wat. Technol.* **9** 495.

Jenkins, D., Ferguson, J. F., and Menar, A. B. (1971) *Wat. Res.* **5** 369.

Jenkins, D., Menar, A. B., and Ferguson, J. F. (1972) *Applications of New Concepts of Physical-Chemical Wastewater Treatment. Progress in Water Technology*, Vol. I, eds. Eckenfelder, W. W., Jr., and Cecil, L. K., Pergamon.

Jenkins, S. H., Sane, M., and Wallbank, T. (1977) *Chem. Ind. (London)* 821.

Jola, M. (1975) *Oberfläch. (Coburg)* **52** 169; 172; 174.

Jørgensen, S. E., Libor, O., and Barkacs, K. (1973) *Wat. Res.* **7** 1885.

Judkins, J. F., Jr., and Hornsby, J. S. (1978) *J. Wat. Pollut. Control Fed.* **50** 2446.

Kavanaugh, M. C., Krejci, V., Weber, T., Eugster, J., and Roberts, P. V. (1978) *J. Wat. Pollut. Control Fed.* **50** 216.

Kelly, R. (1960) *Canad. J. Chem.* **38** 1218.

Keramida, V., and Etzel, J. E. (1982) *Proc. 37th Ind. Waste Conf. Purdue Univ.* 181.

Kisla, T. C., and McKelvey, R. D. (1978) *Environ. Sci. Technol.* **12** 207.

Korchin, S. R. (1979) *Proc. 26th Ontario Ind. Waste Conf.* 301, Ontario Water Resources Commission.

Kunin, R. W. (1986) *Ion Exchange Resins*, Robert E. Krieger.

Kunz, R. G., Giannelli, J. F., and Hensel, H. D. (1976) *J. Wat. Pollut. Control Fed.* **48** 762.

La Mer, V. K. (1966) *Disc. Farad. Soc.* **42** 248.

La Mer, V. K. (1967) *Principles and Applications of Water Chemistry. Proc. 4th Rudolf Conf.*, eds. Faust, S. D., and Hunter, J. V., Wiley.

La Mer, V. K., and Healy, T. W. (1968) *Rev. Pure Appl. Chem.* **13** 112.

Leentvaar, J., Buning, W. G. W., and Koppers, H. M. M. (1978) *Wat. Res.* **12** 35.

Leentvaar, J., Koppers, H. M. M., and Buning, W. G. W. (1979) *J. Wat. Pollut. Control Fed.* **51** 2457.

Le Gendre, G. R., and Runnells, D. D. (1975) *Environ. Sci. Technol.* **9** 744.

Lemlich, R. (ed.) (1972) *Adsorptive Bubble Separation Techniques*, Academic Press.

Lin, S. S., and Carlson, A. D. (1975) *J. Wat. Pollut. Control Fed.* **47** 1978.

Link, W. E., and Rabosky, J. G. (1976) *Proc. 31st Ind. Waste Conf. Purdue Univ.* 485.

Linstedt, K. D., Houck, C. P., and O'Connor, J. J. (1971) *J. Wat. Pollut. Control Fed.* **43** 1507.

Lockwood, R. A., and Chen, K. Y. (1973) *Environ. Sci. Technol.* **7** 1208.

Lyklema, J. (1978) *The Scientific Basis of Flocculation*, ed. Ives, K. J., Sijthoff and Noordhoff.

Matlosz, M., and Newman, J. (1982) *Proc. Electrochem. Soc.* **82** 53.

Mattock, G. (1970) unpublished work.

Mattock, G., and Bryan, C. W. (1983, 1984) unpublished work.

Mattson, J. S., and Mark, H. B., Jr. (1971) *Activated Carbon*, Marcel Dekker.

McGuire, M. J., and Suffet, I. M. (eds.) (1983) *Treatment of Water by Granular Carbon. ACS Advances in Chemistry Series No.* 202.

McIntyre, G., Rodriguez, J. J., Thackston, E. L., and Wilson, D. J. (1983) *J. Wat. Pollut. Control Fed.* **55** 1144.

Merrill, D. T., and Jorden, R. M. (1975) *Progr. Wat. Technol.* **7** 379.

Michaels, A. S. (1954) *Ind. Eng. Chem.* **46** 1485.

Miller, D. G. (1974) *A.I.Ch.E. Symp. Ser. No.* 144 39.

Minton, G. B., and Carlson, D. A. (1972) *J. Wat. Pollut. Control Fed.* **44** 1736.

Minton, G. B., and Carlson, D. A. (1976) *J. Wat. Pollut. Control Fed.* **48** 1697.

Morgan, J. J., and Stumm, W. (1964) *J. Coll. Sci.* **19** 437.

Mysels, K. J. (1959) *Introduction to Colloid Chemistry*, Interscience.

Nakamura, Y., Umehara, A., and Yamada, I. (1973) *U.S. Patent* No. 3, 778, 368.

Nasr, M. S. H., and MacDonald, M. C. (1978) *Can. J. Chem. Eng.* **56** 87.

Netzer, A., and Hughes, D. E. (1984) *Wat. Res.* **18** 927.

Nielsen, A. E. (1964) *Kinetics of Precipitation*, Pergamon.

Novak, J. T., and O'Brian, J. H. (1975) *J. Wat. Pollut. Control Fed.* **47** 2397.

Oda, N., Horie, Y., Idohara, M., and Gomyo, T. (1976) *Jap. Patent* No. 76, 71, 290.

Okuda, T., Sugano, I., and Tsuji, T. (1975) *Filtr. Separn.* **12** 472.

Olthof, M. G., and Eckenfelder, W. W., Jr. (1975) *Wat. Res.* **9** 853.

O'Melia, C. R. (1969) *Public Wks. J.* **100** 87.

O'Melia, C. R. (1972) *Physicochemical Processes for Water Quality Control*, Weber, W. J., Jr., Wiley–Interscience.

O'Melia, C. R. (1978) *The Scientific Basis of Flocculation*, ed. Ives, K. J., Sijthoff and Noordhoff.

O'Shaughnessy, J. C., and Blanc, F. C. (1978) *Proc. 33rd Ind. Waste Conf. Purdue Univ.* 642.

Packham, R. F. (1965) *J. Coll. Sci.* **20** 81.

Panday, K. K., Prasad, G., and Singh, V. N. (1985) *Wat. Res.* **19** 867.

Patron, G., Napoli, D., Ratti, D., and Tubiello, G. (1978) *U.S. Patent* No. 4, 087, 359.

Patterson, J. W. (1975) *Wastewater Treatment Technology*, Ann Arbor Science.

Paulson, E. G. (1977) *Chem. Eng.* **84** (10) 89.

Peniston, O. P., and Johnson, E. L. (1977) *U.S. Patent* No. 4, 018, 678.

Perlmann, G., and Hermann, H. (1938) *Biochem. J.* **32B** 926.

Rebhun, M., and Streit, S. (1974) *Wat. Res.* **8** 195.

Rebhun, M., Narkis, N., and Offer, R. (1975) *Progr. Wat. Technol.* **7** 401.

Roberts, K. L., and Lin, J. G. (1983) *Proc. 38th Ind. Waste Conf. Purdue Univ.* 83.

Rohrer, K. L. (1971) *Ind. Wastes (suppl. to Wat. Sewage Wks. J.)* **118** 36.

Rohrer, K. L. (1974) *Wat. Wastes Eng.* **11** 66.

Rowley, A. G., Husband, F. M., and Cunningham, A. B. (1984) *Wat. Res.* **18** 981.

Rubin, E., and Gaden, E. L., Jr. (1962) *New Chemical Engineering Separation Techniques*, ed. Schoon, H. M., Interscience.

Ruehrwein, R. A., and Ward, D. W. (1952) *Soil Sci.* **73** 485.

Russell, J. M., Cooper, R. N., and Crocombe, B. I. (1984) *Environ. Rechnol. Lett.* **5** 289.

Saito, I. (1984) *Wat. Res.* **18** 319.

Schindler, P. (1959) *Helv. Chim. Acta* **42** 577.

Schindler, P., Althaus, H., Hofer, F., and Minder, W. (1965) *Helv. Chim. Acta* **48** 1204.

Schwoyer, W. L. K. (ed.) (1981) *Polyelectrolytes for Water and Wastewater Treatment*, C.R.C. Press.

Sebba, F. (1962) *Ion Flotation*, Elsevier.

Semmens, M. J., and Martin, W. (1980) *A.I.Ch.E. Symp. Ser. No. 197* 367.

Shelley, M. L., Randall, C. W., and King, P. H. (1976) *J. Wat. Pollut. Control Fed.* **48** 753.

Smith, E. F., MacCarthy, P., Yu, T. C., and Mark, H. B., Jr. (1977) *J. Wat. Pollut. Control Fed.* **49** 633.

Smoluchowski, M. (1917) *Z. Physik. Chem.* **92** 129.

Sparnaay, M. J. (1972) *The Electrical Double Layer*, Pergamon.

Spinelli, J., and Koury, B. (1970) *J. Agric. Food Chem.* **18** 284.

Stahr, R. W., Boepple, C. P., and Knocke, W. R. (1980) *Proc. 35th Ind. Waste Conf. Purdue Univ.* 186.

Stamberg, J. B., Bishop, D. F., Warner, H. P., and Griggs, S. H. (1971) *A.I.Ch.E. Symp. Ser. No. 107* 310.

Stern, O. (1924) *Z. Elekt.* **30** 508.

Stones, T. (1981) *Wat. Pollut. Control* **80** 421.

Stumm, W., and Morgan, J. J. (1981) *Aquatic Chemistry*, Wiley.

Stumm, W., and O'Melia, C. R. (1968) *J. Am. Water Works Assoc.* **60** 514.

Szolnoki, G., and Poloczek, A. (1977) *Ger. Patent* No. 2, 608, 153.

Tan, J. S., and Schneider, R. L. (1975) *J. Phys. Chem.* **79** 1380.

Thackston, E. L., Wilson, D. J., Hanson, J. S., and Miller, D. J., Jr. (1980) *J. Wat. Pollut. Control Fed.* **52** 317.

Thanabalasingam, P., and Pickering, W. F. (1985) *Env. Poll. B.* **10** 115.

Thompson, R., ed. (1986) *Trace Metal Removal from Aqueous Solution Spec. Pub. No. 61* Royal Society of Chemistry.

Toma, W. (1976) *Jap. Patent* No. 76, 68, 965.

Tønseth, H. J., and Berridge, H. B. (1968) *Effl. Wat. Treat. J.* **8** 124.

Tuznik, F., and Kieszkowski, M. (1972) *Electroplat. Metal Finish. J.* **25** 10.

Ulmgren, L. (1975) *Progr. Wat. Technol.* **7** 409.

Verwey, E. J. W., and Overbeek, J. Th. G. (1948) *Theory of the Stability of Lyophobic Colloids,* Elsevier.

Vråle, L. (1978) *Progr. Wat. Technol.* **10** 645.

Van Vuuren, L. R. J., Ross, W. R., and Prinsloo, J. (1976) *Proc. 8th IAWPR Int. Conf.,* ed. Jenkins, S. H., Pergamon.

Van Vuuren, L. R. J., Henzen, M. R., Stander, G. J., and Clayton, A. J. (1970) *Proc. 5th IAWPR Int. Conf.,* ed. Jenkins, S. H., Pergamon.

Walton, A. G. (1967) *The Formation and Properties of Precipitates*, Interscience.

Weber, W. J. (1972) *Physicochemical Processes for Water Quality Control,* Wiley–Interscience.

Wing, R. E. (1981) *Proc. Workshop Inst. Interconnecting and Packaging Electronic Circuits,* Anaheim, May 19–21.

Wing, R. E., and Doane, W. M. (1977) *U.S. Patent* No. 4, 051, 316.

Wing, R. E., and Doane, W. M. (1978) *U.S. Patent* No. 4, 083, 783.

Woodward, F. E., Sproul, O. J., Hall, M. W., and Ghosh, M. M. (1972) *J. Wat. Pollut. Control Fed.* **44** 1909.

Yamaguchi, T., Hatanaka, K., and Ikebe, S. (1976) *Jap. Patent* No. 76, 140, 360.

Zabban, W., and Jewett, H. W. (1967) *Proc. 22nd Ind. Waste Conf. Purdue Univ.* 706.

Zabban, W., and Helwick, R. (1975) *Proc. 30th Ind. Waste Conf. Purdue Univ.* 479.

7

Demulsification

THE NATURE OF EMULSIONS

Introduction

An emulsion is a disperse system in which both the disperse phase and the dispersion medium are liquids, the two being essentially immiscible. The definition can also, for convenience, embrace dispersions of hydrophobic materials that are normally solid or semi-solid at ambient temperatures, such as greases and fats, these being chemically related to oils in being essentially non-polar organic materials. In the present context we are primarily concerned with those systems in which water is the dispersion medium, and with means by which the emulsion can be broken to separate the disperse phase from the carrier water.

One essential difference from many of the sols discussed in Chapter 6 is that whereas in many of the former there is some hydrophilic character, emulsions all involve hydrophobic properties. Further, the droplet sizes in emulsions are generally larger than the sizes of dispersed particles in typical sols as discussed in Chapter 6. In all cases, however, the stability of the system is closely connected to the presence of surface charges and to the magnitudes of interfacial energies. A detailed review of electrokinetic phenomena and of the nature and properties of the electrical double layer surrounding colloid particles has been given in Chapter 6, and since the principles are essentially the same for emulsions, repetition here is unneccesssary. This chapter will therefore concentrate mainly on practical aspects of emulsion breaking, although here again many of the techniques are closely similar to those discussed in Chapter 6.

For reviews of emulsion science, see, for example, Becher (1965), Sherman (1968), and Smith (1974).

Emulsifying agents

Emulsions are generally only stable in the presence of a third substance, an emulsifier, unless the disperse phase is extremely finely distributed as very small droplets—although even here there is an inherent instability in the system. The reason for this is apparent when it is appreciated that colloidal systems depend to a large extent for their stability on the presence of surface charges on the dispersed particles: the essentially non-polar characteristics of oils, greases and fats provide only limited propensity for development of such charges. Emulsifying agents provide mechanisms both for this development and for modification of interfacial energies, both factors contributing to the stability of the emulsions.

The interfacial energy between water and a hydrophobic material (such as oil) that is immiscible with it is usually relatively large with respect to the internal energies of the two phases, and there is therefore an inherent tendency for dispersed droplets to coalesce and form a separate bulk phase: the driving force for this is the reduction in free energy of the system that results. The effect of an emulsifier is probably due to the lowering of interfacial energy that occurs upon adsorption of the emulsifier at the oil–water boundary. With emulsifiers containing potentially ionis-able groups, this also provides a means of establishing a surface charge on the dispersed material, the resultant coulombic forces adding to the stabilizing effect (see Chapter 6).

Emulsifiers fall into classes differentiated according to their chemical nature. For example:

(1) Soaps and detergents constitute a major class of emulsifying agent. These consist of hydrophobic main structures with hydrophilic end or side components, and include such examples as sodium and potassium stearates, naphthenates and cresylates, and salts of sulphonated or sulphated aromatic or polycyclic com-pounds. Emulsion stability with these is usually greater in neutral or alkaline conditions.

(2) Essentially non-ionic hydrophilic materials such as gums, starches and saponins increase the viscosity of an aqueous phase and so reduce the rate at which emulsion droplets, once formed, will aggregate and coalesce. Various syntheti-cally prepared condensates of alcohols or phenols also display this property. Apart from providing a viscosity effect, these materials almost certainly also have an effect on the interfacial energies at the droplet surfaces, and many provide simple physical barriers to coalescence by forming films.

(3) Emulsions may also be stabilised by the presence of finely divided solid particles, such as preparations of carbon, some silicate and aluminosilicate materials, and even fine powders of basic salts such as the sulphates of iron, copper or nickel. Once again it is the effect on the interfacial energies which is important in the stabilisation effect, particularly relevant being the relative surface energies associated with the oil/solid particle and solid particle/water interfaces. Depend-ing on these relative surface energies solid particles may stabilise water-in-oil emulsions (e.g. silica powder, more readily 'wetted' by water than by oil) or the reverse (e.g. carbon powder, which is more readily 'wetted' by oil than by water). The relative volumes of the two phases are also obviously of importance in deciding which type of emulsion is obtained.

Other constituents
Apart from stabilising agents, many emulsions used in industry and which are discharged in waste waters may contain other additives. Notably these may be anti-oxidants (such as nitrites and formates), algicides and bactericides, all introduced to

inhibit other forms of degradation that may occur, and corrosion inhibitors. These materials have to be considered in the overall treatment of emulsions, because they may also need to be degraded or removed as a consequence of being environmentally unacceptable. Treatment procedures for specific components would be according to their chemical class and to the available techniques. Broadly speaking, oxidation is usually the most appropriate, and this would normally be applied to the aqueous phase after demulsification.

The role of demulsifiers

Clearly, to destabilise an emulsion it is necessary to provide some action that will affect the oil/grease–water interfacial film appropriately. Several workers (see King 1941, Becher 1965, Strassner 1968, Neustadter et al. 1975, 1981) have expressed the view that the strength, compactness, viscosity and elasticity of the interfacial film are important factors affecting emulsion stability, so the role of demulsifiers is to displace or modify these films. The specific manner in which a given demulsifying agent works depends, however, on its particular characteristics. It appears that charge reversal or neutralisation alone will not necessarily cause demulsification, because strong interfacial films may still prevent droplet coalescence, and bridging by means of demulsifying agents may additionally be necessary. The conditions for bridging depend on the type of demulsifying agent; and in this respect the comments in the theoretical discussion in Chapter 6 are relevant.

SOURCES AND EFFECTS OF EMULSION DISCHARGES

Sources of waste emulsions

A major source of oily discharges is of course the petroleum industry, particularly in terms of the volumes produced. An important operation for the treatment of such wastes is flotation, and for this the so-called API separator, dissolved air flotation equipment, and various associated improvement devices, such as the tilted plate separator, are widely used as a primary physical means of removing free oil. (These are discussed briefly in the next section.) The aqueous discharge from such separators frequently contains some emulsified as well as free oil, however, and thus some form of subsequent chemical treatment is often needed. These emulsions seem to derive from materials (natural surfactants) present in the oil which provide emulsifying action through being adsorbed at the oil–water interface, with the result that the interfacial film is modified.

Emulsions that have been specially formulated for industrial operations, such as machining of metals, constitute another type of discharge, and may be quite stable. These emulsions may contain emulsifiable oils consisting of the oil together with fat or soap additives, dispersed in water. As an economy measure these fluids are normally recirculated after removal of metal swarf, but eventually need to be disposed of as a consequence of oxidative degradation or build-up of finely divided solids not removed by the recirculation system. In fact the more degraded an emulsion of this type is, frequently the more difficult it is to break it.

Another group of emulsion wastes arises from the use of cleaning solutions, frequently alkaline and containing detergents, for removal of oil and grease from metal surfaces. Discharges here include both spent cleaning solutions and rinse waters used for removing excess cleaner from the work surfaces.

Vegetable, as opposed to mineral, oils are present in discharges from food processing, as for example in extractions from plant seeds such as rape, cotton and soya bean, while animal oils and fats arise from fish processing, in margarine manufacture, in the slaughtering of animals, and in processing of animal by-products. Fats and greases are significant components of such discharges, where emulsification more readily occurs than it does with mineral oils due to their chemical nature. The oily substances may contain, for example, triglycerides in various degrees of chemical unsaturation, which disperse relatively readily. (A review of the uses of emulsions in various industries, and of the component materials used, is given in Lissant (1974).)

Environment problems
In general terms, emulsions present the same potential problems in the environment as do the oils or fats they contain. Breakdown of discharged emulsions in receiving waters will liberate the free non-aqueous component, which will then coat flora and fauna, float on water surfaces, and create a chemical oxygen demand. Minerals oils are even less acceptable than vegetable oils in this last respect, in that they degrade relatively more slowly. Apart from physical coating with consequent inhibition of life processes, oily materials of course also coat equipment that may be used in waterways.

In addition, other components present in emulsions, such as emulsifying agents, anti-oxidants or bactericides, can have specific toxic effects on organisms, and therefore also require attention. For these, where the chemical character can be identified, treatments can be according to processes of oxidation or precipitation as described elsewhere in this book.

DEMULSIFICATION TECHNIQUES
Many publications appear in the literature, detailing procedures for the treatment of particular emulsions. Here we are concerned primarily with the general principles of chemical treatment, from which experimental investigation of a problem can be started. For a review of the many physical as well as chemical procedures, the reader is referred, for example, to a summary by Lohmeyer (1973).

The definition of oil contamination
The effectiveness of any treatment procedure is of course defined by the residues of contaminant that remain after such treatment. Such definition may be in terms of percentage of contaminant removed or as a residual concentration. For most waste water treatment operations this is unambiguous, given an adequate sensitivity, replicability and freedom from interference in the analytical method used. With oils, greases and waxes, and consequently also with emulsions, the situation is not quite so simple, in that individual components are not necessarily identified by the analytical method, with determinations being of a class of materials, and because of possible interferences.

Standard recommended analytical procedures have been published by the API

(1959), EPA (1979), Department of Environment (1983), IP (1985), and APHA/
AWWA/WPCF (1985) that are in principle applicable to distributions of oils and
greases in water. The Department of Environment method involves addition of
sodium chloride, acidification, extraction with carbon tetrachloride or trichlorotrif-
luorethane, drying of the extract with anhydrous magnesium sulphate, and then
extraction of polar substances with 'Florisil' before infra-red absorption measurement. The IP (Institute of Petroleum) method for cutting oils uses acid alone, with
volumetric measurement of the released oil. No specific instruction is given as to the
pH to which the sample is to be brought (a quantity of acid to be added is defined),
and this can have an influence on the degree of extraction of oils from an emulsion if
the sample is strongly alkaline, for example. The APHA/AWWA/WPCF and EPA
methods do specify a pH of 2, but the API method suggests a pH of 5, at which soap
fatty acids may not be liberated completely. This can lead to differences in results
according to the determination method adopted. Interferences are also possible
from various substances such as sulphur compounds and certain organic dyes in some
of the APHA/AWWA/WPCF methods which do not use activated silica for removal
of polar substances. Importantly also, the Department of Environment publication
acknowledges that emulsifying agents may have an influence on the determination,
because there is a lack of well-documented evidence of the conditions under which
emulsified oil can be extracted.

It is clear, therefore, that the definition of oil contamination must be linked
carefully to the method used for determination, and that none of the methods is
specific. For closer characterisation, gas chromatography needs to be applied
(Department of Environment 1982).

Preliminary treatments
Most emulsion discharges contain not only emulsified but also free material in
suspension. To reduce the degree of chemical treatment for the emulsion itself it is
usually sensible to remove the free oil, grease or fat by flotation (and/or sedimentation where solids denser than water are present, as for example with some
machining oil sludges) before the addition of chemicals. If traces only of the free
material are present this is unnecessary, but particularly where the emulsion is not
very stable such physical processes as dissolved air or electroflotation (discussed in
Chapter 5) can be beneficial by effecting some breaking of the emulsion themselves.
For example, the emulsions from some food processing operations, such as slaughtering, can be largely separated by the application of dissolved air flotation (although
some further treatment is usually required). Emulsion wastes from machining and
cleaning operations are more stable, however, and chemical treatment is nearly
always necessary for these before discharge can be made to meet even most sewer
limits. Oily discharges from petroleum refining and associated operations are always
subject initially to flotation, which removes the bulk of the oil contamination but still
leaves emulsified residues demanding some further treatment.

Simple flotation may be used, which for small flows (as from vehicle cleaning
areas, garages, etc.) usually consists of an oil trap arrangement, where the flow
enters a chamber (typically with 20–60 minutes retention time) at the top, then flows
downward under a baffle near the exit, preferably with inlet and outlet weirs to
distribute the flow evenly. The oil or grease collects on the inlet side of the baffle, and

is removed periodically. As with settlement systems, attention should be paid to the rise rate of the free material to determine the maximum underflow rate (which determines the surface area of the tank): the underflow rate, which is given by the volumetric throughput rate divided by the surface area of the flotation zone, should be slower than the rise rate of the oily material.

Such an arrangement may be adequate where subsequent emulsion cracking is to be carried out, or where the flows represent only a small proportion of a total (oil- and grease-free) discharge, which therefore will provide adequate dilution. In other situations, a more sophisticated arrangement may be needed, as for example the API separator (API 1969), extensively used as an oil removal device in the petroleum industry. Fig. 7.1 shows the essentials of such a separator. Experience over many

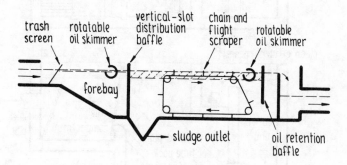

Fig. 7.1 — Schematic diagram of API separator.

years has shown that these devices are adequate for removal of most oil from suspensions, but they are not good enough to meet many present-day discharge limits, and do not affect stable emulsions. Typical performance data have been quoted by Peoples *et al.* (1972) for refinery waste water, indicating an average residual oil content of 60 mg l^{-1}, up to maxima of 180 mg l^{-1}—although for many installations the higher figure is more typical. At the overflow rates normally recommended for the API separator, oil globules of diameter greater than 150 μm are effectively captured: for smaller sized globules much lower flow velocities (an order of magnitude less for removal of globules 50 μm in size) are necessary. This follows directly from Stokes' law:

$$V_s = \frac{(\rho_m - \rho_o)gd^2}{18\,\mu}$$

where ρ_m and ρ_o are the specific gravity of the dispersing medium and oil respectively, d is the globule diameter, μ is the viscosity of the dispersing medium, and V_s is the rise rate.

A substantial improvement in flotation efficiency can be achieved by the introduction of inclined plate separators and related devices using inclined tubes, which effectively provide a very large increase in surface area, and so reduce the overflow rate. The tilted plate separator originally developed by the Shell Oil Co.

(Netherlands) (Brunsmann *et al*. 1962, Kirby 1964), employs parallel plates about 100 mm apart, and tilted at an angle of 45° (different angles can be used). The influent passes down between the plates and rising oil collects on the surfaces of the plates, to move along the plate surfaces and to the top of the liquid in the separator (see Fig. 7.2). Tilted plate separators are now widely used for oil removal and are

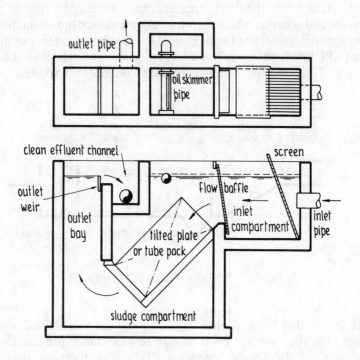

Fig. 7.2 — Typical arrangement of a tilted plate or tube separator for oil flotation (reproduced from Iggleden 1978).

capable of removing efficiently oil particles down to about 60 µm in diameter at nominal overflow rates no slower than used in the API separator. This may correspond in oil removal to residual concentrations of free oil of the order of 20 mg l^{-1}. Iggleden (1978), for example, has claimed at least 95% removal of fuel oil globules of specific gravity 0.9 and diameter down to 60 µm. Performance can, however, be variable, as indeed it is with all gravitationally induced settlement and flotation systems, and much depends on the mode of operation. Furthermore, although tilted plate and related devices (such as tubes) can work well with light oil suspensions, they are not so suitable for separating greasy solids, which tend to block the flow passages.

Flotation can be substantially assisted by the introduction of air into the flotation chamber in the form of very fine bubbles. Various interactions can thus occur between the oil and the air bubbles: oil droplets may be adsorbed on to the air bubbles; the gas bubbles may be entrapped in a flocculated assemblage of rising oil droplets; and bubbles may be absorbed into such an assemblage. The size of the air bubbles is important, and also has an influence on the mode of operation. Thus in so-called dissolved air flotation, influent or recycled effluent is saturated with air to a pressure of some 3–6 bars, and then released at atmospheric pressure into the rising waste water stream with formation of minute air bubbles that are initially 50–100 μm in diameter. Fig. 7.3 illustrates a typical dissolved air flotation system. With induced

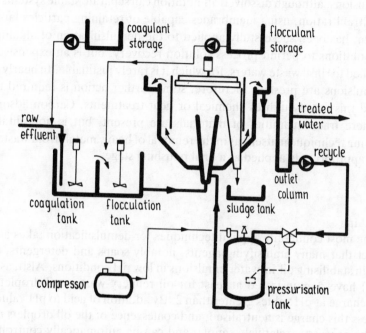

Fig. 7.3 — A typical dissolved air flotation emulsion cracking system (reproduced from Berné (1982), Pergamon Journals Ltd).

air flotation (see Degner and Winter 1979 and Sylvester and Byeseda 1980) the bubbles are larger by an order of magnitude, and air is distributed through the liquid by means of mixers. Both methods improve flotation separation dramatically and are widely used for the separation of oil-in-water suspensions. However, it is important to note that induced air flotation is not suitable for the separation of flocculated emulsions, such as those produced by the techniques discussed in this chapter, because it tends to break up the flocs from the shearing action of the mixer; dissolved air flotation, with its relatively more quiescent conditions, is on the other hand very useful. Flotation techniques, including dissolved air flotation, are reviewed by Vrablik (1959), Klassen (1960), Klassen and Mokrousov (1963), and Lemlich (1972), and the subject is also discussed by Berné (1982) in the context of a review of physical–chemical treatment methods for oil-containing effluents.

Other methods can be used for removing oil from suspensions, such as centrifugation (see Landis 1965) and filtration. Filtration may be applied after initial flotation as a further refining stage, rather as sand filtration is sometimes employed after settlement of solids. Various types of coalescing filters are available commercially, and beds of pebbles or polypropylene crumbs can also be used, as demonstrated by the Warren Springs Research Laboratory (UK). Fibrous bed media have also been used in a filter press construction (Langdon *et al.* 1972) or other pressure chamber (Chambers 1978). However, most of these physical processes have little effect on stable emulsions, although dissolved air flotation can separate stable systems to some extent. Ultrafiltration, using membranes capable of retaining particles larger than 0.1–0.5 μm, has been successfully applied to the demulsification of alkaline metal cleaning solutions to permit process solution recovery, but is an expensive option when applied to rinse waste waters, for which it is rarely justifiable. In nearly all cases where emulsions are present, therefore, some further action is required to break these, and this mostly utilises chemical or heat treatments. Carbon adsorption is useful where trace quantities of materials are present, but is not suitable as a demulsifying technique in itself; as in the removal of heavy metals from waste waters, it is most appropriately applied as a final polishing stage.

Acid cracking

One of the most commonly applied techniques for demulsification takes advantage of the fact that many emulsifying agents, notably soaps and detergents, are least effective in stabilising oil-in-water emulsions in low pH conditions. Also, as Luthy *et al.* (1977) have demonstrated at least for oil refinery wastes, oil droplets have a negative charge at pH values greater than 2. By addition of acid to pH values of 2 or slightly less this charge is neutralised, and coalescence of the oil droplets may then occur. The process is relatively simple, and can be automatically controlled to an empirically determined optimum pH, which helps in maintaining treatment efficiency.

For efficient pH control it is necessary to ensure that the pH electrode not be coated with released oil (oil-in-water emulsions themselves rarely cause much problem by way of coatings provided the solution is well agitated), and so care has to be exercised in the positioning of the pH electrode system in the reactor: it is essential that good turbulence be maintained around the pH glass surface. With continuous flow systems using a reactor with a fixed liquid level this can be easily arranged, but with batch treatment vessels operated on a 'fill-and-empty' principle there is always a risk that the glass surface will be left exposed with a film of oil on it. Attention must therefore be applied to this potential problem, and one procedure is to mount the electrode system in a recirculating side loop that is arranged always to have water-based liquid pumped through it.

One significant drawback in acid cracking is that it gives oil and water residues that are both acidic. Not only does this demand the use of appropriately acid-resistant materials in the construction of subsequent oil separators (such as flotation units), but the oil and waste water both then require neutralisation. The main attraction of acid cracking is therefore where waste acid is available; if such acid contains polyvalent cations, such as Fe^{2+} (e.g. from acid pickling operations) the cations exercise a further charge neutralisation effect, and so promote demulsification, sometimes to the extent that somewhat higher pH values can be used, e.g. 3–4, than with acid alone. Calcium or aluminium salts are also sometimes added as sources of polyvalent cation specifically for charge neutralisation, again usually resulting in a higher pH at which cracking occurs. The amounts of cation needed for charge neutralisation depend on the valency of the cation, as may be expected from the Schülze–Hardy rule. Sodium chloride, which can be used, may be needed at concentrations of the order of 5–10 g l^{-1}. Calcium chloride is usually required at concentrations of an order of magnitude less than this, while with ferric salts a further order of magnitude reduction may be possible. Much depends on the emulsion and the pH, however, and it is impossible to generalise too firmly. Massive doses of reagent may be necessary with some emulsions, although in general if cracking does not occur, or only to a partial extent, on addition of gram per litre levels of salts, it is unlikely that the method will be effective on a plant scale.

Acid and polyvalent metal ion charge neutralisation is useful as a primary demulsification stage where the oil concentrations are relatively high, because it produces a waste material that is not significantly augmented in volume by the reagents introduced. In this respect it can be employed as a separate pre-treatment operation to remove most of the oil present; the residues in the waste water can then be taken out by a polishing stage, such as by inducing interparticle bridging with a coagulation stage. This will occur when the pH is raised in a solution containing a hydrolysable metal cation, and is discussed below.

Acid cracking is not always effective, particularly for example where non-ionic detergents are present as emulsifiers. In these circumstances it is necessary to employ other techniques, as discussed below. Where it does work, the oily residues remaining as emulsion are often of the order of 20–50 mg l^{-1}.

The use of hydrolysable polyvalent cations
Hydrolysable polyvalent cations such as Fe^{3+} can be used to destabilise emulsions, not only by charge neutralisation but also, on hydrolysis, by providing bridges across which emulsified droplets can be linked. The hydrolysed cations also provide additional benefits by presenting surfaces on to which physical adsorption can occur, and, in the process of hydrolysis, by entraining oil and grease materials within the floc structure. The action of hydrolysed metal cations is thus manifold, and therein lies the power of the technique. A theoretical analysis of these processes has already been presented in Chapter 6; the same principles apply here, and so repetition is unnecessary.

As already mentioned, a multiple-stage operation can be used, with charge neutralisation in acid conditions (no hydrolysis of the metal salts) to remove a major part of the oil or fat, followed by adjustment of the pH to cause metal cation hydrolysis and thus adsorption of the remainder of the suspended material. A typical precipitation pH range is 6–8, but the optimum pH has to be decided by trials, and is also dependent to some extent on the cation used. It is customary to follow such demulsification with dissolved air flotation to remove the precipitated solids; this not only removes the entrapped oil, grease or fat, but also aids in removing some emulsifying agents such as detergents, themselves undesirable contaminants in waste waters. Polyelectrolytes, discussed later, can also be used with beneficial effect to flocculate the precipitate formed.

One typical application area for the use of hydrolysable cations is with emulsions formed from refining oil processing. Typically the reagent of choice has been aluminium sulphate, the hydrolysed species from which provide the bridges that can link the oil droplets. Dose levels of 20–30 mg l^{-1} aluminium sulphate appear to be optimal when the waste water is controlled to a pH of 8.0–8.2 (Luthy *et al.* 1978). In a study on some oil shale retort waters, aluminium sulphate was found to be more effective than ferric sulphate as a coagulant (Higgins *et al.* 1982). Performance can be optimised to give residual oil levels of the order of 10–20 mg l^{-1}, or even better. However, demulsification of oil-in-water emulsions of these kinds is nowadays more usually carried out with polyelectrolytes.

Aluminium and ferric salts find further application in the treatment of emulsion cleaners and so-called soluble oils used in machining operations. Sodium aluminate solution has also been used; it offers one advantage in that being alkaline, further addition of alkali is unnecessary. It also appears to act more quickly in some situations than do the other materials.

A typical treatment procedure for machining-oil emulsions consists of a preliminary oil flotation and metal swarf settlement stage. The underflow is then passed to the cracking reactor, where the pH is adjusted to an experimentally determined (acid) value, with simultaneous addition of aluminium or ferric sulphate. Concentration levels of the salts are, for 1–2% oil-in-water emulsions, of the order of 1000–2000 mg l^{-1}, and the pH is usually between 2 and 3. This treatment is often sufficient to cause the oil to float, and can give as much as 90% oil separation. With more stable emulsions, the pH may need to be raised to between 6 and 7 to hydrolyse the cation and produce a floc on to which the liberated oil is adsorbed. The product usually floats, but if for the purpose of pH adjustment alkali is added then the characteristics of the floc will depend on the alkali used: sodium hydroxide tends to make flocs float, while lime may give some settlement. Agitation during reagent admixture should not be too violent, since this tends to break the flocs. Batch treatment is often more efficient in terms of residual oil concentration in the treated water than continuous operations. Dissolved air flotation is helpful in improving the separation but is not usually employed with the relatively small volumes involved in the treatment of machining emulsions.

An interesting variation on the use of heavy metal salts for breaking emulsions has been described by Scholl and Balden (1973), who passed machine shop waste waters containing also some alkaline cleaner and phosphating solution residues through a bed of iron chips, using air agitation and neutral pH conditions. The accumulation of oil and corrosion products on the iron demanded periodic cleaning, but after settlement of solids in the product water from treatment, residual oil levels of less than 30 mg l^{-1} were observed, starting from concentrations in the untreated waters of 1400–14 000 mg l^{-1}. The principal advantage claimed was that the increase in total dissolved salts in the treated water was much smaller than with acid, aluminium sulphate and lime treatment, so the water had higher potential re-use value; but, somewhat surprisingly, the capital costs for air requirements were found to be more expensive than those for the direct chemical treatment by almost 50%.

Emulsion cleaners are used extensively in primary processing operations for the removal of oil, grease, and wax from components before metal finishing (phosphating, electroplating, anodising, etc.). After they have been cleaned, the workpieces are rinsed with water. Two types of effluent thus arise: spent process solution, where the oil or grease level has accumulated to the point where cleaning action is markedly reduced; and rinse waters, containing emulsion and sometimes free oil, at much lower concentrations than in the process solution. Many of these cleaners consist of sodium hydroxide or carbonate solutions, with surfactants (non-ionic as well as ionic), and frequently with phosphate, silicate, or sometimes complexing agents such as EDTA as well. Others may be solvent-based, e.g. with naphtha or an alcohol, often with an anionic surfactant.

Treatment of spent aqueous cleaners may involve an initial oil flotation stage, but this is often bypassed when waste acid is also used for mutual neutralisation. Since such acid from metal finishing pickling operations normally contains dissolved hydrolysable cations, such admixture is helpful in promoting demulsification. The rinse waters are usually adequately treated by similar admixture with acids or other rinse streams containing hydrolysable cations; adjustment of the pH to the optimum precipitation pH for the metal ions (see Chapter 2) provides bridging hydrolysed species that entrap oil and adsorb surfactant in the flocs formed. It is also common to use an anionic polyelectrolyte as well in these operations, as described in Chapter 6. Kieszkowski and Bartkiewicz (1972) have used magnesium sulphate and lime as a preliminary treatment before aluminium sulphate addition to spent aqueous emulsion cleaners, thereby improving removal of anionic surfactants: in trials on several solutions they observed C.O.D. reductions of the order of 60% with 2 g $MgSO_4$ plus 1.8 g $Ca(OH)_2$ per gram anionic surfactant present, extra chemicals improving this performance to 90% C.O.D. reduction. The dose concentration of $Al_2(SO_4)_3$ in the effluent was found to be optimal at 500 mg l^{-1}. Kieszkowski and Bartkiewicz also applied magnesium sulphate/lime treatment to solvent-based cleaners to separate the oil and grease as sludge, and then used aeration to provide foam flotation of residual (non-ionic) surfactants. However, Gomulka and Gomulka (1984) found magnesium salts as well as aluminium sulphate to be less effective than barium chloride as a preliminary reagent for removal of surfactants. Laboratory studies indicated good settlement of insoluble barium complexes, and excess barium was removed as insoluble sulphate by post-treatent with aluminium sulphate, which provided additionally a further reduction in C.O.D. The economics of this approach as a plant procedure would require careful examination, however.

Another industry where aluminium sulphate was proved effective in emulsion-cracking is in the treatment of waste waters containing wool grease. It is common practice to treat wool scouring waste waters with sulphuric acid to pH 4, thereby to recover the grease, but this is not adequately effective when non-ionic surfactants are present: the clarified phase may retain as much as 30% of the original grease content, and the recovered material is of poor quality because of the high fatty acid content

Table 7.1 — Relative performances of demuslification by coagulaton treatment applied with air flotation (after Patterson 1975)

Industrial source	Coagulant	Oil concentration (mg l^{-1})		Per cent removal
		Influent	Effluent	
Refinery	None	125	35	72
Refinery	100 mg l^{-1} aluminium sulphate	100	10	92
Refinery	None	154	40	74
Oil tanker ballast water	100 mg l^{-1} aluminium sulphate (+1 mg l^{-1} polymer)	133	15	89
Paint manufacture	150 mg l^{-1} aluminium sulphate (+1 mg l^{-1} polymer)	1900	0	100
Aircraft maintenance	30 mg l^{-1} aluminium sulphate (+10 mg l^{-1} activated silica)	250–700	20–50	90+
Meat packing	None	3830	270	93
Meat packing	None	4360	170	96
Meat packing	300 mg l^{-1} aluminium sulphate	1944	142	93
Yarn scouring	aluminium chlorohydrate	2300–8160	40–248	89–99
Wool scouring	aluminium chlorohydrate	1650–12260	35–287	82–99+

induced by acidification. Christoe (1977) studied a variety of chemicals to improve the separation, including lime, ferrous and ferric and aluminium salts, and calcium chloride, but found that of these aluminium sulphate is the most effective when used at pH 5.

Mention of recovery points to one disadvantage in using hydrolysable cations, which is that recovered material is inevitably contaminated with the hydrous oxide. This demonstrates a further merit in separating emulsions by physical means as much as possible as a prior stage when the separated material has potential re-use value.

Finally in this section, the relative performances of coagulation treatment of waste waters from different industrial sources can be observed by inspection of Table 7.1, taken from Patterson (1975). It is apparent that percentage removals can be good (in the region of 90%), but the oily residues may still be quite high in concentration terms.

The use of polyelectrolytes

Apart from the disadvantage imposed by use of hydrolysable metal cations for breaking emulsions in reducing the quality of recovered materials, these reagents also generate sludges that have to be disposed of. In both of these respects, water-soluble organic polyelectrolytes offer significant advantages, in that the concentrations needed are only a few milligrams per litre and that they do not generate extra quantities of waste for disposal.

Polyelectrolytes act as demulsifiers in ways already discussed for colloids in Chapter 6, i.e. either as coagulants (destabilisation via charge neutralisation) or as flocculants (by providing inter-particle bridging). It is sometimes possible to realise both functions with the same material, but more frequently the structural requirements for molecules to achieve these separate functions optimally are quite different. For destabilisation via charge neutralisation a high charge density is called for, whereas for effective inter-particle bridging and flocculation a long-chain polymer having low or even zero charge density is indicated. Benefits sometimes also accrue from the use of a hydrolysable metal cation together with a polyelectrolyte. (These principles are discussed in more detail in Chapter 6.)

Cationic polyelectrolytes can be very effective in the treatment of oil emulsions. Luthy *et al.* (1977), in a study of the surface properties of petroleum refinery waste oil emulsions, found that these properties are determined more by the characteristics of the water than by the oil: oil droplets emulsified in distilled water containing sodium chloride show charge reversals at pH 5, but with emulsions arising from refineries with the same oils the isoelectric pH dropped to 2, presumably due to the presence of anionic surfactants. The negative charge present on emulsified oil droplets explains the experimentally observed fact that anionic and non-ionic polyelectrolytes are relatively poor demulsifiers for such systems.

Cationic polyelectrolytes available commercially include polyamines and poly-quaternary salts, such as polydiallyldimethylammonium (PDADMA) chloride. Control of the cationic charge density may be achieved for example by copolymerising acrylamide, which is non-ionic, with PDADMA:

$$\left[\begin{array}{c} -CH_2-CH- \\ \quad | \\ \quad C=O \\ NH_2 \end{array} \right]_x \quad \left[\begin{array}{c} CH_2 \\ -CH \quad CH- \\ \quad | \quad\quad | \\ CH_2 \quad CH_2 \\ \quad N^+ \\ CH_3 \quad CH_3 \end{array} \right]_y$$

The relative proportions of x and y in the resultant molecule dictates the overall positive charge on the polymer. The cationic character of such polymers based on quaternary salts is virtually independent of pH within the pH ranges usually encountered in waste waters. However, the cationicity of polyamines will be strongly pH-dependent in the range of pH near to the pK of the amine groups (typically p$K \pm 2$), with the cationic character increasing with decreasing pH as the amine groups are successively protonated.

Luthy *et al.* (1978) studied the effects of various polyelectrolytes in the separation of emulsified oil assisted by dissolved air flotation, including PDADMA types. They found that the polyelectrolyte dose required to reverse the electrokinetic mobilities of the oil droplets approximated to that which gave the best turbidity removals, this being 10–15 mg l^{-1} for cationic polyelectrolytes. The effective dose was relatively independent of droplet size or oil concentration. Fig. 7.4 summarises graphically the electrokinetic mobilities of oil droplets dosed with various polyelectrolytes in an API separator. As may be expected, the dose of polyelectrolyte required to reverse the charge on the oil droplets decreased with increasing charge density.

The surface activity of demulsifiers in destabilising oil-in-water emulsions can be enhanced by dilution in specific solvents prior to dosing. Graham *et al.* (1983) studied two types of demulsifier in this context: a blend of ethoxylated/propoxylated adducts (molecular weight approximately 2000) with a cationic fatty acid (molecular weight approximately 800), and a material primarily an ethoxylated phenolic resin (molecular weight 1600), both commercially supplied; the solvents used for dilution included methanol, propanol and xylene. Such dilution not only reduces the viscosities of the start demulsifier materials, but can significantly improve their surface activities and enhance emulsion resolution efficiency.

As mentioned earlier, a mixture of emulsion-breaking techniques may be necessary or more effective than one alone. Polyvalent cations hydrolysed to give a floc act as good adsorbents for removing residues as well as breaking emulsions, and these flocs may be improved in character for better flotation by the addition of a polyelectrolyte. Charge reversal with acid and aluminium salts in low pH conditions may be enhanced in its effect by the addition of cationic polyelectrolyte (Harlow *et al.* 1982).

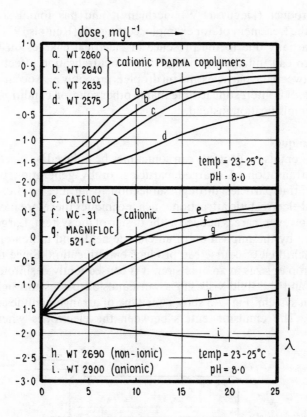

Fig. 7.4 — Electrokinetic mobility of waste oil droplets in effluent from API separator after dosage with polyelectrolytes. λ=electrokinetic mobility, $\mu S^{-1} V^{-1} cm^{-1}$ (After Luthy *et al.* 1978).

Heat treatment

Some emulsions are so stable that they can be broken only extremely slowly, or barely at all, with many demulsifiers. In these circumstances it may be necessary to use heat as well; Harlow *et al.* (1982), for example, augmented the procedure they used for separation of waste oil arising from automotive industry wastes by using a second separation stage in which the temperature was raised to 82°C, and Little and Patterson (1978) showed that shipboard emulsions can be better treated with quaternary ammonium compounds when heat is applied. Obviously the heating of large volumes of emulsion can be expensive, and so is only used as a last resort.

Demulsification with small particles

Emulsions may sometimes be broken by the introduction of very finely divided solids, such as highly dispersed clays and cellulosic materials rendered hydrophobic by treatment with, for example, silicones. A hydrophobic silica is available as a commercial product (Degussa, W. Germany), and has found some successful application in the treatment of, for example, cutting oils (Reufels 1971). The action is presumably partially due to the presence of electrical charges on the very small particles, which can initially provide a charge neutralisation effect on emulsions. Interaction between the particles and oil droplets probably also occurs, possibly with bridging action. It is interesting to note that in other solution conditions fine particles can promote emulsification instead.

Electrical techniques

Coalescence of emulsion droplets can sometimes be induced by electrical methods, either by neutralisation of charged particles, or by generation of a dipole on electrically neutral particles through subjection to an a.c. or d.c. field, thereby causing mutual electrostatic attraction. Such techniques therefore have some apparent interest. However, care is needed in the design of the system. Large droplets tend to be deformed by the imposition of an electric field, and ultimately they may be disrupted, which then of course can produce emulsification. Since there is often a spectrum of droplet sizes in an emulsion, it is consequently desirable to reduce the field strength and the fluid viscosity as an emulsion passes through an electrical demulsification system. Fig. 7.5 shows one type of arrangement designed to realise this objective. The emulsion enters between the electrodes where the field is

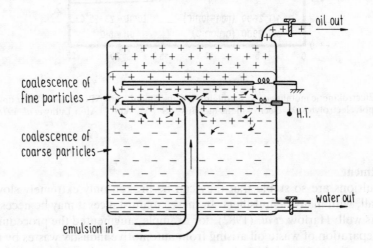

Fig. 7.5 — Electrical demulsification showing electrical field and flow patterns. (After Gopal in Sherman (1978).)

strongest; below the electrodes the larger oil droplets coalesce where the field is weaker.

Ohta (1972) has described a cell system for use on oil emulsions where the cathode is of carbon, iron or lead oxide, and the anode of a noble metal or titanium. Separation is carried out at pH<3, and the aqueous phase is then subjected to a further electrolysis using aluminium electrodes to precipitate surfactants as a sludge. Snyder and Willihnganz (1976) used graphite electrodes, separated into chambers by dividers of filter paper held between plastics screens mounted in plastics frames. Soluble oil emulsions were caused to flow upwards past the faces of the electrodes, and the gases evolved created a turbulence that formed a froth of separated oil. The degree of oil removal was of the order of 80% for a single pass, improving from 75% to 85% by increasing the applied voltage from 10 V to 20 V, but at the expense of a substantial power consumption increase.

Another electrolytically based procedure for oil removal has been described by Ghosh and Brown (1975). Two electrochemically dissimilar materials, e.g. carbon and aluminium or carbon and steel, in granular form are arranged to be in contact with each other, submersed in an aqueous medium. Together they form an electro-chemical cell in which negatively charged oil droplets migrate to the anodic areas, and coalesce there; liberation of metal ions at the anode sites also aids in destabilising the charged oil particles. Ghosh and Brown studied flow-through beds, and found that the residence time is critical and must not be too brief (they state a minimum of 40 seconds). The flow rate, on the other hand, should be fast enough to avoid contamination of the electrode surfaces by by-products, the performance being inversely proportional to the flow rate. It is also inversely related to the degree of porosity of the coalescing bed. It is clear that several variables have to be controlled carefully, and practical plant design would need considerable pre-study.

Electrocoagulation, involving the electrolytic generation of coagulating mater-ials, can be used as an emulsion-breaking technique. For example, Kramer et al. (1978) employed iron electrodes, alternately arranged with wooden spacers Fouling of the electrodes occurred after a period of operation, which was partially removed by polarity reversal, but which after a further 7 hours necessitated cleaning of the electrodes by scraping. This illustrates one of the practical problems arising in this form of treatment.

A magnetic approach has been used by Kaiser et al. (1971), who added an oil-soluble/water-insoluble iron-based fluid to oil-in-water emulsions, and then passed the mixture through packed beds containing magnetic particles to which a magnetic field was applied. These authors claimed virtually complete removal of oil particles 1 μm in diameter with residence times of less than 1 minute.

Electrical techniques are not necessarily attractive, because where the electrical conductivities are low, as with certain oil waste emulsions, the running costs may be high. Furthermore, the more stable emulsions that require bridging action for coagulation are not always destabilised by the application of an electrical field alone.

For further details on electrical demulsification techniques, see the review by Waterman (1965).

THE FINAL NOTE

Most of the reagents used for demulsification are the same as those used for colloid destabilisation, so the descriptions given in Chapter 6 apply here also. It is worth stressing, again, that the optimisation of processes for coagulating colloidal or emulsified particles is, once basic principles have been recognised as a guide to lines of approach that may be taken, very much a matter of empirical study. This is due mainly to the fact that individual waste waters for treatment may have specific components that may modify their behaviour with treatment chemicals quite dramatically, with the result that even laboratory results may not always be reproducible fully on a plant scale.

This last point is indeed representative of much of chemical waste water treatment, as we have sought to demonstrate throughout this book. Before embarkation on a plant design, a full appraisal of the chemical process options and their relative efficacies must be made from underlying theoretical principles. This will provide the basis for establishing the process(es) likely to be most effective; but the best procedure in terms of result and cost will only be found by experimental trial, and this may extend into the full plant system.

REFERENCES

APHA/AWWA/WPCF (1985) *Standard Methods for the Examination of Water and Wastewater*, 16th edn, American Public Health Association, American Water Works Association, Water Pollution Control Federation.

API (1959) *API Analytical Methods No. 733–58*, American Petroleum Institute.

API (1969) *API Manual on Disposal of Refinery Wastes*, American Petroleum Institute.

Becher, P. (1965) *Emulsions, Theory and Practice. ACS Monograph No. 162*, 2nd edn, Reinhold.

Berné, F. (1982) *Wat. Sci. Technol.* **14** 1195.

Brunsmann, J. J., Cornelissen, J., and Eilers, H. (1962) *J. Wat. Pollut. Control Fed.* **34** 44.

Chambers, D. B. (1978) *Chem. Ind. (London)* 834.

Christoe, J. R. (1977) *J. Wat. Pollut. Control Fed.* **49** 848.

Degner, V. W., and Winter, M. K. (1979) *A.I.Ch.E. Symp. Ser. No. 195* 119.

Department of Environment (1982) *Methods for the Examination of Waters and Associated Materials. Gas Chromatographic and Associated Methods for the Characterization of Oils, Fats, Waxes and Tars*, Her Majesty's Stationery Office.

Department of Environment (1983) *Methods for the Examination of Waters and Associated Materials. The Determination of Hydrocarbon Oils in Waters by Solvent Extraction, Infra Red Absorption and Gravimetry*, Her Majesty's Stationery Office.

EPA (1979) *Methods for Chemical Analysis of Water and Wastes*, EPA 600/4-79-020, US Environmental Protection Agency.

Ghosh, M. M., and Brown, W. P. (1975) *J. Wat. Pollut. Control Fed.* **47** 2101.

Gomulka, B., and Gomulka, E. (1984) *Effl. Wat. Treat. J.* **24** 119.

Graham, D. E., Stockwell, A., and Thompson, D. G. (1983) *Chemicals in the Oil Industry, Royal Society of Chemistry Spec. Pub. No. 45* 84.

Harlow, B. D., Hubbell, J. W., and Doran, T. M. (1982) *Proc. 37th Ind. Waste Conf. Purdue Univ.* 197.

Higgins, T. E., McTernan, W. F., Schassberger, L. A., Kacornik, D. J., and Stetzenbach, K. J. (1982) *Proc. 15th Oil Shale Symp.* 451.

Iggleden, G. J. (1978) *Chem. Ind. (London)* 826.

IP (1985) *Methods for Analysis and Testing*, Vol. 1, 1P137/82, Institute of Petroleum.

Kaiser, R., Colton, C. K., Miskolczy, G., and Mir, L. (1971) *A.I.Ch.E. Symp. Ser. No. 124* 115.

Kieszkowski, M., and Bartkiewicz, B. (1972) *Metal Finish. J.* **18** 385.

King, A. (1941) *Trans. Farad. Soc.* **37** 168.

Kirby, A. W. (1964) *Trans. Inst. Chem. Eng.* **4** 76.

Klassen, B. I. (1960) *Proc. Int. Mineral Processing Congress*, Inst. Mining and Metallurgy.

Klassen, V. I., and Mokrousov, V. A. (1963) *Introduction to the Theory of Flotation*, Butterworths.

Kramer, G. R., Buyers, A., and Brownlee, B. (1978) *Proc. 33rd Ind. Waste Conf. Purdue Univ.* 673.

Landis, D. M. (1965) *Chem. Eng. Prog.* **61** 58.

Langdon, W. M., Naik, P. P., and Wasan, D. T. (1972) *Environ. Sci. Technol.* **6** 905.

Lemlich, R. (ed.) (1972) *Adsorptive Bubble Separation Techniques*, Academic Press.

Lissant, K. J. (ed.) (1974) *Emulsions and Emulsion Technology*, Vols. I and II, Marcel Dekker.

Little, R. C., and Patterson, R. L. (1978) *Environ. Sci. Technol.* **12** 584.

Lohmeyer, S. (1973) *Galvanotechnik (Saulgau)* **64** 601; 795; 911.

Luthy, R. G., Selleck, R. E., and Galloway, T. R. (1977) *Environ. Sci. Technol.* **11** 1211.

Luthy, R. G., Selleck, R. E., and Galloway, T. R. (1978) *J. Wat. Pollut. Control Fed.* **50** 331.

Neustadter, E. L., Whittingham, K. F., and Grist, D. M. (1975) *IUPAC Int. Symp. Colloid and Surface Science*, International Union of Pure and Applied Chemistry.

Neustadter, E. L., Whittingham, K. F., and Graham, D. E. (1981) *Surface Phenomena in Enhanced Oil Recovery*, ed. Shah, D. O., 307, Plenum.

Ohta, M. (1972) *Brit. Patent* No. 1, 378, 475.

Patterson, J. W. (1975) *Wastewater Treatment Technology*, 186, Ann Arbor Science.

Peoples, R. F., Krishnanan, P., and Simonsen, R. N. (1972) *J. Wat. Pollut. Control Fed.* **44** 2120.

Reufels, H. (1971) *Oberfläche (Berlin)* **11** 194.

Scholl, E. L., and Balden, A. R. (1973) *Proc. 28th Ind. Waste Conf. Purdue Univ.* 874.

Sherman, P. E. (ed.) (1968) *Emulsion Science*, Academic Press.

Smith, A. L. (ed.) (1974) *Theory and Practice of Emulsion Technology. Proc. Soc. Chem. Ind. Symp.*, Academic Press.

Snyder, D. D., and Willihnganz, R. A. (1976) *Proc. 31st Ind. Waste Conf. Purdue Univ.* 782.

Strassner, J. E. (1968) *J. Petrol. Technol.* **20** 203.

Sylvester, N. D., and Byeseda, J. J. (1980) *Soc. Petrol. Eng.* **20** 579.

Vrablik, E. R. (1959) *Proc. 14th Ind. Waste Conf. Purdue Univ.* 743.

Waterman, L. C. (1965) *Chem. Eng. Progr.* **61** (10) 51.

Appendix 1

Some standard electrode potentials

From Milazzo, G., Caroli, S., and Sharma, V. K. (1978) *Tables of Standard Electrode Potentials*, Copyright © 1978 John Wiley & Sons Ltd. Reprinted by permission of John Wiley & Sons Ltd.

Half-reaction	$E^\ominus$, V
$Ag^+ + e \rightleftharpoons Ag(s)$	$+0.799$
$AgBr(s) + e \rightleftharpoons Ag(s) + Br^-$	$+0.073$
$AgCl(s) + e \rightleftharpoons Ag(s) + Cl^-$	$+0.222$
$Ag(CN)_2^- + e \rightleftharpoons Ag(s) + 2CN^-$	-0.31
$AgI(s) + e \rightleftharpoons Ag(s) + I^-$	-0.151
$Ag(S_2O_3)_2^{3-} + e \rightleftharpoons Ag(s) + 2S_2O_3^{2-}$	$+0.017$
$Al^{3+} + 3e \rightleftharpoons Al(s)$	-1.662
$H_3AsO_4 + 2H^+ + 2e \rightleftharpoons H_3AsO_3 + H_2O$	$+0.559$
$Au^+e \rightleftharpoons Au(s)$	$+1.692$
$Au^{3+} + 2e \rightleftharpoons Au^+$	$+1.401$
$AuCl_4^- + 3e \rightleftharpoons Au(s) + 4Cl^-$	$+1.002$
$Au(CN)_2^- + e \rightleftharpoons Au(s) + 2CN^-$	-0.669
$Ba^{2+} + 2e \rightleftharpoons Ba(s)$	-2.906
$Br_2(l) + 2e \rightleftharpoons 2Br^-$	$+1.065$
$Br_2(aq) + 2e \rightleftharpoons 2Br^-$	$+1.087$
$HBrO + H^+ + e \rightleftharpoons \frac{1}{2}Br_2(g) + H_2O$	$+1.574$
$BrO_3^- + 6H^+ + 5e \rightleftharpoons \frac{1}{2}Br_2(l) + 3H_2O$	$+1.52$
$Ca^{2+} + 2e \rightleftharpoons Ca(s)$	-2.866
$C_6H_4O_2(quinone) + 2H^+ + 2e \rightleftharpoons C_6H_4(OH)_2(hydroquinone)$	$+0.699$
$2CO_2(g) + 2H^+ + 2e \rightleftharpoons H_2C_2O_4$	-0.49
$Cd^{2+} + 2e \rightleftharpoons Cd(s)$	-0.403
$Cl_2(g) + 2e \rightleftharpoons 2Cl^-$	$+1.359$
$HClO + H^+ + e \rightleftharpoons \frac{1}{2}Cl_2(g) + 3H_2O$	-1.63
$ClO_3^- + 6H^+ + 5e \rightleftharpoons \frac{1}{2}Cl_2(g) + 3H_2O$	$+1.47$
$CNO^- + H_2O + 2e \rightleftharpoons CN^- + 2OH^-$	-0.970
$Co^{2+} + 2e \rightleftharpoons Co(s)$	-0.277

Half-reaction $E^\ominus$, V

Half-reaction	$E^\ominus$, V
$Co^{3+} + e \rightleftharpoons Co^{2+}$	$+1.808$
$Cr^{3+} + e \rightleftharpoons Cr^{2+}$	-0.408
$Cr^{3+} + 3e \rightleftharpoons Cr(s)$	-0.744
$Cr_2O_7^{2-} + 14H^+ + 6e \rightleftharpoons 2Cr^{3+} + 7H_2O$	$+1.33$
$Cu^{2+} + 2e \rightleftharpoons Cu(s)$	$+0.337$
$Cu^{2+} + e \rightleftharpoons Cu^+$	$+0.153$
$Cu^+ + e \rightleftharpoons Cu(s)$	$+0.521$
$F_2(g) + 2H^+ + 2e \rightleftharpoons 2HF(aq)$	$+3.06$
$Fe^{2+} + 2e \rightleftharpoons Fe(s)$	-0.440
$Fe^{3+} + e \rightleftharpoons Fe^{2+}$	$+0.771$
$Fe(CN)_6^{3-} + e \rightleftharpoons Fe(CN)_6^{4-}$	$+0.36$
$Fe(OH)_3(s) + e \rightleftharpoons Fe(OH)_2(s) + OH^-$	-0.56
$2H^+ + 2e \rightleftharpoons H_2(g)$	0.000
$Hg_2^{2+} + 2e \rightleftharpoons 2Hg(l)$	$+0.788$
$2Hg^{2+} + 2e \rightleftharpoons Hg_2^{2+}$	$+0.920$
$Hg^{2+} + 2e \rightleftharpoons 2Hg(l)$	$+0.854$
$Hg_2Cl_2(s) + 2e \rightleftharpoons 2Hg(l) + 2Cl^-$	$+0.268$
$Hg_2SO_4(s) + 2e \rightleftharpoons 2Hg(l) + SO_4^{2-}$	$+0.615$
$HO_2 + H_2O + 2e \rightleftharpoons 3OH^-$	$+0.88$
$I_2(s) + 2e \rightleftharpoons 2I^-$	$+0.5355$
$I_3^- + 2e \rightleftharpoons 3I^-$	$+0.536$
$IO_3^- + 6H^+ + 5e \rightleftharpoons \frac{1}{2}I_2(s) + 3H_2O$	$+1.196$
$Mg^{2+} + 2e \rightleftharpoons Mg(s)$	-2.363
$Mn^{2+} + 2e \rightleftharpoons Mn(s)$	-1.180
$MnO_2(s) + 4H^+ + 2e \rightleftharpoons Mn^{2+} + 2H_2O$	$+1.23$
$MnO_4 + 8H^+ + 5e \rightleftharpoons Mn^{2+} + 4H_2O$	$+1.51$
$MnO_4^- + 4H^+ + 3e \rightleftharpoons MnO_2(s) + 2H_2O$	$+1.695$
$MnO_4^- + e \rightleftharpoons MnO_4^{2-}$	$+0.564$
$N_2(g) + 5H^+ + 4e \rightleftharpoons N_2H_5^+$	-0.23
$HNO_2 + H^+ + e \rightleftharpoons NO(g) + H_2O$	$+1.00$
$NO_3^- + 3H^+ + 2e \rightleftharpoons HNO_2 + H_2O$	$+0.94$
$Ni^{2+} + 2e \rightleftharpoons Ni(s)$	-0.250
$H_2O_2 + 2H^+ + 2e \rightleftharpoons 2H_2O$	$+1.776$
$O_2(g) + 4H^+ + 4e \rightleftharpoons 2H_2O$	$+1.229$
$O_2(g) + 2H_2O + 4e \rightleftharpoons 4OH^-$	$+0.401$
$O_2(g) + 2H^+ + 2e \rightleftharpoons H_2O_2$	$+0.682$
$O_3(g) + 2H^+ + 2e \rightleftharpoons O_2(g) + H_2O$	$+2.07$
$Pb^{2+} + 2e \rightleftharpoons Pb(s)$	-0.126
$PbO_2(s) + 4H^+ + 2e \rightleftharpoons Pb^{2+} + 2H_2O$	$+1.455$
$PbSO_4(s) + 2e \rightleftharpoons Pb(s) + SO_4^{2-}$	-0.350
$PtCl_4^{2-} + 2e \rightleftharpoons Pt(s) + 4Cl^-$	$+0.755$
$PtCl_6^{2-} + 2e \rightleftharpoons PtCl_4^{2-} + 2Cl$	$+0.68$
$Pd^{2+} + 2e \rightleftharpoons Pd(s)$	0.987
$S(s) + 2H^+ + 2e \rightleftharpoons H_2S(g)$	$+0.141$
$H_2SO_3 + 4H^+ + 4e \rightleftharpoons S(s) + 3H_2O$	$+0.450$

Half-reaction	$E^\ominus$, V
$2SO_3^{2-} + 2H_2O + 2e \rightleftharpoons S_2O_4^{2-} + 4OH^-$	-1.12
$S_4O_6^{2-} + 2e \rightleftharpoons 2S_2O_3^{2-}$	$+0.08$
$SO_4^{2-} + 4H^+ + 2e \rightleftharpoons H_2SO_3 + H_2O$	$+0.172$
$S_2O_6^{2-} + 2e \rightleftharpoons 2SO_4^{2-}$	$+2.01$
$Sb_2O_5(s) + 6H^+ + 4e \rightleftharpoons 2SbO^+ + 3H_2O$	$+0.581$
$H_2SeO_3 + 4H^+ + 4e \rightleftharpoons Se(s) + 3H_2O$	$+0.740$
$SeO_4^{2-} + 4H^+ + 2e \rightleftharpoons H_2SeO_3 + H_2O$	$+1.15$
$Sn^{2+} + 2e \rightleftharpoons Sn^{2+}$	$+0.154$
$TeO_2.H_2O + 4H^+ + 4e \rightleftharpoons Te(s) + 3H_2O$	$+0.953$
$Ti^3 + e \rightleftharpoons Ti^{2+}$	-0.369
$TiO^{2+} + 2H^+ + e \rightleftharpoons Ti^{3+} + H_2O$	$+0.099$
$Tl^+ + e \rightleftharpoons Tl(s)$	-0.336
$Tl^{3+} + 2e \rightleftharpoons Tl^-$	$+1.25$
$UO_2^{2+} + 4H^+ + 2e \rightleftharpoons U^{4+} + 2H_2O$	$+0.334$
$V^{3+} + e \rightleftharpoons V^{2+}$	-0.256
$VO^{2+} + 2H^+ + e \rightleftharpoons V^{3+} + H_2O$	$+0.359$
$V(OH)_4^+ + 2H^+ + e \rightleftharpoons VO^{2+} + 3H_2O$	$+1.00$
$Zn^{2+} + 2e \rightleftharpoons Zn(s)$	-0.763

Appendix 2

Systematic names and formulae of some chemical substances

(1) Pesticides

Details of the various substances mentioned in the text, together with many others, can be found in Wiswesser, W. J., ed. (1976) *Pesticides Index*, Entomological Society of America. Another useful reference is Büchel, K. H. (ed.) (1983) *Chemistry of Pesticides*, Wiley–Interscience.

(2) Complexing Agents

Names and formulae of complexing agents referred to in the text are as follows:

Citric acid

$$CH_2COOH$$
$$|$$
$$HO \; C \; COOH$$
$$|$$
$$CH_2COOH$$

Ethanolamines

(a) Monoethanolamine

$$HOCH_2CH_2NH_2$$

(b) Diethanolamine

$$HOCH_2CH_2 \diagdown$$
$$\qquad\qquad NH$$
$$HOCH_2CH_2 \diagup$$

(c) Triethanolamine

$$HOCH_2CH_2 \diagdown$$
$$HOCH_2CH_2 — N$$
$$HOCH_2CH_2 \diagup$$

Ethylenediamine

$$H_2NCH_2CH_2NH_2$$

EDTA
(a)   Ethylenediamine-N,N,N',N'-tetra-acetic acid
(b)   (Ethylenenitrilo)tetra-acetic acid

$$HOOC \cdot CH_2 \diagdown \atop HOOC \cdot CH_2 \diagup N - CH_2 - CH_2 - N {\diagup CH_2 \cdot COOH \atop \diagdown CH_2 \cdot COOH}$$

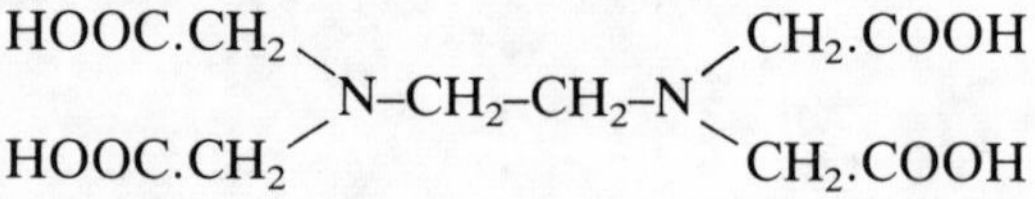

EGTA
(a)   [(Ethylenedioxy)diethylenedinitrilo]tetra-acetic acid
(b)   4,7-Dioxa-1,10-diazadecane-1,1,10,10-tetra-acetic acid

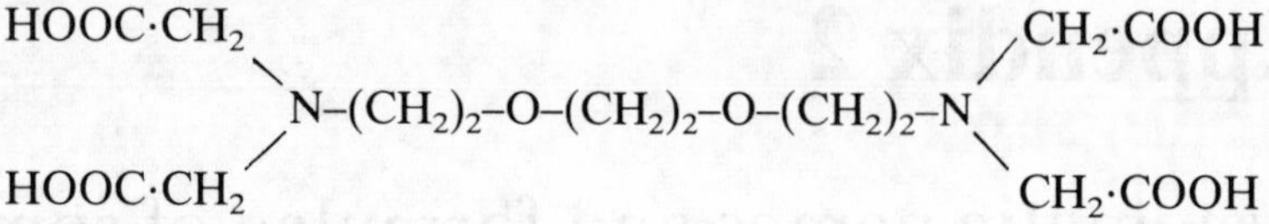

$$HOOC \cdot CH_2 \diagdown \atop HOOC \cdot CH_2 \diagup N - (CH_2)_2 - O - (CH_2)_2 - O - (CH_2)_2 - N {\diagup CH_2 \cdot COOH \atop \diagdown CH_2 \cdot COOH}$$

HEDTA
(a)   Hydroxyethylethylenediamine tetra-acetic acid
(b) N-[N'-(2''-hydroxyethyl)-N'-(carboxymethyl)-2'-aminoethyl]
       iminodiacetic acid

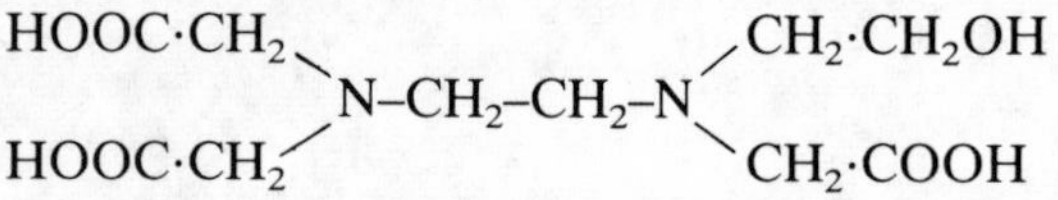

$$HOOC \cdot CH_2 \diagdown \atop HOOC \cdot CH_2 \diagup N - CH_2 - CH_2 - N {\diagup CH_2 \cdot CH_2OH \atop \diagdown CH_2 \cdot COOH}$$

NTA
Nitrilo tri-acetic acid

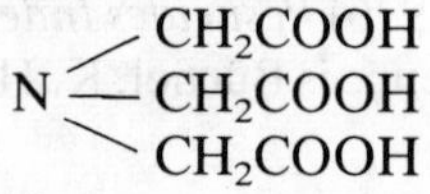

$$N {\diagup CH_2COOH \atop - CH_2COOH \atop \diagdown CH_2COOH}$$

Oxalic acid

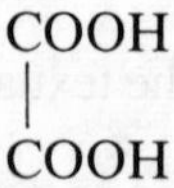

$$COOH \atop | \atop COOH$$

Tartaric acid

$$HOCHCOOH \atop | \atop HOCHCOOH$$

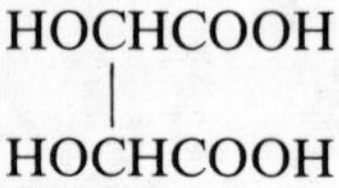

Quadrol (Tetrol)
(a) N,N,N',N'-tetrakis(2-hydroxypropyl)ethylenediamine
(b)   Ethylenedinitrilotetra-1-(2-propanol)

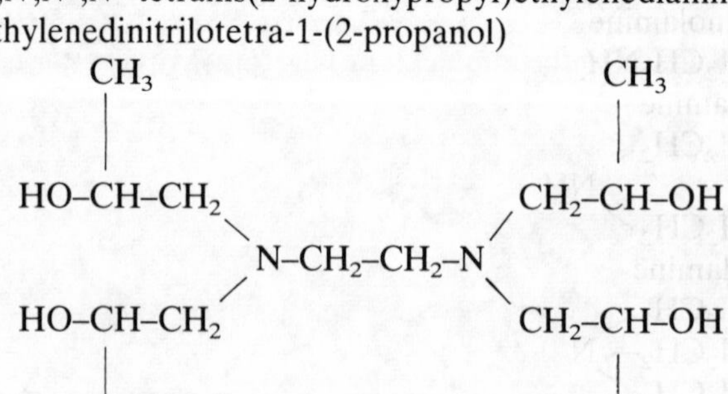

$$\begin{array}{ccc} CH_3 & & CH_3 \\ | & & | \\ HO-CH-CH_2 & & CH_2-CH-OH \\ \diagdown & & \diagup \\ & N-CH_2-CH_2-N & \\ \diagup & & \diagdown \\ HO-CH-CH_2 & & CH_2-CH-OH \\ | & & | \\ CH_3 & & CH_3 \end{array}$$

Appendix 3

Bibliography on industrial waste water treatment

Azad, H. S. (ed). (1976) *Industrial Wastewater Management Handbook*, McGraw-Hill.

Barnes, D., Forster, C. F., and Hruday, S. E. (eds.) (1984) *Surveys in Industrial Wastewater Treatment. Vol. 1. Food and Allied Industries. Vol. 2. Petroleum and Organic Chemicals Industries*, Pitman.

Besselièvre E. B., and Schwartz, M. (1976) *The Treatment of Industrial Wastes*, McGraw-Hill.

Bridgewater, A. V., and Mumford, C. J. (1979) *Water Recycling–Pollution Control Handbook*, George Godwin.

Callely, A. G., Forster, C. F., and Stafford, D. A. (eds.) (1977) *Treatment of Industrial Effluents*, Hodder and Stoughton.

Gurnham, C. F. (1965) *Industrial Waste Treatment and Control*, Academic Press.

Jørgensen, S. E. (1979) *Industrial Waste Water Management*, Elsevier.

Lund, H. F. (ed.) (1971) *Industrial Pollution Control Handbook*, McGraw-Hill.

Nemerov, N. L. (1971) *Liquid Wastes of Industry. Theories, Practices and Treatment*, Addison-Wesley.

Patterson, J. W. (1975) *Wastewater Treatment Technology*, Ann Arbor Science.

Reference can also be made to the annua (June) literature reviews in *J. Wat. Pollut. Control Fed.*

Index

Page numbers indicate the commencement of a reference to the subject, not necessarily the full extent of the reference.